Algorithms in Bioinformatics

Algorithms in Bioinformatics

Theory and Implementation

Paul A. Gagniuc
University Politehnica of Bucharest
Bucharest, Romania

Registered Office
John Wiley & Sons, Inc., 111 River Street, Hoboken, NJ 07030, USA

Editorial Office
111 River Street, Hoboken, NJ 07030, USA

For details of our global editorial offices, customer services, and more information about Wiley products visit us at www.wiley.com.

Wiley also publishes its books in a variety of electronic formats and by print-on-demand. Some content that appears in standard print versions of this book may not be available in other formats.

Library of Congress Cataloging-in-Publication Data

Names: Gagniuc, Paul A., author.
Title: Algorithms in bioinformatics : theory and implementation / Paul A.
   Gagniuc, Polytechnic University of Bucharest, Bucharest, Romania.
Description: First edition. | Hoboken, NJ : Wiley, 2021. | Includes
   bibliographical references and index.
Identifiers: LCCN 2021004386 (print) | LCCN 2021004387 (ebook) | ISBN
   9781119697961 (cloth) | ISBN 9781119697954 (adobe pdf) | ISBN
   9781119697992 (epub)
Subjects: LCSH: Bioinformatics. | Algorithms.
Classification: LCC QH324.2 .G34 2021 (print) | LCC QH324.2 (ebook) | DDC
   570.285–dc23
LC record available at https://lccn.loc.gov/2021004386
LC ebook record available at https://lccn.loc.gov/2021004387

Cover Design: Wiley
Cover Image: © Science Photo Library/Alamy Stock Photo

Set in 9.5/12.5pt STIXTwoText by Straive, Chennai, India

I dedicate this book to my family,
my children Nichita and Ana,
my beautiful wife Elvira,
and to my mother-in-law Anastasia.

Contents

Preface

Algorithms in Bioinformatics: Theory and Implementation is a concise yet comprehensive textbook of bioinformatics, which describes some of the main algorithms that are used to elucidate biological functions and relationships. This unique guide to *Algorithms in Bioinformatics* approaches the subject along the four convergent lines of mathematics, implementation, simulation, experimentation, and can be ideal for upper-undergraduate bioinformatics courses, researchers, doctoral students, and sociologists or engineers charged with data analysis. This work first begins with a general introduction to biology, which is meant to bring a more concrete understanding of the molecular processes concerning the field of bioinformatics. Following this introduction, an in-detail look is made to subjects like sequence alignment, forced alignment, detection of motifs, sequence logos, Markov chains, or information entropy. Other novel approaches are also described, such as self-sequence alignment, objective digital stains (ODSs) or spectral forecast, and the discrete probability detector (DPD) algorithm. This work also contains thorough step-by-step explanations regarding the meaning of the background models in bioinformatics from several angles. More importantly, it introduces the readers to the art of algorithms, shows how to design computer implementations, and provides extensive worked examples with detailed case studies. The implementations presented here point out how native programming in Javascript can broaden the horizons of possibilities in bioinformatics by considering the might of modern Internet browsers. Graphical illustrations are used for technical details on computational algorithms to aid an in-depth understanding of their inner workings. Moreover, this work brings to the reader's attention more than 100 open-source implementations and 33 Powerpoint presentations.

About the Companion Website

This book is accompanied by a companion website:

www.wiley.com/go/gagniuc/algorithmsinbioinformatics

The website includes the following:
- Algorithms in JavaScript
- Charts for visualization
- PowerPoint presentations
- Sequence alignment (Jupiter bioinformatics application)
- HTML/JavaScript files

1

The Tree of Life (I)

1.1 Introduction

This chapter provides an overview of life and draws near some important questions: When did life on earth begin? What is life? How is it organized? When did multicellular organisms appear and why? How many species exist on Earth? Notions of biology related to the emergence and classification of life are discussed in connection with different strategies on organism formation. Some ultrastructural images (electron microscopy) are presented as examples for reference. The lower and upper physical dimensions of eukaryotic and prokaryotic organisms are explored in detail. The same exploration is made for viruses that interact within the kingdoms of life (Animalia, Plantae, Fungi, Protista, [Archaea and Bacteria] or Monera). Moreover, a discussion closely debates the reference system and the requirements for life; with special considerations for the "spark of metabolism." Next, an introduction is made on some concrete topics, namely: The origins of eukaryotic cells, the endosymbiosis theory, the origins of organelles, the notion of reductive evolution, and the importance of horizontal gene transfer (HGT). Toward the end of the chapter, the main hypotheses regarding the origin of eukaryotic multicellularity are explored using the behavior observed in current species.

1.2 Emergence of Life

The Earth is believed to be ~4.5 billion years old. Geological evidence shows that liquid water, continental crust, and a rudimentary atmosphere existed on Earth just 100 million years later (4.4 billion years ago) [1]. The planetary atmosphere consisted of water vapor, carbon dioxide, methane, and ammonium [2]. It is unknown exactly when or how life began on Earth [3]. It is considered that life began on the early Earth soon after conditions became favorable for a chain of consecutive, yet undetermined chemical reactions [4]. The field of

Algorithms in Bioinformatics: Theory and Implementation, First Edition. Paul A. Gagniuc.
© 2021 John Wiley & Sons, Inc. Published 2021 by John Wiley & Sons, Inc.
Companion website: www.wiley.com/go/gagniuc/algorithmsinbioinformatics

prebiotic chemistry tries to explain how organic compounds formed in the absence of biology and how these simple molecules self-assembled to ignite life on Earth and possibly on other planets. The oldest fossils of single-celled organisms date around 3.5 billion years ago [5, 6]. Nonetheless, only organisms with a dense biological structure would have resisted the intense metamorphism experienced by crustal rocks for more than 3.5 billion years. In turn, a dense biological structure may indicate high organism complexity. Thus, the earliest known microfossils could actually indicate the presence of structurally complex unicellular organisms [7]. It stands to reason that those "first" organisms must have required a long time to develop their complexity. Evidence for life on Earth before 3.8 billion years ago has been proposed in the past [7]. Preserved carbon, potentially of biogenic nature, pushes the origin of life on Earth to 4.1 billion years [8]. This indicates that life may have occurred fairly quickly after the formation of the planet (4.5 − 4.1 = 0.4 or 4.5 − 3.8 = 0.7). That is, 400 million to 700 million years after the formation of the planet. Moreover, the observation has important implications for our beliefs about how fast life ignites on other planets with similar conditions. In the next important event, life brings chemical mod-·ifications to the planetary atmosphere. An oxygen-containing atmosphere and evidence of cyanobacteria and photosynthesis date around 2.4–2.2 billion years ago [9, 10]. Large colonial organisms with coordinated growth in oxygenated environments have been found as far as 2.1 billion years ago [11]. Life made a gradual step toward eukaryotic unicellular organisms ~2 billion years ago. Eukaryotes divide into three main groups around 1.5 billion years ago, namely in the unicellular ancestors of modern plants, fungi, and animals [12, 13]. The appearance of multicellular eukaryotes is an unclear period. During evolution, gain or loss of multicellularity often occurred until a stable multicellular state was reached [14]. Knowledge of the complexity and size of current single-celled eukaryotic organisms calls into question many more complex fossils. It is difficult to investigate whether some macroscopic-sized fossils indicate multicellular or unicellular macroscopic organisms. Some certainty appears in the fossil record with the rise of bilaterians (bilateral symmetry in organisms) over 550–600 million years ago [15]. Nevertheless, even in the case of these fossils, some uncertainty overshadows interpretation. The macroscopic dimensions and the observed bilateral symmetry still cannot indicate with certainty the multicellular nature of these extinct organisms. Again, many of these fossils can be interpreted as multicellular organisms or as unicellular organisms (e.g. giant protists) [16]. More clear evidence suggests that multicellular organisms may have been present around 635 million years ago [17]. Recent molecular clock analyses estimate that animals started to evolve ~650 million years ago [18]. Bilaterian metazoans (animals with bilateral symmetry) first appeared around 600 million years [12]. Moreover, the evolutionary origins of the blood vascular system date around the same period [19]. Later, bilaterians split into the protostomes and deuterostomes [12].

Protostomes give rise to bring about all the arthropods (e.g. insects, spiders, crabs, shrimp, and so on). Deuterostomes eventually give rise to all vertebrates [12]. Perhaps the most important leap made in the evolution of life was the appearance of motility in multicellular organisms. Fossilized trails of bilaterian animals suggest that eukaryotic multicellular organisms have acquired motility around 551–539 million years ago [20]. A "few" million years later, the first true vertebrate with a backbone appears in the fossil record (545–490 million years ago) [21]. Moreover, fossil evidence shows that animals were exploring the land for the first time around 500 million years ago (544–457) [22]. Plants begin colonizing the land around the same time (470 million years ago) [22]. The first four-legged animals (tetrapods) explored the land 385–359 million years ago and gave rise to all amphibians, reptiles, birds, and mammals [22, 23]. The oldest fossilized tree also dates from this period [22]. Important diversifications in eukaryotic species appear after this period, both on land and in water. The first mammal-like forms appear in the fossil record around 225 million years ago [24]. Much closer periods bring many wonders. For instance, the largest eukaryotes in Earth's history have been observed around 100 million years ago, namely the cretaceous dinosaurs (e.g. *Argentinosaurus*; length: 22–35 m; estimated mass: 50 000–100 000 kg) [25]. The last extinction event (Cretaceous–Tertiary extinction), which occurred over 66–65 million years ago, allowed for a relaxed evolution of mammals [26]. Our own story begins of course at the origin, of life. However, more distinguished developments start with the origins of the first primates around 55 million years ago [27]. Note that the timeline of past events is detailed in the literature and here only some general points were reached.

1.2.1 Timeline Disagreements

Microfossils (the imprint left by an organism in stone), stromatolites (layered rocks derived from photosynthetic cyanobacteria remains sedimented over time), sedimentary carbon isotope ratios or molecular fossils derived from cellular and membrane lipids ("biomarkers"), are used for estimations of the origin and diversification of life in the distant past [28]. Data expressed in billions and hundreds of millions of years are particularly subjective and can lead to variations in the literature up to plus or minus half a billion years. These issues are known and must be taken at face value. While timeline estimates may vary, the order of events is particularly objective. Note that timeline disagreements in the paleontology literature rather indicate that evolution has no milestones but trial periods that overlap; some trials more successful and others that we will probably never know about. Nevertheless, the closer the events get to the present, the more reliable the numbers become. Although relative, timeline estimations in paleontology represent a reliable reference system for important past events on our planet.

1.3 Classifications and Mechanisms

Life on Earth was classified by us into three major domains, namely Bacteria, Archaea, and Eukarya (Figure 1.1) [29]. Bacteria (Greek – bakterion, "small stick") and Archaea (Greek – Arkhē, "origin") are prokaryotes (Greek – pro, "before"; karyon, "kernel" or "core"). Prokaryotes are single-cell organisms (unicellular) without a "core," namely without a nucleus, and are considered similar or close in sophistication to the first living organisms on Earth. Eukarya (Greek – eu, "well" or "true") includes all unicellular and multicellular organisms with cells that contain nuclei, and it refers to animals (including us), plants, fungi, and single-celled protists.

All contemporary forms of life store information in DNA molecules. DNA molecules are polymers consisting of four types of organic molecules linked together by phosphate groups, namely: adenine (A), thymine (T), cytosine (C), and guanine (G). Indirectly, all cellular processes are orchestrated by the information contained in the DNA molecule. Cellular processes store and use energy in the form of discrete packets (adenosine triphosphate molecules or ATP). In prokaryotes, DNA shows a double-stranded circular (usually) form

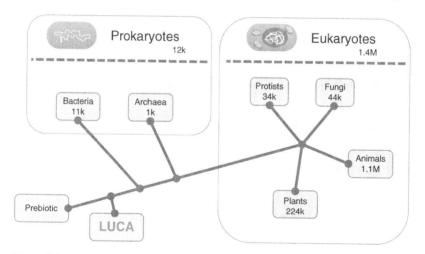

Figure 1.1 The tree of life – basic diagram. The prebiotic period shown on the bottom-left represents the formation of primordial chemical molecules necessary for the ignition of life. Next, the diagram indicates the appearance of LUCA (last universal common ancestor), the first "rudimentary" form of life. The first prokaryotes appear later based on the evolution of LUCA, namely bacteria and archaea. Eukaryotes appear next in the evolutionary chain. Eukaryotes divide the tree of life into four other main subdivisions (eukaryotic kingdoms), namely: protists, fungi, animals, and plants. Note that the approximate number of known species is presented for each subdivision. Source: Refs. [29, 74, 252, 253].

and it is located in the internal environment of the cell (cytoplasm). Cells of eukaryotic organisms contain a double-stranded linear (usually) DNA folded inside a membrane-bound organelle, named the nucleus (from Latin – nucleus, "kernel" or "seed"; pl. nuclei). The nuclear membrane is a controlled barrier that separates the DNA molecules from the cytoplasm (Figure 1.2). Naturally, images based on electron microscopy can best show the classic structure and the inner "frozen" dynamics of eukaryotic cells (Figure 1.2). In double-stranded DNA, a cytosine molecule from one strand and a guanine molecule from the other strand, form three hydrogen bonds while adenine and thymine form two hydrogen bonds. The successive alternation of these simple hydrogen bonds along the double-stranded DNA molecule dictates the energy required to separate the two strands and establishes the local stability of the duplex. In both eukaryotes and prokaryotes, the order of the four types of nucleotides defines the information structure throughout a DNA molecule. These structures include the well-known "genes" (Greek – geneá, "generation"). Genes are regions of different lengths, found along the DNA molecule. Broadly, gene regions are in turn accompanied by regulatory structures, such as gene promoters and enhancers. Genes are involved in transcription, namely in the synthesis of RNA transcripts. Note that RNA molecules are also polymers consisting of four types of organic molecules, namely: adenine (A), uracil (U), cytosine (C), guanine (G). The RNA transcript is a single-stranded nucleotide sequence that is complementary to the DNA strand harboring the gene. In turn, the information on the RNA transcript dictates whether the transcript becomes a functional molecule within the cell or whether it becomes a template for protein synthesis.

1.4 Chromatin Structure

In multicellular organisms, every cell type usually contains the same DNA information; however, it exhibits a different phenotype. What determines this behavior ? Double-stranded DNA molecules of eukaryotic organisms are folded and distributed into chromosomes, which take the form of chromatin (DNA, histone proteins, and non-histone proteins). The basic organization of chromatin consists of filaments made of repetitive units called nucleosomes. Each nucleosome consists of eight histone proteins (i.e. type H2A, H2B, H3, and H4) that wrap ~146 base pairs (bp) of double-stranded DNA (Figure 1.3a). Nucleosomes are connected to each other by 10–80 bp of DNA associated with linker histone H1 that wraps another 20 bp [30]. This basic form of chromatin self-assembles into higher orders of organization. Inside the cell nucleus, these higher orders of chromatin organization include the chromatin fibers, the fractal globules, and reach a final level of organization, namely the chromosomal territories [31, 32].

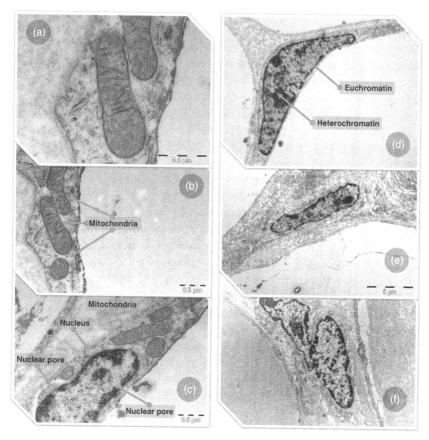

Figure 1.2 Ultrastructural images of adipocyte cells from Bos taurus (Cattles).
Adipocytes especially show how tolerant and adaptable cellular organelles are to various
constant mechanical stresses. (a, b) Shows mitochondria in adipocytes. The right side of
homogeneous light gray content represents the lipid droplet. (c) Shows a few
mitochondria in proximity to the cell nucleus. (d, e) Shows the shape of the cell nucleus
in different mechanical constraints induced by the size of the lipid droplets. Again, the
homogeneous light gray content represents the lipid droplets from the surrounding cells.
(f) Shows two adipocytes with adjacent nuclei. Within each nucleus (c–f), the genetic
material can be observed in different states of activity. Inside each nucleus, the dark gray
(to almost black) areas represent heterochromatin and the normal gray areas represent
euchromatin. In short, euchromatin contains a specific and dynamic set of active genes
that is expressed only in adipocytes, while areas of heterochromatin contain the
remaining unexpressed genes. At the edge of the nuclear membrane, nuclear pores can
be observed. Interruptions with a light gray hue can be seen along the perimeter of the
nuclear membrane. Those are the nuclear pores. Note that each image shows only a
small fraction of the actual size of an adipocyte. Source: Courtesy Dr. Elvira Gagniuc,
Department of Pathology, Faculty of Veterinary Medicine, University of Agronomic
Sciences and Veterinary Medicine, Bucharest.

Chromatin is a highly dynamic three-dimensional structure, which self-arranges differently from one cell type to another, thus, establishing and maintaining the cell identity [31, 33, 34]. But what determines these chromatin self-arrangements? The predispositions for self-arrangements are determined in advance by chromatin remodelers [35]. Among the chromatin remodelers, two main families of enzymes are heavily involved in the global chromatin organization during the cell cycle, namely acetyltransferases and deacetylases. These enzymes make changes to the histone tails of the nucleosomes (histone tails are amino acid chains that extend from the nucleosomal core – Figure 1.3a). Histone tail acetylation on the nucleosomes leads to a relaxed form of chromatin, which subsequently allows transcription factors (TF) to gain access to their target genes (euchromatin). Histone deacetylation leads to a higher-order folding of nucleosomal arrays, which in turn form a dense chromatin structure that is inaccessible to the transcription machinery (heterochromatin) [36]. Thus, patterns of histone acetylation and deacetylation along the chromatin filaments dictate the initial chromatin folding and unfolding inside the cell nucleus and consequently the activity of a specific subset of genes [37]. These acetylation–deacetylation mechanisms are combined (among others) with DNA methylation mechanisms, which lead to a gradual and stable inactivation of certain genes over several cell generations (a topic discussed in other chapters). Nevertheless, the global distribution of chromatin is established immediately after the cell division and exposes a subset of genes specific to the cell type. The DNA regions that contain the genes that are part of the cell type subset are positioned more toward the center of the cell nucleus in the relaxed volume of the chromatin (called euchromatin) [31]. In contrast, genes outside the cell-type subset are distributed in the condensed volumes of chromatin (called heterochromatin) that are usually positioned toward the inner part of the nuclear membrane (Figure 1.2c–f).

Nevertheless, chromatin also undergoes partial changes while in G0 phase (the resting stage of the cell cycle – e.g. many neuronal cells are always in this state). These partial and continuous changes of the chromatin structure are meant to silence or activate certain genes from the main subset of genes. Such facultative heterochromatin areas are present on the outer surface of the heterochromatin landscape (the euchromatin – heterochromatin borders), and their condensation state depends on successive interactions between different gene products of the subset [31]. Note that DNA molecules are present not only in the area of the cell nucleus but also in other organelles (e.g. mitochondria and chloroplasts). Much like in prokaryotes, the circular double-stranded DNA molecules found in these eukaryotic organelles show their own type of organization called a nucleoid (meaning nucleus-like; it is an infrequent DNA–protein assembly) [38, 39].

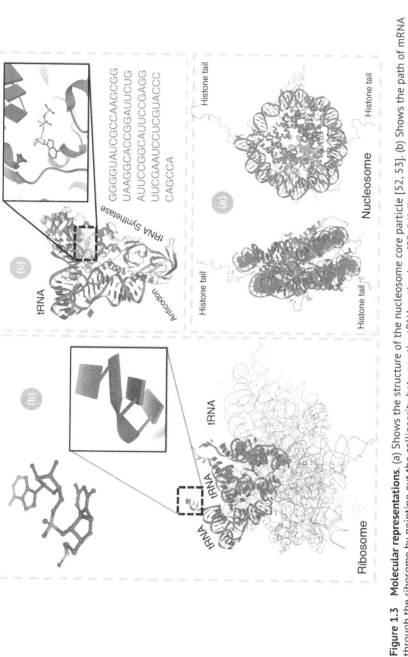

Figure 1.3   Molecular representations. (a) Shows the structure of the nucleosome core particle [52, 53]. (b) Shows the path of mRNA through the ribosome by pointing out the collinearity between the tRNA anticodons [52, 53]. The window highlights the binding region between an amino acid and a tRNA. (c) Shows the *Escherichia coli* glutaminyl transfer RNA synthetase complexed with transfer RNA(Gln) and ATP [53, 55]. The tRNA sequence is presented next to this ribonucleoprotein particle. The last letters in the sequence correspond in reverse order to the region in the tRNA highlighted by the dotted line window (i.e. "ACCG …"). The position of the tRNA anticodon is also highlighted here. Source: Refs. [52, 53, 55].

1.5 Molecular Mechanisms

Eukaryotes and prokaryotes prefer different strategies for synthesizing multiple proteins from a single DNA region (a transcription unit). In prokaryotes, several protein-coding areas (genes) are arranged linearly in a region called an operon, which is usually regulated by a single promoter. An operon is a cluster of coregulated genes with related functions [40]. Thus, operon expression leads to a number of proteins equal to the number of coding areas (genes) in the operon. All the genes in the operon are transcribed into a continuous RNA molecule, which is almost simultaneously translated into proteins. However, functional gene clustering (operon-like) has been reported in eukaryotes (i.e. fungi, plants, and animals) [40]. Eukaryotes, on the other hand, primarily use a single coding area interrupted by noncoding areas (introns). Different combinations between smaller fragments (exons) of the coding area lead to several types of RNAs and consequently to several types of proteins. The protein versions that originate from to a single gene are called "protein isoforms." Note that protein isoforms are not necessarily functionally related [41].

1.5.1 Precursor Messenger RNA

Promoter and enhancer regions regulate the transcription of nearby genes. The initiation of transcription is conditioned by various regulatory proteins that bind to the regulatory regions of DNA (the promoter and enhancer regions). In eukaryotes, the regulatory proteins facilitate changes in the local chromatin structure to allow proper recruitment and binding of RNA polymerase to one of the DNA strands. Thus, the local chromatin structure either promotes or inhibits RNA polymerase and TF binding. Transcription begins once the RNA polymerase enzyme binds to the promoter region of the gene. Regulatory proteins in conjunction with different combinations of TF dictate the frequency of synthesis for pre-messenger RNA (mRNA) molecules (how many copies per unit of time). For instance, different combinations of TFs lead to different three-dimensional macromolecular conformations (the transcription mediator complex) [42]. These temporary macromolecular constructions (made of TFs and other proteins) and their interaction with chromatin, allow the access of RNA polymerase to the DNA sequence to a greater or lesser extent. The difficulty of recruitment imposes a probability distribution for binding. In turn, this binding probability of RNA polymerase sets the frequency of synthesis for pre-mRNAs. As a rule of thumb, a more open chromatin structure is associated with active gene transcription events, while a more compact chromatin structure indicates transcriptional inactivity (no expression).

1.5.2 Precursor Messenger RNA to Messenger RNA

The DNA region of the gene can be subdivided into other types of regions. Especially in eukaryotes, many genes are organized into coding (exons) and noncoding regions (spliceosomal introns). Both exons and introns are transcribed into a continuous pre-mRNA fragment. While the pre-mRNA is being transcribed/synthesized, the intronic regions are removed by spliceosomes (a ribonucleoprotein complex) and the ligation of the exon regions forms the mRNA. The process of intron removal is called splicing. Exon ligation in the same order, in which these regions appear in a gene, is called constitutive splicing. Thus, constitutive splicing allows for a "one gene–one protein" model (or, one pre-mRNA – one mRNA). When exon ligation does not follow the order observed in the gene (i.e. certain exons are skipped), several mRNA variants are produced from a single pre-mRNA variant. If the gene encodes for proteins, then each mRNA variant will generate a different type of protein (protein isoforms). This process is known as alternative splicing, and is responsible for the exaggerated abundance of protein types in the eukaryotic proteome.

1.5.3 Classes of Introns

Introns are regions that interrupt the coding region of functional RNA or protein-coding genes. There are four known classes of introns: Group I introns, Group II Introns, nuclear pre-mRNA introns (Spliceosomal introns), and transfer RNA (tRNA) introns. Group I introns are self-splicing introns and are found in some ribosomal RNA (rRNA) genes [43]. Group II introns are mobile ribozymes that self-splice from precursor RNAs (pre-RNAs) and are found in bacterial genomes and organellar genomes, suggesting that catalytic RNAs, as informational structures, predate the origin of eukaryotes and perhaps the origin of cellular life [44, 45]. Nuclear pre-mRNA introns are found in protein-coding genes and require a ribonucleoprotein complex (spliceosomes) for splicing. The tRNA introns are found in various tRNA genes in all the three kingdoms of life, and require certain enzymes for splicing [46].

1.5.4 Messenger RNA

The stochastic behavior of brownian motion (random walk) and the concentration gradients, scatter the mRNA molecules from the origin of synthesis to other locations with a lower concentration. Thus, the mRNA scattering (as is the case with many other molecules of different sizes and shapes) is done naturally on the least frictional paths. In the case of eukaryotes, the origin of pre-mRNA synthesis and processing is the inner space of the cell nucleus (high mRNA concentration) and the mRNA molecules diffuse through the nuclear pores into the relaxed environment of the cytoplasm (low mRNA concentration). In the case of prokaryotes, the mRNAs diffuse from the origin of synthesis (which is close to the DNA molecule floating directly in the cytoplasm) into the rest of the cytoplasm. The information from some mRNAs allows for protein synthesis, whereas the information from other RNAs provides direct biological functionality. Moreover, some mRNA molecules become functional by themselves due to a self-complementary between different regions of the same molecule. Other RNA molecules become functional after they are processed by different proteins (or vice versa).

1.5.5 mRNA to Proteins

In both eukaryotes and prokaryotes, mRNA molecules, which contain the information structure for protein synthesis, are stochastically encountered by two ribosomal subunits that initiate the translation step. Once bound to an mRNA transcript, the two subunits form the ribosome. The ribosome is a ribonucleoprotein (made of RNA and proteins) organelle that facilitates the formation of chemical bonds between amino acids in the order specified by the information encoded in the mRNA molecule. Life evolved a molecular scheme for translation, known as the "genetic code" [47]. In this scheme, groups of three nucleotides are associated with different amino acids used for polypeptide synthesis. Each set of consecutive and nonoverlapping nucleotide triplets on the mRNA transcript is known as a codon. Polypeptide synthesis begins from a start codon, which initiates the position of the reading frame. Usually, the start codon is represented by the "AUG" triplet (representation with the highest frequency across all life). However, other triplet combinations (non-AUG start codons) can take the role of a start codon (with a lower frequency) [48]. Post initialization, the mRNA transcript slides in between the two ribosomal subunits by one codon at a time following the reading frame set by the start codon [49, 50]. Different versions of tRNAs present in various concentrations in the cytoplasm are each linked to an amino acid. The type of amino acid connected to a tRNA is associated with an anticodon, a special nucleotide triplet region from the tRNA destined for a temporary bind to an mRNA transcript. Thus, tRNAs are the temporary links between the mRNA transcript and the nascent amino acid chain. An assembled ribosome contains three "openings" (A, P, and E sites) for tRNA–mRNA interactions (Figure 1.3.b). The smaller subunit of the ribosome allows for a complementary between three nucleotides (the codon) on the mRNA transcript and three nucleotides (anticodon) of a tRNA molecule (Figure 1.3.b). Once the mRNA–tRNA binding has been facilitated by the smaller subunit, the amino acid transfer from a tRNA to the nascent amino acid chain is facilitated by the larger subunit of the ribosome [51]. The tRNA molecules with appropriate anticodons come into contact through complementary with the mRNA transcript.

The amino acid chain is passed from the previous tRNA to the amino acid of the next incoming tRNA, increasing the growing peptide by one amino acid on each switch. Thus, the amino acid chain remains attached to the most recently bound tRNA and is not released until a termination codon appears in the mRNA transcript (UAA, UAG, UGA) [56]. Since it is an evolved/evolving scheme, small variations of the genetic code exist above different kingdoms of life, and these variations are central to the ultimate goals of bioinformatics (i.e. how life works).

1.5.6 Transfer RNA

On the other side of the translation, an ancient group of enzymes set the rules of the genetic code [57]. The aminoacyl–tRNA synthetase (tRNA-ligase) represents a group of enzymes. The function of these enzymes is to attach an appropriate amino acid to a corresponding tRNA (Figure 1.3.c). Many of these enzymes recognize their tRNA molecules using the anticodon [58]. Consequently, there is one tRNA-ligase for each tRNA–amino acid pair. For instance, in humans there are twenty different types of aminoacyl–tRNA synthetases, one for each amino acid of the genetic code [59]. Some organisms lack the genes needed for all twenty aminoacyl–tRNA synthetases. However, such organisms use all twenty amino acids for protein synthesis. In such cases, a tradeoff is made in the complexity of a tRNA-ligase, such that one enzyme associates more than one pair [60, 61]. Thus, the tRNA matching with an amino acid is based on additional properties exhibited by the tRNA, such as the geometry (shape) of the molecule, specific nucleotide positions along the tRNA chain, and so on [62].

1.5.7 Small RNA

RNAs have multiple and versatile roles across all biological systems and one of the roles is mRNA silencing and post-transcriptional regulation of gene expression. Small RNAs are short (~18–30 nucleotides), noncoding RNA molecules that can regulate gene expression in both the cytoplasm and the nucleus. A few classes of small RNAs have been defined, such as microRNAs (miRNAs), small interfering RNAs (siRNAs), and Piwi-interacting RNAs (piRNAs) [63]. For instance, miRNAs are small noncoding RNA molecules (~21–25 nucleotides in length) that play an important regulatory role in animals and plants by targeting specific mRNAs for degradation or translation repression [64, 65]. It appears that an imperfect complementary between miRNAs and different mRNA targets has the potential to regulate several genes simultaneously. Moreover, miRNAs cross the boundary of a single cell. To add to the complexity of these processes, some miRNAs are secreted into exosomes or microvesicles and may have the ability to move through circulation to other distant cells or tissues [66–68]. Without question, the fine-grained regulation that underlies the complexity of eukaryotes is found in these short RNA molecules.

1.5.8   The Transcriptome

The set of all RNA molecules produced by a given organism is known as the "transcriptome." This includes, of course, the mRNA transcripts but also the RNA molecules mentioned above (i.e. mRNAs, tRNAs, rRNAs, siRNAs, miRNAs, piRNAs, and so on) as well as other uncharacterized noncoding RNA molecules. When expressed, genes produce mRNAs in different quantities, which are then detectable [69]. Currently, two main techniques are representative for capturing gene expression, namely: RNA-Seq and microarrays [70, 71]. RNA-Seq (RNA sequencing) allows for full sequencing of all RNA molecules present in a sample, whereas microarrays target <u>known</u> transcripts of different genes through hybridization (complementary) [72]. Thus, RNA-Seq experiments can estimate the subset of genes expressed in a cell type or in different tissues (several cell types) at any one time by an alignment of the sequenced RNAs to the reference genome (the DNA of the organism) [73]. However, the transcriptome can be seen as an ideal set, because the complete set of possible RNAs cannot be fully detected. Reasoning dictates that each state of a cell shows a specific subset of RNAs from the transcriptome. Of the total number of states that a cell can exhibit, only a few states can be induced and captured by RNA-Seq. Thus, a small subset of RNAs from the transcriptome may remain undetectable. At the tissue level, there are a number of cell types, each with a specific set of active genes. Often, the analysis of the pattern of gene expression is performed at the tissue level, i.e. on several cell types at the same time. From a global perspective, this leads to a union between the sets of genes expressed in each of the cell types that make up the tissue. Furthermore, genes that are expressed in several cell types (such as housekeeping genes) may show the highest amounts of mRNA, while genes that are only expressed in certain cell types can show lower amounts of mRNA.

1.5.9   Gene Networks and Information Processing

The mRNA and/or the protein products encoded by one gene often regulate the expression of other genes. In multicellular eukaryotes, the set of genes that are expressed in a specific cell type forms an "open" gene network. Each gene network is a self-orchestrated feedback loop constantly adapting to different inputs from the environment. The dynamics of a gene network may be deduced in practice from the gene expression levels. The RNA-Seq technique shows the set of genes, and their expression levels (amount of mRNA) at the time of cell/tissue sampling. Repeated sampling at different time intervals can complete a puzzle related to the functional relationship between the genes of the set. Direct or indirect activation of a gene promoter by the product of other genes (mRNA or proteins) is done with a relative delay and largely depends on the frequency by which the gene product is synthesized. The frequency of synthesis impacts the time of accumulation of the gene product (mRNA or proteins) in the cell as well as its stochastic diffusion toward other promoters and macromolecules with which it can interact. Note that the environment can be represented by a number of factors: the current set of molecules inside the cell, the signal molecules synthesized by other cells (other gene networks) or the amount of nutrients, pressure, temperature, and so on.

1.5.10 Eukaryotic vs. Prokaryotic Regulation

In prokaryotes, gene expression is primarily regulated at the level of transcription. Moreover, transcription and translation occur almost simultaneously in the cell cytoplasm. Eukaryotic regulation of gene expression is dynamically orchestrated at several levels, such as epigenetics (chromatin and TF), transcription, post-transcription, translation, and post-translation (further processing of the amino acid polypeptides from a primary structure to more complex, secondary, tertiary structure, and so on). Eukaryotic gene expression occurs with a delay when compared to prokaryotes, as transcription takes place within the nucleus and translation occurs outside the nucleus within the cytoplasm.

1.5.11 What Is Life?

The information in DNA molecules supports a continuous biochemical feedback inside the cell, which self-regulates according to external and internal stimuli (i.e. nutrients, signal molecules, pressure, electromagnetic radiation, and so on). Through energy consumption, this continuous process is maintained in a permanent imbalance above the "inanimate" background. From our reference system, this dynamic self-regulating biochemical feedback is considered life.

1.6 Known Species

How many species really? Unfortunately, life is not as diverse as previously believed [74, 75]. For more than 250 years, our species has cataloged all the other land and water species at our disposal, and continue to do so. Based on this census, the tree of life contains a total of 1.43 million known species, from which almost 1.42 million are eukaryotes and 12k species are prokaryotes (Table 1.1 and Figure 1.1) [74]. Most of the time, real census ruins the "feng shui" of predictions. In total, even the most enthusiastic predictions forecast between 8 million and 11 million species in existence [74, 76, 77].

However, under the heavy umbrella of uncertainty, predictions and census rarely match when it comes to the total number of species on earth. Among the cataloged species for the tree of life, animal species constitute 78% and plant species represent 15%. Out of a total of 1.4 million species, about 1.2 million species live on land and 19k live in the aquatic environment (Table 1.1).

Table 1.1 The total number of known species.

Kingdoms	Land	Water	Total
Eukaryotes			
Animals	953k	171k	1125k
Fungi	43k	1k	44k
Plants	216k	9k	224k
Protists	21k	13k	34k
Prokaryotes			
Bacteria	10k	1k	11k
Archaea	1k	0k	1k
Total	1244k	194k	1439k

The table shows the total number of known species on earth.
The data are divided into two major categories on rows,
namely prokaryotes and eukaryotes; and two major
categories on columns two and three, namely the
environment. Note that 1k means 1000 species.
Source: Refs. [74, 283].

1.7 Approaches for Compartmentalization

Compartmentalization is not a condition for life; however, a closed environment for biochemical containment might be. Most likely, unicellular species that evolved in water had fewer survival issues related to mechanics and gravity. To obtain a specialization, some of these species evolved a cooperation between internal biochemical processes rather than between individual cells. Thus, their volume and shape increased to the point where a single cell began to resemble a multicellular organism. Such unicellular organisms contain multiple nuclei, and for historical reasons are called coenocytes (Greek – coeno, "common"; cyte, "box"). Nevertheless, the virtual "cells" of these unicellular organisms are not defined by a physical barrier. This virtualization may be achieved only through controlled biochemical interactions between concentration gradients of different types of molecules (i.e. from gradients of simple nucleic acids, amino acids, fatty acids and sugars, up to RNA and proteins). Spatially spaced point sources of such chemical and biochemical gradients can form a well-organized virtual structure in these unicellular organisms. Moreover, the biochemical versatility can continue up to the point of inclusion of smaller unicellular species to form a biochemical symbiosis.

1.7.1 Two Main Approaches for Organism Formation

Life shows two main approaches for organism formation. The first approach forced a cooperation between biochemical processes (virtual cells in one physical boundary). The second approach forced a cooperation for gradual specialization among individual cells of the same species (or even between species). Interestingly, the second approach shows a lower entropy than the first. However, cooperation between individual cells did not rule out further specialization between biochemical processes inside individual cells.

1.7.2 Size and Metabolism

Competition and gravity preclude the emergence of unicellular organisms over a certain size. Moreover, gradient-based biochemical signaling and interactions would be inefficient on long distances inside large unicellular organisms. Multicellular organisms seem to have found a balance between the speed of response and the size of the cells. Small cells have a larger surface area relative to their volume. Each unit of volume can exchange gases and nutrients at a higher rate compared to larger cells. Note that the principle is equivalent to smaller salt granules that dissolve faster in water than large ones. Cooperation for development of cell specialization in the direction of a circulatory system formation ensured an optimal exchange with the outside environment and a fast response for the entire organism. In the case of very large unicellular organisms, the response time for any stimulus may be dictated by distances inside the cell and the metabolic rate. For instance, a biochemical interaction between two points in the cytoplasm of such an organism would require time and high amounts of messenger molecules to diffuse in a large volume until the target is stochastically encountered. In other words, "time contracts" for giant unicellular organisms. It is likely that giant single-celled organisms have existed in the distant past. However, competition with smaller unicellular organisms with higher response times may have eliminated them from the evolutionary chain.

1.8 Sizes in Eukaryotes

Physical sizes of different organisms provide an intuitive assessment over their complexity (Table 1.2.). The comparison between the extremes of eukaryotes can be particularly important for a good view of nonlinearity that exists between entropy and size, of which, we will discuss about later. This subchapter briefly discusses the physical sizes of different unicellular and multicellular eukaryotic organisms.

Table 1.2 Extreme sizes of unicellular organisms.

Unicellular organisms	Eukaryotes (μm)	Prokaryotes (μm)
Min	0.8	0.15
Max	300 000	1400

The table shows the minimum and maximum physical dimensions of unicellular organisms in both eukaryotes and prokaryotes. The values represent averages of the measurements published in the scientific literature and are presented in micrometers.

1.8.1 Sizes in Unicellular Eukaryotes

Eukaryotic unicellular organisms provide important clues to the schism that led to multicellularity. Also, single-celled eukaryotic organisms are particularly important biological models that can lead to a deeper understanding of the fundamental mechanisms behind the evolutionary process.

Marine life shows both the maximum and minimum sizes for unicellular organisms. For instance, a member of the green algae, *Caulerpa taxifolia*, is a unicellular organism of 30 centimeters in length, or more [78]. The *Syringammina fragilissima* is another example of a unicellular organism, which reaches ~20–25 cm in diameter or *Ventricaria ventricosa*, which is a cell of 2–4 cm in diameter [79, 80]. On the other hand, the smallest unicellular eukaryote appears to be *Ostreococcus tauri*, a marine green alga with a diameter of about 0.8 μm [81, 82].

1.8.2 Sizes in Multicellular Eukaryotes

Through cooperation, eukaryotic multicellular organisms have been able to evolve large dimensions. In water, buoyancy counterbalances gravity and it allowed for evolution of the largest organisms on the planet. For instance, *Balaenoptera musculus* (the blue whale) is a marine mammal of 27–30 m and around 170–200 tones [83]. It may be the largest contemporary organism on the planet. On land, *Loxodonta africana* (the African savanna elephant) is the largest living land animal [84]. Among birds, *Struthio camelus* (the common ostrich) can reach 2.8 m in height and weigh over 150 kg [85].

1.9 Sizes in Prokaryotes

On the other hand, prokaryotes from *Mycoplasma* species show some of the smallest possible dimensions for life (~100 species). For instance, bacteria

Mycoplasma gallicepticum and *Mycoplasma genitalium* are likely two of the smallest self-replicating forms of life, with a diameter of ~0.0002 mm (0.2 μm or 200 nm) [86, 87]. This small size is 2 up to four times smaller than the wavelength of a photon of light from the visible spectrum (700–400 nm). However, the largest species of bacterium found among prokaryotes are *Thiomargarita namibiensis* and *Epulopiscium fishelsoni* (between 0.5 and 0.7 mm), which are comparable in size to some unicellular eukaryotes [88, 89].

1.10 Virus Sizes

Small viruses are predominant and may be round, or rod shaped, or a combination of the two in the case of prokaryotes. A capsid is the protein shell of a virus, which can remind us of the idea of Platonic solids. For some small viruses, usually the self-assembly process is dictated by their nucleic acids sequence (motif seeds) [90]. This protein shell seals their genetic material from the environment. However, many capsid proteins can self-assemble with no additional help. Viral proteins have structural properties that allow regular and repetitive interactions among them. Identical protein subunits are distributed with helical symmetry for rod-shaped viruses and polyhedron symmetry for round viruses. Because they have small genomes, viral genes must repeat protein subunits. Each subunit has identical bonding contacts with the neighbors. Repeated interaction with chemically complementary surfaces at the subunit interfaces, naturally leads to a symmetric arrangement (3D patterns). The bonding contacts are usually noncovalent. This ensures error-free self-assembly and reversibility. Thus, if the gene responsible for the viral protein is inserted into another cell for expression, that cell will produce viral proteins that will self-assemble into shells – fake viruses with no genome inside. Depending on the species, the 3D shape and the repetitive interactions of viral proteins allow for different types of bonding patterns, which in turn lead to different configurations and capsid sizes (Table 1.3.).

Thus, some viruses can be comparable in size with certain life forms (Table 1.4.). For a degree of comparison, *M. gallicepticum* is 3 up to 10 times smaller than the

Table 1.3 Extreme sizes in viruses.

Viruses	Eukaryotic viruses (μm)	Prokaryotic viruses (μm)
Min	0.017	0.03
Max	1.5	0.2

The table shows the minimum and maximum physical dimensions of viruses found in both eukaryotes and prokaryotes. The values represent averages of the measurements published in the scientific literature and are presented in micrometers.

Table 1.4 Single-celled organisms vs. viruses.

	Eukaryotes		Prokaryotes		Eukaryotic viruses		Prokaryotic viruses	
	Species	Val (µm)	Species	Val (µm)	Species	Val (µm)	Species	Val (µm)
Min	*Prasinophyte algae*	0.8	*Mycoplasma genitalium*	0.15	*Porcine circovirus*	0.017	Phages	0.03
Max	*Caulerpa taxifolia*	300 000	*Thiomargarita namibiensis*	1400	*Pithovirus sibericum*	1.5	Phages	0.2

The table shows a comparison between extreme microscopic sizes of viruses and unicellular organisms, that covers both eukaryotes and prokaryotes.

diameter of the largest giant viruses. Giant viruses that infect single-celled eukaryotes like amoebas (i.e. *Acanthamoeba castellanii*), such as *Pithovirus sibericum*, *Pandoravirus salinus*, or *Pandoravirus dulcis*, are about 1–1.5 μm (1000–1500 nm) in length [91, 92]. Other more well-known giant viruses are *Megavirus chilensis* (400 nm) or *Acanthamoeba polyphaga Mimivirus* (390 nm), each with considerable dimensions over the size of certain prokaryotes [91]. In terms of physical size and genome complexity, giant viruses closed a significant gap between the realms of viruses and the prokaryotic/eukaryotic unicellular organisms [91].

On the other scale, *Porcine circovirus* is the smallest virus (17 nm) and is found in multicellular eukaryotes [93, 94]. Almost all isolated viruses from prokaryotes show ranges between 30 and 60 nm. Giant prokaryotic viruses with capsids diameters ranging from 200 to more than 700 nm have been reported [95]. Nevertheless, these comparisons between virus sizes in prokaryotes and eukaryotes can be misleading as more specialized life forms can lead to more extreme variations in size, complexity, and methods of infection.

1.10.1 Viruses vs. the Spark of Metabolism

How can *P. sibericum* be so big yet lifeless? There are several reasons for which viruses are not considered alive nor do they become alive from our perspective. More robust viral species of considerable size have a reasonable probability to incorporate parts of biochemical mechanisms from the infected cells (inside their capsid). Thus, although giants viruses may incorporate functional metabolic pathways of a cell, those functional parts will have nothing to consume since the capsid does not allow the proper exchange of molecules between the interior of the capsid and the outside environment. Those metabolic pathways that can consume component parts inside the capsid may inactivate the virus. Even assuming that there can be a possibility for a primitive metabolism, capsid proteins hinder replication of a possible "new life form." This is the likely reason why a virus of considerable size lacks the spark of metabolism. But are viruses alive? The virus environment is the cell. Without this environment, viruses become inactive until different stochastic processes lead to reactivation. For cells, the environment is represented by molecules that can be metabolized. Without these substances, cells either decay in simpler macromolecules or enter a hibernation state like viruses do. Therefore, the answer is relative and dependent on our reference system.

1.11 The Diffusion Coefficient

But why a discussion about the size of organisms? Mass diffusivity (diffusion coefficient) is a physical constant that impacts the way an organism can evolve.

The cell volume must be balanced with the cell surface; otherwise, the exchange with the external environment becomes inefficient. This exchange consists of metabolites that must exit the cell per unit time or nutrients that must enter the cell per unit time. Multicellularity allows an organism to exceed the size limits normally imposed by diffusion. On the other hand, unicellular organisms with increased size have a decreased surface-to-volume ratio and may have difficulty in absorbing and transporting sufficient nutrients throughout the cell. As a counterbalance, unicellular eukaryotic organisms have among the most varied shapes and sizes observed in nature. Both unicellular and multicellular organisms can achieve a high surface-to-volume ratio by favoring DNA mutations that lead to a convoluted surface. For instance, to increase their surface area, choanocyte organisms can take many forms, such as *C. taxifolia,* which resembles a kind of "pine leaf" or *S. fragilissima,* which has a convoluted surface.

1.12 The Origins of Eukaryotic Cells

All eukaryotic cells contain membrane-bound organelles (e.g. the nucleus, mitochondria, chloroplasts, and so on). The complexity of the eukaryotic cell is given by the presence and the interaction of organelles. The origin of organelles has always been a mystery difficult to explain. However, the endosymbiotic theory is the leading evolutionary theory for the origin of eukaryotic cells. The idea of endosymbiosis was first proposed by Konstantin Mereschkowski in 1905 [96, 97]. According to the theory of endosymbiosis, the eukaryotic cell is like a Matryoshka (Russian doll). A symbiotic relationship where one organism lives inside the other is known as endosymbiosis. The term "primary endosymbiosis" refers to the engulfment and retention of a prokaryote organism by another prokaryote or eukaryote organism. The term "secondary endosymbiosis" refers to one eukaryote organism having engulfed and retained another eukaryote organism with an organelle already obtained by primary endosymbiosis. Note that today the endosymbiotic behavior is most beautifully observed in protists (e.g. *Paramecium bursaria*).

1.12.1 Endosymbiosis Theory

The last universal common ancestor (LUCA) was likely a population of unicellular organisms that led to the emergence of two domains in prokaryotes: Bacteria and Archaea. It seems that LUCA were complex unicellular organisms and not the immediate descendants of primeval cells. Rumor "has it" that prokaryotes are the descendants of LUCA by reductive evolution [98]. Nevertheless, evidence shows that about 2 billion years ago, eukaryotic cells may have evolved from a merger between the two prokaryotic domains. Endosymbiosis theory suggests a scenario in which an archaeal cell engulfed a bacterial cell. This kind of merger was repeated independently many times and eventually evolved to form all the membrane-bound organelles, including the mitochondria and chloroplasts. The last eukaryotic common ancestor (LECA) was likely a population of unicellular organisms that eventually (i.e. within 300 million years of LECA) led to a diversification of eukaryotes in supergroups (around 1.5–2 billion years ago) [99, 100]. Today, existing prokaryotic groups reveal the intermediate steps in the eukaryotic-cell evolution [101]. Moreover, complex archaea that bridge the gap between prokaryotes and eukaryotes have been found [102].

1.12.2   DNA and Organelles

Prokaryotic organisms of the distant past are perhaps the ancestors of almost all membrane-bound organelles that are found today inside the eukaryotic cells. Among the membrane-bound organelles, some contain their own genome and others lost their genome throughout evolution [103]. The adaptation to the intra-cellular environment led to the loss of many of the original genes accumulated for environmental survival. Other important genes from the organelles have been transferred to the nucleus over the evolutionary timeline. Organellar DNA transfer to the nucleus is a known process by which, during evolution, some critical genes of the organelles are moved for preservation and synchronization of cell division [104, 105]. But why preservation ? The DNA mutation rate is lower in the nucleus. In some important organelles, high concentrations of reactive oxygen species (ROS) can lead to oxidative stress and DNA damage [106, 107]. Thus, the relocation of a gene from the organelle to the nucleus enables a more secure conservation over time. Moreover, the transfer process also leads to an obligate codependency. The relocated genes control the division of the organelles (synchronization) and encode products that interact with organelle-encoded proteins. In turn, the genes still present in the organellar genome encode proteins that interact with nuclear proteins [108]. Ultimately, the organelle interacts with its own evolved genes physically present in the nucleus of the cell. Nevertheless, some of these DNA containing organelles still are organisms in their own right; genetically equipped for the environment imposed by the cytoplasm of the host cell. Intracellular signaling pathways that coordinate gene expression between organellar and nuclear genomes are highly complex; toward almost complicated [109]. Moreover, additional signaling pathways exist between different organellar genomes [109]. These complications may be one of the reasons for the reductive evolution of the organellar genomes. However, many organelles retained a large part from their ancestral genome. Thus, different equilibrium states between organellar and nuclear genomes must exist. Moreover, contrary to certain sedimented expectations, prokaryotic organisms also may contain organelles with special arrangements; for instance organelles such as magnetosomes, chlorosomes, pirellulosomes, anammoxosomes, carboxysomes, and so on [110–112].

1.12.3 Membrane-bound Organelles with DNA

Mitochondria arose approximately 2.3–1.8 billion years ago from a unicellular organism related to modern α-proteobacteria. Bacterial species *Rickettsia prowazekii* of the genus *Rickettsia* is probably the closest phylogenetic relative to the mitochondria [113, 114]. Also, ATP production in *Rickettsia* is the same as that in mitochondria. On the other hand, plastids (i.e. chloroplasts and chloroplast-like organelles) arose 1.6–1.5 billion years ago from the ancestors of cyanobacteria [115]. That is, a mitochondriate eukaryote became host to a cyanobacterium-like prokaryote. Organelles with their own genomes, such as plastids and mitochondria, are found in most eukaryotic cells [116]. A multicellular eukaryote contains hundreds to thousands of these organelles in each cell. The number of organelles is specific to each cell type and may vary depending on the state/metabolic needs of the cell.

1.12.4 Membrane-bound Organelles Without DNA

How much of the genome can be transferred to the nucleus or can be permanently lost in evolution? The complete loss of DNA in an organelle is a possibility. Among the organelles that have lost their genome in evolution is the hydrogenosome. Hydrogenosomes are cell organelles that have a double membrane and synthesize ATP via hydrogen-producing fermentations [117]. Hydrogenosomes were once mitochondria and are a classic example of complete mitochondrial genome loss [118]. Considering the hydrogenosome example, it is reasonable to assume that all eukaryotes may contain an organelle of mitochondrial ancestry [119]. However, DNA in the hydrogenosomes of some anaerobic ciliates has been detected [120]. Ciliates are protists with hair-like organelles (i.e. cilia) used for propulsion and adherence inside liquid media. Cases of genome-containing hydrogenosomes show that these organelles are somewhere toward the end of their reductive evolution period. It can also mean that the loss of the genome is not just a one-way street, which would have important implications for elucidating the occurrence of life on Earth. A basic question arises when considering the above: Are the genome-less hydrogenosomes organisms in their own way? This is an interesting question because it shows the versatility of life and the ideas discussed in the "*Philosophical transactions*" chapter or in the "*Viruses vs. the spark of metabolism*" subchapter.

1.12.5 Control and Division of Organelles

Both genome and genome-less organelles divide. But how ? Regular bacterial fission (division) uses a dynamin-related protein (DRP) to constrict the membrane at its inner face [121]. However, DRPs are also essential for mitochondrial and peroxisomal fission [122]. Fission is required to provide a population of organelles for daughter cells during mitosis. In contrast to bacterial fission, mitochondria use dynamin and DPRs to constrict the outer membrane (a ring) from the cytosolic face [121]. For instance, the unicellular eukaryote *Trichomonas vaginalis* is a parasite that uses hydrogenosomes instead of regular mitochondria. The role of DRPs in the division of the hydrogenosome is similar to that described for peroxisomes and mitochondria [123]. Moreover, plastids use similar mechanisms and a plant-specific DPR [124, 125]. In eukaryotes, dynamin and DPRs have their genes stored in the nuclear genome. Thus, control over the division of membrane-bound organelles is held by the nuclear genome. Genes encoding DPRs were once present in the symbiotic α-proteobacterium ancestors of mitochondria or in the symbiotic cyanobacterium ancestors of the plastids. In other words, the alternative self-assembly of a complementary dynamin constriction mechanism on the outer membrane, allowed a transition of the organellar fission genes to the nuclear genome for synchronization of cell division. Moreover, this complementary mechanism allowed a reductive evolution up to the point of complete genome elimination in some organelles, such as in the case of hydrogenosomes. But how do organelle genes physically get into the nucleus to recombine with chromosomal DNA? Insertion of DNA from organelle genomes into the nucleus is DNA-mediated (RNA-mediated insertions may also occur) through a process called HGT [126]. As a side note, peroxisomes are single-membrane organelles that catalyze the breakdown of very-long-chain fatty acids through beta-oxidation [127]. Peroxisomes are a hybrid of mitochondrial and ER-derived (ER – endoplasmic reticulum) pre-peroxisomes [128].

1.12.6 The Horizontal Gene Transfer

The symbiosis between organisms is not possible without the HGT. The HGT is the weak "glue" that unites all species and it has important evolutionary implications [129]. In the future, these implications may very well undo the classification discussed above for the tree of life and how we understand the origin of life on Earth. The HGT refers to the transmission of DNA between different genomes, whereas the vertical gene transfer (VGT) is made between generations by sexual or asexual reproduction. The way in which the classification for the tree of life works is largely based on the VGT concept; thus, one can imagine the issue. HGT was first observed as a phenomenon in *Streptococcus pneumoniae* species by Frederick Griffith in 1928 [130]. The main observation made by

Frederick Griffith was that virulence (pathogenicity) in this species of bacterium is transmitted by contact or proximity. This was an important revelation for the later field of genetics. Since then, increasing evidence shows that DNA fragments of different sizes may be exchanged between the kingdoms of life, to a greater or lesser extent [129]. Not long ago, the transfer of genetic information from the members of the *Agrobacterium* genus to eukaryote cells was seen as an extraordinary and rare process [131, 132]. Today, evidence indicates clearly that transfer of genetic information between species and inside different cell compartments is a common process, which takes place over the evolutionary time. For instance, bacteria have acquired genetic material from eukaryotic hosts and vice versa [133]. Viruses contain genes derived from their eukaryotic hosts and vice versa [134]. In plants, for instance, the HGT between genomes takes place through intracellular transfer of DNA among the nuclear, mitochondrial, and plastid genomes. The transfer of mitochondrial genes to the nucleus is known to be an ongoing evolutionary process. However, evidence also shows a HGT of mitochondrial DNA to the plastid genome [135]. Moreover, expression of a transferred nuclear gene in a mitochondrial genome was also observed [136]. For example, the orf164 gene in the mitochondrial genome of Arabidopsis is derived from the nuclear ARF17 gene that codes for an auxin-responsive protein [136]. Thus, the transfer of DNA segments from any location to any other location seems to be a rule across all life. However, HGT is most frequent between closely related species with similar genome features and less frequent otherwise [137]. In other words, HGT is a process that occurs at different frequencies between prokaryotes, between eukaryotes, between prokaryotes and eukaryotes and vice versa [138]. Perhaps, the importance of HGT goes as far as the emergence of new species (speciation) [139, 140].

1.12.7 On the Mechanisms of Horizontal Gene Transfer

Understanding of the mechanisms and vectors underpinning HGT across the kingdoms of life is still limited. Mobile genetic elements (MGEs) represent the main known vectors for HGT [137]. Well-known HGT events often include, but are not limited to, transposable elements (TE), plasmids or bacteriophage elements [141]. The behavior of a MGE has a certain degree of stochasticity and may incorporate a complete gene(s) or may include only sections of a gene, or with a high probability none of the two. Sections of genes transferred by MGEs decay in time and are recognized in bioinformatic analyzes as pseudogenes (nonfunctional genes) [126]. Among the MGEs, TE can best show the level of complexity that a DNA fragment can exhibit. The TE were first observed in *Zea mays* (corn) by Barbara McClintock in 1950 [142]. The main observation made by Barbara McClintock was that the genetic material can jump from one place to another within a genome. The insertion of TEs into the coding pigment-genes was responsible for unstable phenotypes on the kernels of a maize ear (kernels of different colors). Note that each kernel is an embryo produced from an individual fertilization and one ear of corn contains around 800 kernels positioned in 16 rows.

1.13 Origins of Eukaryotic Multicellularity

Above the evolutionary time, cells of multicellular organisms evolved a series of states (cell types). The mechanisms that lead to the formation of such states are unknown. Biology includes several competing hypotheses on the origin of eukaryotic multicellularity; all of them based on observations made on the behavior of current species. These hypotheses suggest multiple pathways that can lead to multicellular organisms; some pathways more successful than others. Moreover, these competing hypotheses may all be valid. Note that only a few general notions are mentioned here.

1.13.1 Colonies Inside an Early Unicellular Common Ancestor

One of the hypotheses for multicellularity suggests a repeated division of the nucleus within the same unicellular organism and a subsequent formation of membranes in between the nuclei. A reminiscent coenocytic behavior can be seen in multicellular eukaryotic organisms, for instance, in the eggs (0.51 ± 0.003 mm) laid by the well-known *Drosophila melanogaster* (vinegar fly). The initial stages of the vinegar fly eggs contain multiple nuclei in a common cytoplasmic space (the entire volume of the egg) [143]. Only a few stages of development later, the cell membranes around the floating nuclei start to appear almost simultaneously to constitute the initial cells of the larva [143].

1.13.2 Colonies of Early Unicellular Common Ancestors

A second hypothesis for the origin of multicellularity proposes that unicellular organisms may aggregate to form unitary colonies that can achieve multicellularity and cell specialization over time. According to this theory, multicellularity emerged from cooperation between unicellular organisms. Examples of cooperation among organisms have been observed in nature at different scales and in various forms. One of the simplest integrated multicellular organisms is *Tetrabaena socialis* in which four identical cells constitute the individual [144]. The nuclear genome of *T. socialis* dictates the number of cells in the colony [145]. Another example is the choanoflagellate (Greek and Latin – khoánē, "funnel"; flagellate, "flagellum") *Salpingoeca rosetta,* which can exist as a unicellular organism or it can switch to form multicellular spherical colonies called rosettes (form bridges between cells by incomplete cytokinesis), showing a primitive level of cell differentiation and specialization. Formation of multicellular colonies is induced by different signal molecules. The source of such signal molecules can originate from individuals of the same species (i.e. slime molds) or from individuals of different species (i.e. bacterium species) [146]. In the case of *S. rosetta*, the signal molecules for colony formation originates from the food source, namely the *Algoriphagus machipongonensis* bacterium (phylum *Bacteroidetes*) [147, 148]. Choanoflagellates, sponges and algae of the genus Volvox are more complex examples of first evolutionary stages that indicate the border between colonial organisms and multicellular organisms. Choanoflagellates are the closest relative of metazoans (all animals composed of cells differentiated into tissues and organs) [149, 150]. Some genes required for multicellularity in animals, such as genes for adhesion, genes for signaling, and genes for extracellular matrix formation, are also found in choanoflagellates [151]. This suggests that these genes may have evolved in a common ancestor before the transition to multicellularity in animals [152]. Sponges are one of the oldest primitive multicellular organisms in the fossil record. Choanoflagellates are small single-celled protists, partially similar in shape and function with some of the sponges cells (choanocytes) [153]. Many associations have been made in the past between choanocytes and choanoflagellates. However, the transcriptome of sponge choanocytes is the least similar to the transcriptomes of choanoflagellates and is significantly enriched in genes unique to either animals or sponges alone [154]. Slime molds are also interesting examples, which can indicate how some multicellular organisms formed. Slime molds are unrelated eukaryotic organisms that can live as single cells. In certain conditions (i.e. starvation), single cells of the same species can aggregate to form multicellular reproductive structures [155]. For instance, the multicellular aggregate (a slug-like mass of a few thousand cells called a *grex*) of amoebae *Dictyostelium discoideum* can show cellular adhesion, cellular specialization, tissue organization, and coordination that allows for mechanical movement [156, 157]. Although the behavior of *D. discoideum* is not necessarily a close example of the process that led to multicellular organisms, it can certainly serve as a clue for detailed research on the emergence of multicellularity.

1.13.3 Colonies of Inseparable Early Unicellular Common Ancestors

A third hypothesis for the origin of multicellularity suggests an early unicellular organism that underwent repeated divisions with incomplete separations between generations, which further led to a forced cooperation and specialization for a primitive tissue formation. Plant embryos and animal embryos adhere to this behavior. More to the point, the eukaryotic microorganism *Saccharomyces cerevisiae* (the budding yeast) is an ideal example for this hypothesis [158]. Yeast colonies growing on solid media show specific structural patterns in their three-dimensional multicellular organization. These structural patterns are specific to each yeast strain [159]. Moreover, variations in the multicellular organization appear to be dependent on the environmental parameters, such as the position of surrounding cells, nutrient gradients, temperature, and so on [160].

1.13.4   Chimerism and Mosaicism

The cooperation of eukaryotic cells is best observed in the case of two phenomena, namely chimerism and mosaicism. Mosaicism is represented by two or more cell populations in different tissues originating from one fertilized egg. Namely, one cell population with the original genotype (usually representing the majority) and other cell populations with slight variations of the original genotype. One of the mechanisms that lead to mosaicism is represented by transposomes [161]. With embryonic development, the genotype of an organism can undergo various types of mutations, including transposome-induced mutations above different cell lines. These mutations can occur late in embriogenesis, leading to marginal effects at the organism level, or they can occur early in the embryonic development of an organism with more pronounced/noticeable effects [162]. Transposome-induced mutations represent a normal and nonrandom variability in multicellular organisms, leading to different phenotypic characteristics [161]. The classic example, however, is represented by the experiment performed by Barbara McClintock on corn kernels, where the transposomes inactivate the gene for the pigment protein and the phenotype is easily recognizable (please see the "horizontal gene transfer" subchapter from above). Mosaicism can also be represented by other types of mutations. For example, in humans, Down syndrome is characterized by an additional copy of chromosome 21, which is frequently attached to chromosome 14 (Trisomy 21). The extra copy of chromosome 21 slightly changes the chromosomal territories in the cell nucleus and the way heterochromatin and euchromatin are distributed. This leads to unusual variations in the expression of certain genes, especially those present on the extra chromosome 21 and those present on the neighboring chromosomal territories. Trisomy 21 occurs at the beginning of embryogenesis. However, such a mutation may appear late in embryogenesis, which results in mosaic tissues, part with normal cells and part with cells with an extra copy of chromosome 21 [163, 164]. Such a mosaic can be clinically unnoticeable unless genetic analysis is made on different tissues of the organism. Moreover, it is believed that a combination of germinal and somatic Trisomy 21 mosaicism may be reasonably common in the general population [163]. In development, cells with different genotypes compete in the tissue population. Such a competition can lead to the possibility (especially for mosaicism that appeared late in embryogenesis) in which cells with the original genotype completely marginalize the function of mosaic genotypes or vice versa, depending on which is more fit for a specific function. On the other hand, chimerism is represented by fusion of more than one fertilized zygote, namely cells of different organisms that are orchestrated by common molecular signals to form a single body [165]. Chimerism can be observed in all multicellular species to a greater or lesser extent and may be accompanied by genetic mosaicism in any of the genotypes that form the composite organism.

1.14 Conclusions

Biological literature is probably the most sophisticated among all sciences and can be particularly overwhelming. An introduction was made to some important concepts that can provide an overview on living organisms, such as the emergence of life, classification, number of species, the origins of eukaryotic cells, the endosymbiosis theory, organelles, reductive evolution, the importance of HGT, and the main hypotheses regarding the origin of eukaryotic multicellularity. Among the biological concepts described here, some have wider implications. Examples of genome-less organelles, such as hydrogenosomes, or processes such as the HGT, question life as we understand it. Endosymbionts best explain the significance of the environment and also explain the distribution of life in a blurry, nonunitary context. In other words, endosymbiosis widens the threshold of life and shows how difficult it is to place a border between how much life resides inside or outside the cell. Moreover, the HGT appears to connect all the species on earth to a greater or lesser extent. Much evidence shows that some of these ancient processes (e.g. catalytic RNAs) are likely adding or subtracting innovative mechanisms for continuous adaptations among different species (if not all).

2

Tree of Life: Genomes (II)

2.1 Introduction

An insight into the context of biological information is of utmost importance for different approaches in bioinformatics. The first part of the chapter discusses the units of measurement and explains the meaning of some notations used here. A few interesting unit conversions, with accompanying algorithms, are shown in addition to the subject. Next, eukaryotic and prokaryotic organisms with the largest/smallest genomes are presented in detail. Moreover, different computations performed for this chapter show the average genome size above the major kingdoms of life, including the average genome size of different organelles, plasmids, and viruses. Toward the end of the chapter, a comparative analysis is made between the average number of genes and the average number of proteins above the main kingdoms of life. This informative analysis highlights the frequency of a process called alternative splicing, which allows certain eukaryotic genes to encode for several types of proteins.

2.2 Rules of Engagement

Genome size refers to the amount of DNA contained in a haploid genome (a single set of chromosomes). The genome size is expressed in terms of base pairs (bp) and the related transformations: kilo base pairs (1 kbp = 1000 bp), or mega base pairs (1 Mbp = 1 000 000 bp), or giga base pairs (1000 Mbp = 1 Gbp), and so on. By excellence, base pairs are discrete units. Nonetheless, these units of measurement are also used to express averages. For single-stranded DNA (ssDNA)/RNA sequences, the unit of measurement is the nucleotide (nt) and is written as: 1 000 000 nt, 1000 Knt, 1 Mnt, 0.001 Gnt, and so on. However, most often than not, base pairs are written as simple bases when the context is understood (e.g. 1 000 000 b, 1000 kb, 1 Mb, 0.001 Gb, and so on).

Algorithms in Bioinformatics: Theory and Implementation, First Edition. Paul A. Gagniuc.
© 2021 John Wiley & Sons, Inc. Published 2021 by John Wiley & Sons, Inc.
Companion website: www.wiley.com/go/gagniuc/algorithmsinbioinformatics

For instance, the notations "b," "kb," "Mb," "Gb" are used when referring to DNA/RNA sequences in text format. FASTA files contain nucleic acid sequences in the 5′–3′ direction. Technically, all nucleic acids represented as FASTA are single-stranded; however, through complementarity, the reference can be considered as double-stranded. In this chapter, the CG% content is mentioned as an intuitive parameter for the overall composition of the genomes of different species. Note that the (C+G)% or GC% content represents the percentage of guanine and cytosine along a DNA or RNA sequence (e.g. a DNA/RNA fragment, a gene, an entire genome).

2.3 Genome Sizes in the Tree of Life

There is no direct correlation between the genome size of a species and the complexity of its phenotype. In any case, the intellectual curiosity regarding the size of genomes still remains. Determination of genome size based on DNA sequencing data is one of the most accurate methods to date. To observe the lack of correlation between genome size and phenotype, upper-bound extremes can be considered here. As expected in an intuitive manner, eukaryotes show the largest genomes. In animals, the amphibian *Ambystoma mexicanum* (the Mexican Axolotl) shows the largest (sequenced) genome observed in nature to date. *A. mexicanum* shows a genome size of 32 396 Mbp (32 Gb) and a physical length that can reach up to 30 cm [166]. In plants, the record is held by *Pinus lambertiana* (27 603 Mbp) and *Sequoia sempervirens* (26 537 Mbp). *P. lambertiana* is the tallest and most massive pine tree [167, 168]. *S. sempervirens* species includes the tallest living trees on Earth (115.5 m in height or 379 ft) [169]. Among the prokaryotes, *Minicystis rosea* and *Sorangium cellulosum So0157-2* show the largest genomes. The bacterial genome of *M. rosea* contains 16 Mbp of DNA (GC%: 69.1) and shows the maximum genome size found in prokaryotes [170]. Secondary to this species is the bacterial genome of *S. cellulosum So0157-2*, with 14.78 Mbp of DNA (GC%: 72.1) [171]. As discussed in the previous chapter, endosymbiosis challenges the notion of the smallest genome necessary for life. The smallest prokaryotic genomes were found in different obligate symbionts. One such case is *Nasuia deltocephalinicola* with a genome of 112 kbp (0.11 Mbp) [172, 173]. The eukaryotes with the smallest nuclear genome necessary for life are found in the kingdom of fungi. The spore-forming unicellular parasite *Encephalitozoon intestinalis* shows a genome size of ~2.3 Mbp and a total of 1.8k protein-coding genes [174]. Nonetheless, the smallest free-living eukaryote is *Ostreococcus tauri*, a marine green alga with a diameter of about 0.8 μm and a genome size of 12.6 Mbp (8.2k protein-coding genes) [175].

2.3.1 Alternative Methods

The data mentioned above were determined by DNA sequencing approaches made so far. DNA sequencing is an ongoing process for several decades and the species chosen for sequencing are usually either of economic or research importance (or even of historical significance). There are many species that have not yet been sequenced, either due to their minor importance to humans or due to large genomes that cannot be easily managed. Usually, the size of the genetic material can be estimated by methods other than sequencing. One of these methods is flow cytometry, which estimates the weight of the genetic material [176]. This weight, expressed in picograms (pg), can then be converted to base pairs. One picogram is equal to 978 megabase pairs (1 pg = 978 Mbp) [177]. For instance, *Paris japonica* (flower) shows a genome weight of 152.23 pg, which suggests a genome size of 148 880 Mbp (152.23 pg × 978 Mbp = 149 Gbp) [178].

2.3.2 The Weaving of Scales

To get a sense of genome size closer to our reference system, some transformations can express the mega base pairs as physical lengths. The linear length of a double-stranded DNA (dsDNA) molecule can be calculated by multiplying the average distance between bases (~3.4 angstrom = 0.34 nm [179, 180]; 1 angstrom = 0.1 nm) by the total number of base pairs in a genome. Here, genomes are expressed in mega base pairs. Since 1Mbp is equal to one million base pairs, the size of a genome can be multiplied by one million and then multiplied further by the average distance between bases (0.34 nm). One meter is equal to 1 000 000 000 nanometers (1×10^9). Thus, the result expressed in nanometers is divided by 1×10^9 for conversion to meters.

$$dsDNA = \frac{(0.34\,nm \times 1\,000\,000 \times genome\ size\ Mbp)}{1\,000\,000\,000}$$

Depending on the organism, cells of different tissues can be characterized based on the number of sets of chromosomes present: monoploid (one set of chromosomes), diploid (two sets), triploid (three sets), tetraploid (four sets), pentaploid (five sets), and so on. For instance, the human genome contains 3.1 Gbp (3100 Mbp). Thus, in a human haploid (or monoploid) cell (e.g. a single set of chromosomes found in a gamete), the unfolded length of a single set of chromosomes, arranged linearly one after the other, would show an approximate length of:

$$dsDNA\ haploid\ cell = \frac{(0.34\,nm \times 1\,000\,000 \times genome\ size\ Mbp)}{1\,000\,000\,000}$$
$$= \frac{(0.34\,nm \times 1\,000\,000 \times 3100\,Mbp)}{1\,000\,000\,000}$$
$$= \frac{1\,054\,000\,000}{1\,000\,000\,000} = {\sim}1.054\ m$$

Thus, a single set of human chromosomes ($n = 23$ Chr) can theoretically unfold up to 1 m. However, the human body is constituted mainly of somatic cells (diploid cells – two sets of chromosomes/cell). For a diploid cell ($2n = 46$ Chr), the linear length of all 46 dsDNA molecules is calculated as above and the result in multiplied by two:

$$\text{dsDNA diploid cell} = 2 \times \frac{(0.34 \text{ nm} \times 1\,000\,000 \times \text{genome size Mbp})}{1\,000\,000\,000}$$

$$= 2 \times \frac{(0.34 \text{ nm} \times 1\,000\,000 \times 3100 \text{ Mbp})}{1\,000\,000\,000}$$

$$= 2 \times \frac{1\,054\,000\,000}{1\,000\,000\,000} = {\sim}2.108 \text{ m}$$

Therefore, the two sets ($2n = 46$ Chr) of human chromosomes found inside a somatic cell can theoretically unfold up to 2.1 m. The linear length of dsDNA molecules from all chromosomes of a somatic cell and the estimated average number of somatic cells in the human body, can be used for various mental experiments (e.g. comparisons between DNA lengths and cosmic distances). These calculations can be empirically extended for ssDNA molecules placed linearly one after the other. For instance, the 2.1 m of dsDNA from a somatic cell, of course, doubles if the ssDNA approach is considered (2.1 m × 2 DNA strands = 4.2 m of ssDNA). The implementation found in Additional algorithm 2.1 uses the above formula to convert the number of bases of a genome to physical length expressed in meters. **Important**: For convenience, from this point on all notations "b", "kb", "Mb", "Gb" will refer to dsDNA (double stranded DNA).

Additional algorithm 2.1 **Note that the source code is in context and works with copy/paste.**

```
<script>

document.write('Homo sapiens (3100 Mb): <br>');

document.write('DNA in a haploid cell nucleus: ');
document.write(f(3100) + ' meters<br>');

document.write('DNA in a somatic cell nucleus: ');
document.write((2 * f(3100)) + ' meters<br>');

function f(Mb){return (0.34 * 1000000 * Mb)/1000000000;}

</script>
```

```
Output:

Homo sapiens (3100 Mb):

DNA in a haploid cell nucleus: 1.054 meters

DNA in a somatic cell nucleus: 2.108 meters
```

Above, the example is given on *Homo sapiens* and the result shows the calculated total length of unfolded chromosomes for both haploid cells and diploid (somatic) cells. This computation can be applied to all genomes mentioned so far by calling function *f* repeatedly. Thus, Additional algorithm 2.1 is extended to perform this calculation for an arbitrary number of species (Additional algorithm 2.2).

Additional algorithm 2.2 Note that the source code is in context and works with copy/paste.

```
<script>

// DNA to meters

var a = 'Ambystoma mexicanum|32396Mb' +
        'Pinus lambertiana|27603Mb' +
        'Sequoia sempervirens|26537Mb' +
        'Minicystis rosea|16Mb' +
        'Sorangium cellulosum So0157-2|14.78Mb' +
        'Escherichia coli|4.9Mb' +
        'Encephalitozoon intestinalis|2.3Mb' +
        'Ostreococcus tauri|12.6Mb' +
        'Homo sapiens|3100Mb';

var t = a.split('Mb');

for (var u=0; u<t.length-1; u ++)
{
  var r = t[u].split('|');
  document.write(r[0] + ' (' + r[1] + ' Mb) = ');
  document.write(f(r[1]) + ' meters<br>');
}

function f(Mb){return (0.34 * 1000000 * Mb)/1000000000;}

</script>
```

```
Output:

Ambystoma mexicanum (32396 Mb) = 11.01464 meters

Pinus lambertiana (27603 Mb) = 9.38502 meters

Sequoia sempervirens (26537 Mb) = 9.02258 meters

Minicystis rosea (16 Mb) = 0.00544 meters

Sorangium cellulosum So0157-2 (14.78 Mb) = 0.0050252 meters

Escherichia coli (4.9 Mb) = 0.0016660000000000002 meters
```

(Continued)

Additional algorithm 2.2 (Continued)

```
Encephalitozoon intestinalis (2.3 Mb) = 0.0007819999999999999 meters

Ostreococcus tauri (12.6 Mb) = 0.004284 meters

Homo sapiens (3100 Mb) = 1.054 meters
```

To call function f repeatedly, a parsing-based method is used. Above, variable a contains a series of records. The structure of these records is based on two delimiters, namely: "|" and "Mb." Delimiter "|" separates the species name ($r[0]$) from the size of the genome ($r[1]$), while the "Mb" delimiter separates the records from each other ($t[u]$). Please note that 0.001 m equals 1 mm. For instance, the output of Additional algorithm 2.2 shows that *Escherichia coli* contains a genome of ~1.6 mm in length (0.0016 m), or that *E. intestinalis* contains a genome of 0.78 mm in length (0.00078 m).

2.3.3 Computations on the Average Genome Size

A series of computations show the average genome size observed for each division in the tree of life, as well as the average size of viral genomes and the average DNA length of plasmids (Figure 2.1 and Table 2.1). These values were calculated from the raw data extracted from the file transfer protocol (FTP) of the National Center for Biotechnology Information (NCBI). The NCBI section for *Genome Information by Organism* contains general data in relation to each branch from the tree of life: eukaryotes (13k); prokaryotes (265k); viruses (41k); plasmids (23k); organelles (17k). These categories amount to ~359k DNA/RNA sequences of different assembly levels of readiness, of which 341k sequence samples of assembly level "complete" were used to calculate the averages presented here. Thus, filters were used to obtain a clean data set. For instance, only levels for "complete chromosomes" or "complete genomes" were considered for these calculations.

Moreover, the maximum values presented in the main text were extracted from these data and checked against the literature. The files containing the raw data can be found in the additional materials online. **Important note**: The number of samples shown on the last row of Table 1.4 can be misleading. Table 1.4 shows 252k prokaryote samples, whereas the cataloged prokaryotes in Table 1.1 show a total of 12k species. In the NCBI database, prokaryotes have more than one reference or representative genome per species. According to NCBI filters, around 3.2k of the prokaryote genomes are representative.

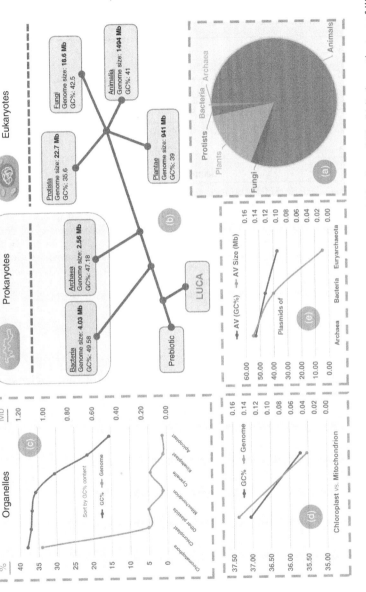

Figure 2.1 **The average genome size.** (a) Shows the proportion of known species in each kingdom of life. (b) It shows the tree of life with data on the main kingdoms of life. Each kingdom is labeled with the average genome size and the average GC% content. (c) Shows the average organellar genome for a number of organelles investigated to date. Here, the organelles are sorted by GC%. (d) It shows a comparison between mitochondria and chloroplasts. (e) Shows a comparison between plasmids from bacteria, archaea, and eukaryotes. For each chart (c–e), the left axis indicates the GC% percentage and the right axis indicates the average size of the genome expressed in mega base pairs (written here as Mb instead of Mbp, for ease).

Table 2.1 The average genome size in the tree of life.

Genome size average (Mb)				
Eukaryotes (Mb)	Prokaryotes (Mb)	Plasmids (Mb)	Organelles (Mb)	Viruses (Mb)
AV 433.92	3.74	0.11	0.07	0.04
SD ±1160.87	±1.81	±0.23	±0.39	±0.43
Average GC% content				
Eukaryotes (%)	Prokaryotes (%)	Plasmids (%)	Organelles (%)	Viruses (%)
AV 41.92	48.72	45.91	36.05	45.34
SD ±10.90	±11.87	±11.32	±7.92	±9.27
Samples				
Total 12 039	252 029	21 801	16 388	38 431

The table shows the average genome size and the average GC% content in: Eukaryotes, prokaryotes, plasmids, organelles, and viruses (eukaryotic and prokaryotic). Note that smaller standard deviation (SD) values indicate that more of the data are clustered about the mean while a larger SD value indicates the data are more spread out (larger variation in the data). The unit of length for DNA is shown in mega bases (Mb). For instance, DNA fragments equal to 1 million nucleotides (1 000 000 b) are 1 mega base in length (1 Mb) or 1000 kilo bases (1000 kb) in length. The last row (samples) indicates how many sequenced genomes have been used for these computations.

2.3.4 Observations on Data

Eukaryotic organisms show an average genome size of 434 Mb and prokaryotic organisms show an average genome size of 3.7 Mb. DNA-containing organelles (70 kb) and viruses (40 kb) show mildly close values for the average genome size. On the other hand, plasmids (110 kb) contain almost twice as much genetic material when compared to the average genome size of organelles and viruses. Out of curiosity, a calculation can be made here on the reductive evolution of organelles. Considering the ancestry of the organelles, the average genome of prokaryotes (3.74 Mb) was taken as the reference system in this approach:

$$AOG\% = \frac{100}{ARG} \times AOG = \frac{100}{3.74 \text{ Mb}} \times 0.07 \text{ Mb} = 1.87\%$$

where average prokaryote genome (*ARG*) is the average size of the reference genome and average organellar genome (*AOG*) is the average size of the organelle genome. *AOG%* represents the size of *AOG* expressed as a percentage from *ARG*. The *AOG* represents 1.8% from the *ARG*. Thus, the reductive evolution is then represented by the reference (the average genome of prokaryotes) percentage value of 100% minus 1.8%:

$$REv = 100\% - AOG\% = 100\% - 1.87\% = 98.12\%$$

where REv represents the reduction of the AOG since first endosymbiosis occurred (2 billion – 1.5 billion years ago). Thus, during this period, the AOG underwent a reductive evolution of 98%. Note that genomes fluctuate in size over long periods of time and the reductive evolution is not necessarily a "one-way street" [181]. The average GC% content was also calculated. The average GC% shows a fairly large difference between prokaryotes and eukaryotes. Plasmids and viruses show a close GC% average of ~ 45% (Table 2.1).

This observation is not entirely surprising since plasmid-to-virus transition scenarios have been proposed in the past [182]. Prokaryotic and eukaryotic ssDNA viruses have their origin in bacterial and archaeal plasmids [183]. Plasmid propagation by virus-like particles was observed in the saline waters of cold environments. For instance, a plasmid from an *Antarctic haloarchaeon* uses specialized membrane vesicles to disseminate and infect plasmid-free cells [184]. The average GC% of organellar genomes is somewhat close to the average GC% value observed in the eukaryotic genomes, but very far from the GC% average value observed in the prokaryotic genomes. In prokaryotes, the average genome size and the average GC% content was also calculated separately for bacteria and archaea (Table 2.2).

The archaeal genomes show an average size and a GC% much lower than what it was observed in bacterial genomes (Table 2.2). The same computations were made for DNA-containing organelles, plasmids, and viruses, and the results will be discussed further.

Table 2.2 The average genome size in prokaryotes.

Prokaryotes	Archaea		Bacteria	
	Size (Mb)	GC%	Size (Mb)	GC%
AV	2.56	47.18	4.03	49.58
SD	±0.98	±11.55	±1.79	±12.57

The table shows the average genome size and the average GC% content in bacteria and archaea. Note that the unit of length for DNA is shown in mega bases (Mb). For instance, DNA fragments equal to 1 million nucleotides (1 000 000 b) are 1 mega base in length (1 Mb) or 1000 kilo bases (1000 kb) in length.

2.4 Organellar Genomes

Chloroplasts were once free-living cyanobacteria, while mitochondria were once free-living proteobacteria. Both have preserved remnants of eubacterial genomes. The average genome size of eukaryotic organelles was calculated at 0.07 Mb (70 kb). With a few exceptions, the average values from each type of organelle show a uniformity regarding the genome size (Table 2.3). Assuming a somewhat constant reductive evolution, this observation may indicate the occurrence of primary and secondary endosymbiosis during the same period for most of the organelles.

Depending on the species, both chloroplasts and mitochondria have evolved slightly differently, however, sometimes even radically different by accelerated reductive evolution (e.g. hydrogenosomes). Many known membrane-bound organelles are derived from either cyanobacteria or proteobacteria lineages. Nonetheless, there are organelles that show a more recent history, different from that of chloroplasts or mitochondria. One such organelle is the chromatophore. About chromatophores and other plastid-like or mitochondria-like organelles, we will discuss further. However, discussions will include only the organelles currently considered by the NCBI database. Note that there are other variations of DNA-containing organelles that have been cataloged as "other plastids" and others who are just waiting to be discovered.

2.4.1 Chloroplasts

Chloroplasts are organelles found in plant cells. Chloroplasts sustain all life on earth by converting solar energy (photons of different frequencies) to carbohydrates through the process of photosynthesis and oxygen release. Like many other organelles, chloroplasts originated from endosymbiotic photosynthetic organisms and retain their own unique DNA [185]. The average length of the chloroplast genome (cpDNA) is 0.153 Mb (153 kb). The chloroplast genome encodes many key proteins that are involved in photosynthesis and other metabolic processes [186]. The chloroplast genome contains between 30 and 50 different RNA genes and a number of protein-coding genes, which range from about 100 in land plants and green algae to 150–200 in nongreen algae [187].

2.4.2 Apicoplasts

Apicoplasts are organelles discovered in parasites [188]. The apicoplast is a vestigial plastid homologous to the chloroplasts of algae and plants. Apicoplasts are present in the majority of parasites from Phylum *apicomplexa* [189]. The apicoplast is of secondary endosymbiotic origin (please see the "The origins of

Table 2.3 The average genome size in different eukaryotic organelles.

	Genome size average (Mb)						
	Apicoplast	Chloroplast	Chromatophore	Cyanelle	Kinetoplast	Mitochondrion	Other plastids
AV	0.035	0.152	1.022	0.136	0.031	0.034	0.15
SD	±0.008	±0.028	—	—	±0.01	±0.463	±0.038
Average GC% content							
	Apicoplast	Chloroplast	Chromatophore	Cyanelle	Kinetoplast	Mitochondrion	Other plastids
AV	15.5	37.04	37.99	30.47	21.58	35.71	36.74
SD	±3.09	±2.61	-	-	±2.96	±9.28	±2.95
Samples	54	4247	1	1	3	11 144	938

Note that smaller standard deviation (SD) values indicate that more of the data are clustered about the mean, while a larger SD value indicates the data are more spread out (larger variation in the data). The unit of length for DNA is shown in mega bases (Mb). For instance, DNA fragments equal to 1 million nucleotides (1 000 000 b) are 1 mega base in length (1 Mb) or 1000 kilo bases (1000 kb) in length. For instance, an average genome size of 0.035 Mb is 35 kb. The last row (samples) indicates how many sequenced genomes have been used for these computations.
Source: Refs. [190, 198].

eukaryotic cells" subchapter) and contains a circular genome around 0.035 Mb (35 kb; Table 2.3) [190]. The plastid encodes RNAs and proteins [191].

2.4.3 Chromatophores

Chromatophores are photosynthetic organelles discovered in species of freshwater amoeboids from *Paulinella* genus. Evidence shows that primary endosymbiosis of chromatophores occurred relatively recently (90–140 million years ago) compared to the origin of canonical Archaeplastida plastids (>1.5 billion years ago) [192]. *Paulinella* lineage gained the chromatophores independently. Thus, chromatophores are considered a valuable model for recent endosymbiosis and a perfect example of organellogenesis. The chromatophore genome is 1 Mb in length and encompasses ~800 protein-coding genes [193]. A uniformity exists between the genome size of different organelles, except for chromatophores (Table 2.2). *Synechococcus* (WH5701) is a unicellular cyanobacterium found in the marine environment and the closest relative of the chromatophore. Bioinformatic evidence shows that chromatophores experienced a genome reduction compared to *Synechococcus*, from which about 74% of the genes were lost. Consequently, these reductions led to the loss of essential functions and made the chromatophore totally dependent on the host for growth and survival [193].

2.4.4 Cyanelles

Cyanelles (or muroplasts) are photosynthetic organelles found in glaucocysto-phyte algae [194]. The *Cyanophora paradoxa* is a representative member of the glaucocystophyceae and is used as a biological model for the study of these plastids [195]. Cyanelles are surrounded by a peptidoglycan-like envelope (a peptidoglycan wall) [196, 197]. These organelles show close morphological and biochemical resemblance to endosymbiotic cyanobacteria. The cyanelle genome is 0.136 Mb (136 kb; Table 2.3) in length [198].

2.4.5 Kinetoplasts

Kinetoplasts are networks of circular DNA molecules (kDNA), found in the mitochondria of kinetoplastids (unicellular eukaryotic organisms capable of self-propulsion) [199]. The mitochondrial genome (kinetoplast) of these flagellated protozoans is considered of the highest complexity encountered among all known organelles. The kinetoplast is physically connected to the flagellar basal bodies of these organisms [200]. The kDNA isolated for electron microscopy may span on a 2D grid of 10 by 15 μm [201]. Depending on the species, the kDNA net-work can exist in many configurations. kDNA is composed of two types of linked

DNA rings. Variations include 5000 up to 10 000 minicircles and 25–50 maxicircles. The minicircles are approximately 0.5–10 kb in size and the maxicircles range from 20 to 40 kb [201]. kDNA represents approximately 30% of the total DNA of these protists. It is worth mentioning here that studies done on kinetoplastids have helped our understanding of the RNA-editing mechanisms [202, 203].

2.4.6 Mitochondria

Mitochondria are the most investigated organelle (Figure 1.2a–c). Mitochondria arose about 2.3–1.6 billion years ago from an α-proteobacterial endosymbiont [204, 205]. Energy production is the main function of this organelle. Potential energy is created by oxidation of glucose and the release of adenosine triphosphate (ATP). ATP is generated by the mitochondrial ATP synthase from adenosine diphosphate (ADP) and phosphate ions (P_i). In turn, ATP hydrolysis leads to ADP and energy release ($ATP + H_2O \rightleftarrows ADP + P_i$), which drives all the fundamental cell functions in most eukaryotes. To put things in perspective, the human body uses an average of 50 kg of ATP per day [206]. Depending on the species and type of tissue, mitochondria ranges from hundreds to thousands of copies per cell [207]. Mitochondria contain their own genome [208]. The average length of the mitochondrial genome is 31 kb (Table 2.3).

2.5 Plasmids

Plasmids are circular or linear dsDNA molecules found in all kingdoms of life [209]. Each plasmid carries only a few genes and is capable of replicating autonomously in the host environment. Their importance is crucial, as these are among the main vectors of horizontal gene transfer [210]. Plasmids naturally exist in prokaryotes, where they were first described [211]. In eukaryotes, plasmids are most common among fungi and higher plants [212]. The average length of plasmid DNA is 0.11 Mb (±0.23; that is 110 kb) (Table 2.1). Note that bacterial plasmids contain the largest number of samples and weigh the most on the main average shown in Table 2.1. Nevertheless, plasmids vary in size and some of the largest can reach the size of a bacterial chromosome (e.g. megaplasmids). For instance, *Ralstonia solanacearum* is a plant pathogen that contains one of the largest megaplasmids (2.1 Mb) [213]. Another example is *Streptomyces clavuligerus*, a bacterium that contains a linear megaplasmid of 1.8 Mb [214]. Of course, these observations can quickly lead to hypotheses regarding speciation and the origin of chromosomes (not discussed here). Eukaryotic plasmids show an average DNA size of about 0.01 Mb (10 kb) and an average GC% close to that of organelles (37%) (Table 2.4).

Table 2.4 The average DNA length of different plasmids.

Plasmids	Archaea		Bacteria		Euryarchaeota	
	Size (Mb)	GC%	Size (Mb)	GC%	Size (Mb)	GC%
AV	0.15	53.07	0.11	45.87	0.01	37.14
SD	±0.17	±9.77	±0.23	±11.31	±0.04	±5.71
Samples	256	256	21426	21426	118	118

Note that the unit of length for DNA is shown in mega bases (Mb). DNA fragments equal to 1 million nucleotides (1 000 000 b) are 1 mega base in length (1 Mb) or 1000 kilo bases (1000 kb) in length. For instance, 0.15 Mb is 150 kb. The last row (samples) indicates how many sequenced plasmids have been used for these computations.

It was previously mentioned that archaeal genomes showed an average size and a GC% much lower than what it was observed in bacterial genomes (Table 2.2). However, the situation seems to be reversed in the case of plasmids. The bacterial plasmids show an average size and a GC% much lower than what it was observed in archaeal plasmids (Table 2.4).

2.6 Virus Genomes

Some viruses contain a RNA-based genome and others contain a DNA-based genome. Among the DNA-based viral genomes, some species contain dsDNA and other species show a ssDNA. The same is true for RNA-based viruses; some species contain double-stranded RNA (dsRNA) and other species show a single-stranded RNA (ssRNA) [215]. Prokaryotic and eukaryotic viruses, taken together, show an average genome size of ∼0.04 Mb (40 kb) (Tables 2.1 and 2.5). Eukaryotes contain both the smallest and largest viral dimensions, and the smallest and largest viral genome sizes. Viruses with RNA genomes dominate the eukaryotic world [215]. RNA viruses without DNA replication intermediates are called riboviruses. Some famous riboviruses are influenza, SARS, COVID-19, hepatitis C, hepatitis E, Ebola, rabies, polio, and so on. RNA viruses that include DNA intermediates are called retroviruses. The most famous retroviruses are the human immuno-deficiency viruses (HIV-1 and HIV-2) that cause the acquired immuno-deficiency syndrome (AIDS). But what are DNA replication intermediates? Retroviruses use their own reverse transcriptase enzymes to produce a DNA copy of their RNA genome. The new DNA fragment is then incorporated into the genome of the host cell by an integrase enzyme. Post incorporation, the cell

Table 2.5 The average genome size of different eukaryotic and prokaryotic viruses.

Viruses	Average genome size (Mb)	GC%
AV	0.0339	45.3970
SD	±0.0652	±9.2474
Samples	37962	37962

Note that smaller standard deviation (SD) values indicate that more of the data are clustered about the mean while a larger SD value indicates the data are more spread out (larger variation in the data). The unit of length for DNA is shown in mega bases (Mb). DNA fragments equal to 1 million nucleotides (1 000 000 b) are 1 mega base in length (1 Mb) or 1000 kilo bases (1000 kb) in length. For instance, an average genome size of 0.0339 Mb is 33.9 kb. The last row (samples) indicates how many sequenced genomes were used for this calculation.

transcribes and translates its own genes and the viral genes needed to assemble new copies of the virus. It is worth mentioning here that mutation rates in RNA viruses are up to a million times higher than their hosts [216].

The physical size of organisms and the size of their genomes lack any proportionality or correlation. But the relationship between DNA quantity and physical size is partially different in the case of viruses. Interestingly, the largest viruses also contain the largest genomes and the smallest viruses contain the smallest genomes [91]. However, these extremes are occupied by virus species with a DNA-based genome. For instance, *Pandoravirus salinus* is among the largest virus species (1 μm long) and contains 2.5 Mb of dsDNA packed in particles of bacterium-like shapes [217]. Their large size is explained by DNA transposons that have colonized the genome of the giant virus *P. salinus* over long periods of time [218]. On the other hand, *Porcine circovirus* is the smallest virus, with a capsid diameter of 17 nm and a ssDNA-based genome size of ~1.7 kb [219, 220]. RNA-based viral genomes are also among the smallest. For instance, hepatitis delta virus (HDV) contains a 36-nm virion (virus particle) and an ssRNA molecule around ~1.7 kb [221–223]. As previously stated, plasmids, and viruses show a close GC% average of ~45% (Tables 2.1 and 2.5). Evidence suggests that prokaryotic and eukaryotic ssDNA viruses have their origin in bacterial and archaeal plasmids [183]. Furthermore, as mentioned earlier, giant viruses overlap the cellular world. For instance, DNA methylation contributes to various regulations in all domains of life. Genes of giant dsDNA viruses encode DNA methyltransferases, which make use of this mechanism [224].

2.7   Viroids and Their Implications

Discussions about viruses and their simplicity or complexity form bridges that were once hard to imagine. Large viruses partially overlap with cellular mechanisms and their upper limit appears to be life. But which is the lower limit for viruses? The smallest viruses discussed here are less representative for the lower limit of infectious mechanisms. The lower limit is represented by different RNA fragments or different proteins such as prions. Prions are misfolded proteins with the ability to transmit their misfolded shape onto correctly folded proteins of the same type (please see the mad cow disease). Prion mechanisms are, perhaps, less relevant to the occurrence of life on Earth and will not be discussed here. However, the mechanisms related to self-replicating proteins represent one of the competing hypotheses for the preorigins of life. For instance, amyloid fibers arise spontaneously from amino acids under prebiotic conditions. Thus, amyloid catalysts may have played an important role in prebiotic molecular evolution [225]. In the RNA world, the current bet for the origin of life on Earth is represented by catalytic RNAs. Examples of short RNA fragments with different properties are found in many varied and distant cases throughout the scientific literature. For instance, RNA fragments of several hundred nucleotides called "viroids" are the smallest infectious pathogens [226]. Viroids were first observed in the roots of *Solanum tuberosum* (potatoes) by Theodor Otto Diener in 1971 [226]. The ssRNA circular structure of viroids or viroid-like satellite RNAs lacks the presence of any genes and stands somewhere in between "nothing" and RNA viruses [216]. Apparently, RNAs are the only biological macromolecules that can function both as genotype and phenotype [227]. Some viroids and viroid-like RNAs exhibit catalytic properties that allow self-cleavage and ligation [228]. This catalytic property links the opportunistic RNAs to self-splicing introns (Group I introns). Group I introns are found in protein coding genes of bacteria and their phages, nuclear ribosomal RNA (rRNA) genes, mitochondrial mRNA and rRNA genes, chloroplast transfer RNA (tRNA) genes, and so on [229–231]. In 1981, Theodor Otto Diener asked the question: Are viroids escaped introns? [232]. A small fraction of the nuclear group I introns have the potential of being mobile elements [233]. Of course, today one can ask a complementary question: Are introns some distant viroid-like RNAs introduced into the genome of different organisms through DNA intermediates? It is likely that noncoding RNAs were the indirect source for all introns [227]. These speculations place the early opportunistic catalytic RNAs at the point of origin for the eukaryotic proteome diversity. In conclusion, viroid-like molecules could have been directly implicated in the occurrence of life on Earth. It is reasonable to believe that an intersection between self-replicating proteins and catalytic RNAs has probably led to some truly rudimentary precellular forms of life. Thus, it can be speculated here that in the prebiotic period there could have been two rudimentary life forms, which gradually merged to form the Last Universal Common Ancestor (LUCA) population. Please note that "viroids" are short ssRNAs and "virions" are virus particles.

2.8   Genes vs. Proteins in the Tree of Life

Throughout different organisms, the proteome may be smaller, equal to (hardly ever), or larger than the genome. In eukaryotic species in particular, one gene may encode for more than one protein via a process known as alternative splicing. Note that RNA-splicing mechanisms are discussed in detail in Chapter 8. A comparative analysis between the average number of genes and the average number of proteins is shown in Table 2.6. Based on the values shown in this table, various rough estimates can be made on the frequency of alternative splicing in different kingdoms of life. A general equation can be formulated by assuming a "one gene–one protein" correspondence. Given that an equality between the number of proteins and the number of genes means 100%, everything that is above this threshold is a surplus that can be attributed to alternative splicing and protein splicing. Thus, the average number of genes divides the unity (a value of 1 – it can also be 100 for simplicity) and the result is multiplied by the average number of proteins. To find the average protein surplus (S), the unity is deduced

Table 2.6   Genes vs. proteins in the tree of life.

Eukaryotes	Size (Mb)	Genes	Proteins	GC%
Animals	1493.6	27 075.8	39 140.1	41.0
Fungi	18.6	7707.5	6951.2	42.5
Plants	940.8	39 140.1	45 405.0	38.7
Protists	22.7	5915.1	5628.1	35.6
Other	45.6	7546.3	7354.8	43.4
Prokaryotes	Size (Mb)	Genes	Proteins	GC%
Bacteria & Archaea	4.0	3829.0	3598.4	49.5

The table shows a comparison between the average genome size, the average GC% content, the average number of genes, and the average number of resulting proteins. Note that the unit of length for DNA is shown in mega bases (Mb). DNA fragments equal to 1 million nucleotides (1 000 000 b) are 1000 kilo bases in length (1000 kb) or 1 mega bases in length (1 Mb), or 0.001 giga bases in length (0.001 Gb). For instance, an average genome size of 1493.6 Mb is 1.4936 Gb (~1.4 Gb).

from this result only if the proteome is larger than the genome, as follows:

$$S = \frac{1}{genes} \times proteins - 1 = \frac{1}{27\,075} \times 39\,140 - 1 = 0.45 = \sim 45\%$$

The average animal proteome is 45% more diverse when compared to the average number of animal genes. By using the same formula from above, the average plant proteome is 16% more diverse than the average number of known genes. The fungi, protest, and prokaryote average proteome is moderately undersized. As before, the average number of genes divides unity (a value of 1) and the result is multiplied by the average number of proteins. However, since in this case the proteome is smaller than the genome, the final result is subtracted from 1 (unity), as follows:

$$S = 1 - \frac{1}{genes} \times proteins = 1 - \frac{1}{7707} \times 6951 = 0.098 = \sim 10\%$$

where S in this case is the part of the proteome that should exist, assuming a "one gene – one protein" correspondence. Thus, the fungal average proteome indicates ~10% fewer types of proteins than the average number of genes. This suggest that 10% of the fungal genome encodes for functional RNAs. The situation is similar for protists. In protists, about 5% of genes could encode for functional RNAs and the remaining 95% encodes for proteins. In the case of prokaryotes, about 93% of archaeal and bacterial genes encode for proteins and the remaining 7% could encode for functional RNAs.

Although informative, note that an undersized proteome does not rule out the possibility of alternative splicing or protein splicing in any of these kingdoms. Animals and plants show the most diverse proteomes, well above the average number of genes (Table 2.6). Individually, some species may show a particularly high proteome diversity compared to these averages. For instance, in plants, *Triticum durum* (macaroni wheat) contains ~63k of genes and a proteome of ~190k. Following the same reasoning as above, the proteome of *T. durum* is ~197% more diverse when compared to the number of genes. In animals, a significant difference can also be found. Current NCBI data shows that the human genome contains ~60k of genes (the list of annotated features includes protein-coding genes, noncoding genes, and pseudogenes) and a proteome of ~120k (*H. sapiens* GRCh38.p13). The proteome of *H. sapiens* is ~95% more diverse when compared to the number of genes. Note: However, when it comes to the human genome and the proteome, a discussion can be almost dangerous over time. In literature, the number of genes and proteins for *H. sapiens* can vary depending on different agreements or/and advances in bioinformatics [234–236]. But why all this uncertainty related to the number of genes or proteins? All genes are predicted by using bioinformatic means. Many predictions are then verified by alignment of sequenced mRNAs against a reference genome. However, many genes express themselves only in special conditions or over certain periods of time, or only once in a life time. Thus, their mRNAs cannot be detected and sequenced to further confirm the bioinformatic predictions. To add to this matter, many genes may overlap and often gene promoters can show bidirectional activity [237–239]. It stands to reason that such elusive genes are difficult to locate with certainty and other genes will prove difficult to detect in the future. Moreover, many results derived from large-scale experiments (e.g. genome studies) are directly under the umbrella of chaos theory. Small changes in the initial parameters of different algorithms can lead to huge variations in the final predictions. This has already been evident over time in the case of the human genome [234–236].

2.9 Conclusions

Bioinformatics is the field that will perhaps lead to a clearer understanding of both the origins and the current mechanisms of life. Here, an introduction provided the necessary context for a better understanding of different approaches used in the field of bioinformatics and, possibly, for new ideas "just waiting" to be implemented in the future. In a first stage, the chapter described the units of measurement used throughout the book, and presented a series of useful conversions, followed by discussions regarding organisms with the largest/smallest genomes. In a second stage, the average values related to the genome size in different kingdoms of life were calculated and discussed. In this context, the global features of viral genomes, plasmid DNA and various genome-containing organelles found in different eukaryotic organisms have been described in brief. Furthermore, viroids have been mentioned in connection with the properties shown by catalytic RNAs. Toward the end of the chapter, discussions were gradually switched from catalytic RNAs to RNA splicing. The frequency of RNA splicing was further pointed out by a comparison between the average number of genes and proteins in the main kingdoms of life.

3

Sequence Alignment (I)

3.1 Introduction

The order of molecules within biological polymers dictates their macromolecular predisposition for a certain structure and allows for a number of biochemical properties, which ultimately reflect on different biological functions. The polymers in question may represent a DNA, RNA, or a protein molecule. The significance of the order of molecules in polymers of biological origin is partially unknown. Different regions from these polymers, whose function is already known, can be regarded as a reference system for detection of similar biological functions within the same organism or between organisms. Evolutionary relationships between different biological sequences are another area of great interest that indicates how species diverged (or converged) in the direction of certain phenotypes. Thus, the issue becomes fully immersed in the field of computer science. The functional regions of different sequences, whose role has been partially deduced, can be searched in new sequences by using (among others) a special class of algorithms, namely the sequence alignment algorithms. An alignment is a method that highlights regions of similarity between two or more sequences. On computers, such sequences are represented by strings of symbols (letters, digits, or other types of characters). Therefore, this chapter explores the technical aspects of the main algorithms used for sequence alignment.

3.2 Style and Visualization

To start an implementation for sequence alignment, two initial steps are required. The first step is related to the way the letters are placed relative to each other and the second step allows a visualization of the content of a matrix or vector throughout the chapter. The monospace font-family allows any symbol to be framed in a

Algorithms in Bioinformatics: Theory and Implementation, First Edition. Paul A. Gagniuc.
© 2021 John Wiley & Sons, Inc. Published 2021 by John Wiley & Sons, Inc.
Companion website: www.wiley.com/go/gagniuc/algorithmsinbioinformatics

constant horizontal space. In other words, any letter shows a constant length. The declaration of a monospace is especially important for aligning text sequences one above the other in a correct frame. The monospace style for the HTML format is stated below and accompanies the majority of algorithms throughout the chapter (Additional algorithm 3.1).

Additional algorithm 3.1 Note that the source code is out of context and is intended for explanation of the method.

```
<style>
    body {
        padding: 1rem;
        font-family: monospace;
        font-size: 18px;
        font-style: normal;
        font-variant: normal;
        line-height: 20px;
    }
</style>

All letters have equal width.<br>

||||||||||||||||||||||||||||||||||||
```

Output:

```
All letters have equal width.

||||||||||||||||||||||||||||||||||||
```

The output of such a statement shows that two strings constructed of different symbols occupy exactly the same length. Next, a special function is needed to observe the results in HTML format. A general-purpose function is defined for visualization of the contents of any matrix or vector. The function below is called *show matrix content* (SMC) and will be used throughout this chapter to view changes made to these mathematical structures (Additional algorithm 3.2):

Additional algorithm 3.2 Note that the source code is out of context and is intended for explanation of the method.

```
// SHOW MATRIX CONTENT

function SMC(m) {
    var r = "<table border=1>";
    for(var i=0; i<m.length; i++) {
        r += "<tr>";
        for(var j=0; j<m[i].length; j++){
            r += "<td>"+m[i][j]+"</td>";
        }
        r += "</tr>";
    }
    r += "</table>";

    return r;
}
```

The above function builds a table in HTML format in which it associates the cells of a table with the elements of an array. The *SMC* function receives an array *m*, which it traverses element by element. The variable *i* represents each row and the variable *j* represents each column in the matrix *m*. **Note**: by default, matrices and vectors start from index 0. Next, *m.length* represents the total number of rows in the matrix *m* and *m[i].length* represents the total number of columns in the matrix *m* in row *i*. Variable *r* is a string variable. The table is built in variable *r* by using the HTML format for a table structure. In HTML format, a table starts with a "<table>" tag and ends with the "</table>" tag. The "border" property indicates the thickness of the line that separates the cells in the table. The "<tr>" tag means a new row inside the table, while the "<td>" tag represents a new horizontal cell. The *SMC* function returns the contents of the variable *r*. The content of the variable *r* is handed over directly to the browser for interpretation by declaring "*document.write(SMC(m));*". The browser parses the tags built by the *SMC* function and shows them to the user as a simple table. **Note**: all the names of the variables inside the function are independent of the names of the variables outside the function. Thus, variables like *m, i, j, r* or other variables inside the *SMC* function may have other meanings outside the function. The functions described (such as *SMC*) should be seen as separate "boxes" from the main algorithm.

3.3 Initialization of the Score Matrix

The construction of a score matrix starts with the initialization, in which the size of the matrix and its format are established (Additional algorithm 3.3). A score matrix can, of course, be constructed in several ways. For example, such an array may contain only numeric values or it may contain both symbols and numeric values. If an array is initialized only with numeric values, an implementation can match the rows and columns in the array with the positions from the two sequences that participate in the alignment. First, the mixed approach is favored, in which sequences are included in the matrix on the first line and the first column. The initialization stage is valid for both global and local sequence alignment. The implementation starts by declaring the two sequences used for the alignment and the main parameters (Additional algorithm 3.3). First, two sequences are considered for an alignment, namely:

> $s0$: "TGAATTCAGTTA"
> $s1$: "TGGATCGA"

The variables for penalty and reward are the main parameters of the classical sequence alignment algorithms, namely: Match: $+2$; Mismatch: -1; and gap: -2. Next, the score matrix (m) and the traceback matrix (t) are declared. Note that the traceback matrix is purely optional and is used throughout the chapter to highlight the alignment path. The two sequences (s0 and s1) are inserted letter by letter in a two-dimensional array using the split function (s[0] and s[1]). This strategy allows the increment variables (i and j) to traverse and access the symbols in any of the two sequences. The length of the two sequences is stored in dedicated variables (n_0 and n_1). All possible pairs of letters from the two sequences are considered, which gives rise to a 2D matrix representation. Both a score matrix (m) and a traceback matrix (t) are built in parallel using a nested loop. A "nested" loop is a loop within a loop (an inner loop within the body of an outer loop). Of course, the total number of steps taken by the two loops is dictated by the length (n_0 and n_1) of the two sequences involved in this process (steps = n_0 × n_1). Variable i corresponds to the columns, whereas variable j corresponds to the lines of a matrix. Inside the nested loop, all elements of the two matrices are filled by default with zero values (m[i][j]=0; and t[i][j]=0;). The nested loop also contains two sets of logical conditions (Additional algorithm 3.3). The first set of logical conditions check if the increment variables (i and j) point to an element in the first row or column and rewrite the value of zero if necessary. For instance, if the i, j coordinates indicate an element in the first column, then the value zero is replaced by a letter corresponding to the $(i - 2)$ position in the $s0$ sequence (s[0][i-2];). In contrast, if the i, j coordinates indicate an element in the first row of the matrix, then the value zero is replaced by a letter corresponding to the

($j - 2$) position in the *s1* sequence (`s[1][j-2];`). Thus, the first row and the first column in the matrix hold the two sequences. Note: to accommodate the two sequences and the adjacent values, the rows and columns on the matrix are two positions longer. Thus, the two positions are always subtracted from variables i and j (i.e. `s[0][i-2];` or `s[1][j-2];`), to allow a direct correspondence between the positions of the letters in the two sequences and the structure/format of the matrix. The second set of logical conditions check if the two loops point to an element in the second row or column and rewrite the value of zero with decreasing values that are multiples of the gap penalty.

Additional algorithm 3.3 Note that the source code is in context and works with copy/paste.

```
<script>

var s0 = 'TGAATTCAGTTA';
var s1 = 'TGGATCGA';

var Match = +2;
var Mismatch = -1;
var gap = -2;

var m = [];
var t = [];
var s = [];

s[0] = [] = s0.split(");
s[1] = [] = s1.split(");

var n_0 = s[0].length + 1;
var n_1 = s[1].length + 1;

for(var i=0; i<=n_0; i++) {

    m[i]=[];
    t[i]=[];

    for(var j=0; j<=n_1; j++) {
```

(Continued)

Additional algorithm 3.3 (Continued)

```
            m[i][j]=0;
            t[i][j]=0;

            if (i>1) {m[i][0]=s[0][i-2];}
            if (j>1) {m[0][j]=s[1][j-2];}

            if (i==1 && j>1) {m[i][j]=m[i][j-1] + gap;}
            if (j==1 && i>1) {m[i][j]=m[i-1][j] + gap;}
        }
    }

document.write(SMC(m));

// SHOW MATRIX CONTENT
function SMC(m) {
    var r = "<table border=1>";
    for(var i=0; i<m.length; i++) {
        r += "<tr>";
        for(var j=0; j<m[i].length; j++){
            r += "<td>"+m[i][j]+"</td>";
        }
        r += "</tr>";
    }
    r += "</table>";

    return r;
}
</script>
```

Output:

0	0	T	G	G	A	T	C	G	A
0	0	-2	-4	-6	-8	-10	-12	-14	-16
T	-2	0	0	0	0	0	0	0	0
G	-4	0	0	0	0	0	0	0	0
A	-6	0	0	0	0	0	0	0	0
A	-8	0	0	0	0	0	0	0	0
T	-10	0	0	0	0	0	0	0	0

T	-12	0	0	0	0	0	0	0	0
C	-14	0	0	0	0	0	0	0	0
A	-16	0	0	0	0	0	0	0	0
G	-18	0	0	0	0	0	0	0	0
T	-20	0	0	0	0	0	0	0	0
T	-22	0	0	0	0	0	0	0	0
A	-24	0	0	0	0	0	0	0	0

The successive values on the second column are obtained by adding the gap value to the value present in the $i-1$ element ($m[i-1][j] + gap$). In contrast, the successive values from the second row are obtained by adding the gap value to the value present in the j-1element of a matrix ($m[i][j-1] + gap$). In this manner, the second row and the second column in the matrix hold decreasing values that are multiples of the gap penalty (i.e. -2, -4, -6, -8, ...). Note: In this step, the score matrix (m) and the traceback matrix (t) are exactly the same. However, the t matrix remains untouched until the traceback phase.

3.4 Calculation of Scores

Up to this point, matrices m and t contain the two sequences ($s0$ on the elements from the first column and $s1$ on the elements from the first line), the values multiple of the gap penalty and a value of zero in the other elements. In a second phase of initialization, the zero values are replaced with score values (Figure 3.1).

The score values are computed according to different rules for global and local alignment, which initially take into account the numbers on row two and column two (i.e. the values multiple of gap). However, the order in which these scores are calculated over the matrix is universal for both global and local alignment. Score values are calculated (in this case) starting from the third element of the third row (i.e. $m_{2,2}$; note that the first column and first row in the matrix starts at the $m_{0,0}$ element) and the computations continue successively to the right until the last element on the row. This process continues row by row in the same manner until the last row of the matrix is finally reached. On the last row of the matrix, the computation of scores ends in the lower right corner of the matrix.

3.4.1 Initialization of the Score Matrix for Global Alignment

Although the order in which the scores are calculated is universal for global and local alignment, the method by which the score value is computed differs

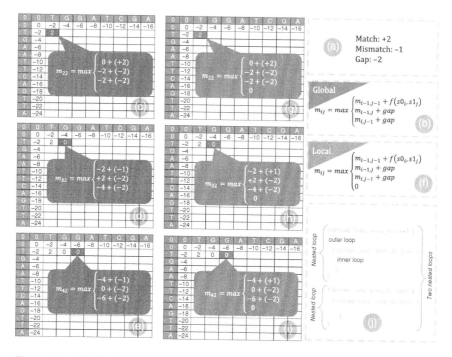

Figure 3.1   Initialization of the score matrix. (a) Shows the values of the parameters used in the experiments from this chapter. (b) Shows the main expressions that compete in the maximization function in the case of global alignment. (c–e) Shows the first three steps in the score matrix initialization for global alignment. (f) Show the expressions that compete in the maximization function in case of local alignment. (g–i) Shows the first three steps in the score matrix initialization for local alignment. (j) Shows what a nested loop means.

slightly (Figure 3.1c–e,b). In the global alignment, the computation of a score value involves a repeated maximization between three expressions, which take into account the parameter values (match; mismatch; and gap) and the values in the neighboring cells (diagonal, left, and top):

$$m_{ij} = max \begin{cases} m_{i-1,j-1} + f\left(s0_i, s1_j\right) \\ m_{i-1,j} + gap \\ m_{i,j-1} + gap \end{cases}$$

where f is a matching function:

$$f\left(s0_i, s1_j\right) = \begin{cases} \alpha, & s0_i = s1_j \\ \beta, & s0_i \neq s1_j \end{cases}$$

where α represents the value of the match parameter and β represents the value of the mismatch parameter. Also, the i_{th} character in sequence $s0$ is $s0_i$ and the j_{th}

character of sequence *s1* is $s1_j$. The implementation below traverses the *m* matrix by using a nested loop, in the same manner as described for the implementation of Additional algorithm 3.3. The current cell (m[i][j]) is the cell for which a new value is calculated. The seeds of these values have their origin in the numbers found on the second row (m[i][1]) or column (m[1][j]) in the score matrix. The nested loop allows three variables to compete inside a maximization function, namely *A*, *B*, and *C* (Additional algorithm 3.4).

Additional algorithm 3.4 Note that the source code is out of context and is intended for explanation of the method.

```
for(var i=0; i<=n_0; i++) {

    for(var j=0; j<=n_1; j++) {

        if(i>1 && j>1){

            var A = m[i-1][j-1] + f(m[i][0],m[0][j]); // \
            var B = m[i-1][j] + gap;                   // -
            var C = m[i][j-1] + gap;                   // |

            m[i][j] = Math.max(A, B, C);

        }
    }
}
```

The value of *A* is calculated by adding the value from the diagonal of the current cell to the result returned by a function *f* (Additional algorithm 3.5). Function *f* is a matching function that returns the value of the *Match* variable (+2), only if the letter from the first row (top of the current cell, m[i][0]) is the same as the letter from the first column (left of the current cell, m[0][j]).

Additional algorithm 3.5 Note that the source code is out of context and is intended for explanation of the method.

```
function f(a1, a2) {

    if(a1 === a2){return Match;} else {return Mismatch;}

}
```

If the letters differ, then function f will return the value of the *Mismatch* variable (-1). Note: since the two sequences are stored on the first row and the first column of the m matrix, the s[0][i-2] reference is equivalent to m[i][0] and the s[1][j-2] reference is equivalent to m[0][j]. Therefore, a comparison is possible either directly between the two sequences, for instance:

```
var A = m[i-1][j-1] + f(s[0][i-2], s[1][j-2]);
```

or by using the two sequences stored in the m matrix:

```
var A = m[i-1][j-1] + f(m[i][0], m[0][j]);
```

The values calculated for variables B and C have a more direct approach. For example, the value of the B variable is calculated by adding the value found on the left of the current cell and the *gap* value (m[i-1][j] + gap). Likewise, the value of the C variable is calculated by adding the value found above the current cell to the value of the *gap* variable (m[i][j-1] + gap). The highest value between A, B, and C is written in the current element/cell of the m matrix. For a complete view, Additional algorithm 3.3 can be concatenated with Additional algorithm 3.4 to test what has been discussed so far (Additional algorithm 3.6):

Additional algorithm 3.6 Note that the source code is in context and works with copy/paste.

```
<script>

var s0 = 'TGAATTCAGTTA';
var s1 = 'TGGATCGA';

var Match = +2;
var Mismatch = -1;
var gap = -2;

var m = [];
var t = [];
var s = [];

s[0] = [] = s0.split('');
s[1] = [] = s1.split('');

var n_0 = s[0].length + 1;
var n_1 = s[1].length + 1;

for(var i=0; i<=n_0; i++) {

    m[i]=[];
    t[i]=[];

    for(var j=0; j<=n_1; j++) {
```

```
        m[i][j]=0;
        t[i][j]=0;

        if (i>1) {m[i][0]=s[0][i-2];}
        if (j>1) {m[0][j]=s[1][j-2];}

        if (i==1 && j>1) {m[i][j]=m[i][j-1] + gap;}
        if (j==1 && i>1) {m[i][j]=m[i-1][j] + gap;}
    }
}

for(var i=0; i<=n_0; i++) {

    for(var j=0; j<=n_1; j++) {

        if(i>1 && j>1){

            var A = m[i-1][j-1] + f(m[i][0],m[0][j]); // \
            var B = m[i-1][j] + gap;                  // -
            var C = m[i][j-1] + gap;                  // |

            m[i][j] = Math.max(A, B, C);
        }
    }
}

function f(a1, a2) {
    if(a1 === a2){return Match;} else {return Mismatch;}
}

document.write(SMC(m));

// SHOW MATRIX CONTENT
function SMC(m) {
    var r = "<table border=1>";
    for(var i=0; i<m.length; i++) {
        r += "<tr>";
        for(var j=0; j<m[i].length; j++){
            r += "<td>"+m[i][j]+"</td>";
        }
        r += "</tr>";
    }
    r += "</table>";

    return r;
}
</script>
```

(Continued)

Additional algorithm 3.6 (Continued)

Output:

	0	0	T	G	G	A	T	C	G	A
0	0	0	-2	-4	-6	-8	-10	-12	-14	-16
T	-2	2	0	-2	-4	-6	-8	-10	-12
G	-4	0	4	2	0	-2	-4	-6	-8
A	-6	-2	2	3	4	2	0	-2	-4
A	-8	-4	0	1	5	3	1	-1	0
T	-10	-6	-2	-1	3	7	5	3	1
T	-12	-8	-4	-3	1	5	6	4	2
C	-14	-10	-6	-5	-1	3	7	5	3
A	-16	-12	-8	-7	-3	1	5	6	7
G	-18	-14	-10	-6	-5	-1	3	7	5
T	-20	-16	-12	-8	-7	-3	1	5	6
T	-22	-18	-14	-10	-9	-5	-1	3	4
A	-24	-20	-16	-12	-8	-7	-3	1	5

Depending on how the parameters are set, a score matrix for global alignment can show either positive or negative values. In the case of global alignment, the two phases of the score matrix initialization are complete (Additional algorithm 3.6).

3.4.2 Initialization of the Score Matrix for Local Alignment

All discussions from the global alignment are valid here as well. However, in the case of the local sequence alignment algorithm, the second phase in the initialization of the score matrix brings a new rule (Figure 3.1g–i,f). All the negative values are eliminated from the m matrix as follows:

$$m_{ij} = max \begin{cases} m_{i-1,j-1} + f\left(s0_i, s1_j\right) \\ m_{i-1,j} + gap \\ m_{i,j-1} + gap \\ 0 \end{cases}$$

And f is the matching function:

$$f\left(s0_i, s1_j\right) = \begin{cases} \alpha, & s0_i = s1_j \\ \beta, & s0_i \neq s1_j \end{cases}$$

where α represents the value of the match parameter and β represents the value of the mismatch parameter. Again, the i_{th} character in sequence $s0$ is $s0_i$ and the j_{th} character of sequence $s1$ is $s1_j$. On the implementation side, all negative values are eliminated from the m matrix by the appearance of constant D. Thus, the nested loop allows four values to compete inside a maximization function, namely A, B, C, and D (Additional algorithm 3.7).

Additional algorithm 3.7 Note that the source code is out of context and is intended for explanation of the method.

```
for(var i=0; i<=n_0; i++) {

    for(var j=0; j<=n_1; j++) {

        if(i>1 && j>1){

            var A = m[i-1][j-1] + f(m[i][0],m[0][j]); // \
            var B = m[i-1][j] + gap;                   // -
            var C = m[i][j-1] + gap;                   // |
            var D = 0;

            m[i][j] = Math.max(A, B, C, D);


        }
    }
}
```

For instance, if the maximum value shown by *A*, *B*, or *C* is below zero, then the value of *D* becomes the maximum value. For a complete view, Additional algorithm 3.3 can be concatenated with Additional algorithm 3.7 to test what has been discussed so far (Additional algorithm 3.8):

Additional algorithm 3.8 Note that the source code is in context and works with copy/paste.

```
<script>

var s0 = 'TGAATTCAGTTA';
var s1 = 'TGGATCGA';

var Match = +2;
var Mismatch = -1;
var gap = -2;

var m = [];
var t = [];
var s = [];

s[0] = [] = s0.split('');
s[1] = [] = s1.split('');
```

(Continued)

Additional algorithm 3.8 (Continued)

```
var n_0 = s[0].length + 1;
var n_1 = s[1].length + 1;

for(var i=0; i<=n_0; i++) {

    m[i]=[];
    t[i]=[];

    for(var j=0; j<=n_1; j++) {

        m[i][j]=0;
        t[i][j]=0;

        if (i>1) {m[i][0]=s[0][i-2];}
        if (j>1) {m[0][j]=s[1][j-2];}

        if (i==1 && j>1) {m[i][j]=m[i][j-1] + gap;}
        if (j==1 && i>1) {m[i][j]=m[i-1][j] + gap;}
    }
}

for(var i=0; i<=n_0; i++) {
    for(var j=0; j<=n_1; j++) {

        if(i>1 && j>1){

            var A = m[i-1][j-1] + f(m[i][0],m[0][j]); // \
            var B = m[i-1][j] + gap;                  // -
            var C = m[i][j-1] + gap;                  // |
            var D = 0;

            m[i][j] = Math.max(A, B, C, D);
        }
    }
}

function f(a1, a2) {
    if(a1 === a2){return Match;} else {return Mismatch;}
}

document.write(SMC(m));


// SHOW MATRIX CONTENT

function SMC(m) {
    var r = "<table border=1>";
```

```
    for(var i=0; i<m.length; i++) {
        r += "<tr>";
        for(var j=0; j<m[i].length; j++){
            r += "<td>"+m[i][j]+"</td>";
        }
        r += "</tr>";
    }
    r += "</table>";

    return r;
}

</script>
```

Output:

0	0	T	G	G	A	T	C	G	A
0	0	-2	-4	-6	-8	-10	-12	-14	-16
T	-2	2	0	0	0	0	0	0	0
G	-4	0	4	2	0	0	0	2	0
A	-6	0	2	3	4	2	0	0	4
A	-8	0	0	1	5	3	1	0	2
T	-10	0	0	0	3	7	5	3	1
T	-12	0	0	0	1	5	6	4	2
C	-14	0	0	0	0	3	7	5	3
A	-16	0	0	0	2	1	5	6	7
G	-18	0	2	2	0	1	3	7	5
T	-20	0	0	1	1	2	1	5	6
T	-22	0	0	0	0	3	1	3	4
A	-24	0	0	0	2	1	2	1	5

Thus, a score matrix for local alignment can only display values above zero, regardless of how the parameters are set. Up to this stage, the initialization methods for both global and local alignment have been addressed in detail. To conclude, the difference between the initialization of the score matrix for global alignment and local alignment consists in the number of variables that compete inside the maximizing function (global alignment: A, B, C; local alignment: A, B, C, and D).

3.4.3 Optimization of the Initialization Steps

Notice that Additional algorithm 3.8 contains two pairs of nested loops (four loops in total). Both pairs traverse the m matrix (and t) in exactly the same manner. Thus, the two stages of initialization can be merged and the four loops can be reduced to two loops. The entire implementation is reduced to (Additional algorithm 3.9):

Additional algorithm 3.9 Note that the source code is out of context and is intended for explanation of the method.

```
for(var i=0; i<=n_0; i++) {

    m[i]=[];
    t[i]=[];

    for(var j=0; j<=n_1; j++) {

        m[i][j]=0;
        t[i][j]=0;

        if (i>1) {m[i][0]=s[0][i-2];}
        if (j>1) {m[0][j]=s[1][j-2];}

        if (i==1 && j>1) {m[i][j]=m[i][j-1] + gap;}
        if (j==1 && i>1) {m[i][j]=m[i-1][j] + gap;}

        if(i>1 && j>1){

            var A = m[i-1][j-1] + f(m[i][0],m[0][j]); // \
            var B = m[i-1][j] + gap;                  // -
            var C = m[i][j-1] + gap;                  // |
            var D = 0;

            m[i][j] = Math.max(A, B, C, D);
        }
    }
}
```

The output of Additional algorithm 3.9 from above provides the same result as the output of Additional algorithm 3.8. Notice that the contents of the nested loop of the second initialization step were simply moved to the nested loop of the first initialization step without additional changes. Thus, four loops were merged into two loops and the processing time was reduced.

3.4.4 Curiosities

The reduction process can be pushed even further. In the nested loop approach, a matrix can be traversed in two ways. The first option involves going through the matrix on columns (*i*) with an incrementation of the row (*j*) at the end of each column. The second option involves traversing the matrix on rows (*j*) with an incrementation of the column (*i*) at the end of each row. Nonetheless, the two remaining loops can be merged into a single loop. But how can a single loop traverse a matrix? A matrix can be regarded as a vector whose content successively holds each line from the matrix. Such a hypothetical vector can be traversed continuously from one end to the other with a single loop. But how to distinguish between rows and columns? To traverse a matrix using a single loop, the implementation below allows for two approaches (Additional algorithm 3.10). Note that the output is too long and is not shown here (please run the algorithm from the additional materials online).

Additional algorithm 3.10 Note that the source code is out of context and is intended for explanation of the method.

```
<script>

var s0 = 'TAGCCCTATCGGTCA';
var s1 = 'TACGGG';

var m = [];
var s = [];

s[0] = [] = s0.split('');
s[1] = [] = s1.split('');

var n_0 = s[0].length+1;
var n_1 = s[1].length+1;

document.write('Traversing a matrix on columns by using one loop:<br>');

var i = 0;
var j = 0;

for(var v=0; v<=(n_0*n_1)-1; v++) {

    i = v % n_0;

    if(i==0 && v!=0 && j<n_1 && v!=(n_0*n_1)){j+=1;}

    document.write('m[' + i + '-' + j + ']<br>');
}


document.write('Traversing a matrix on rows by using one loop:<br>');

var i = 0;
var j = 0;

for(var v=0; v<=(n_0*n_1); v++) {
```

(Continued)

Additional algorithm 3.10 (Continued)

```
    j = v % n_1;

    if(j==0 && v!=0 && i<n_0 && v!=(n_0*n_1)){i += 1;}

    document.write('m[' + i + '-' + j + ']<br>');
}

</script>
```

Since the rows of a matrix have an equal length, this simple property can be exploited here for the one loop strategy. For example, consider the first option (traversing the matrix on columns). One solution is to use the modulo operator. The modulo operator returns the remainder of a division. Thus, consider a loop between *0* and $n\_0 \times n\_1$ (the length of the first sequence multiplied by the length of the second sequence). At each step in the loop, i holds the remainder of n_0 divided by v ($i = v$ % n_0). Each time i is equal to zero it means that n_0 is a multiple of v. In other words, the increase of i follows an abrupt reset each time n_0 divided by v is zero. Consequently, a value of zero ($i = 0$) signifies that the end of a line has been reached and a j variable can be incremented by one ($j+=1$; a new row begins). Thus, j denotes the number of rows in the matrix, whereas i denotes the columns in the same matrix. These two variables (i and j) can therefore be used in the same manner as before (i.e. as in the nested loop). In contrast, the second option (traversing the matrix on rows) makes a switch between variables i and j and uses the length of the second sequence (n_1) in the modulo operator approach (i.e. $j = v$ % n_1). A complete adaptation that uses a single-loop initialization is presented in Additional algorithm 3.11. The approach is similar to Additional algorithm 3.8; however, it uses a single loop, which traverses the m matrix on columns to perform the score matrix initialization, specifically for local alignment.

Additional algorithm 3.11 Note that the source code is in context and works with copy/paste.

```
<script>

var s0 = 'TGAATTCAGTTA';
var s1 = 'TGGATCGA';

var Match = +2;
var Mismatch = -1;
var gap = -2;
```

```
var m = [];
var s = [];

s[0] = [] = s0.split('');
s[1] = [] = s1.split('');

var n_0 = s[0].length + 1;
var n_1 = s[1].length + 1;

var nm = n_0*n_1;

var i = 0;
var j = 0;

m[0]=[];

for(var v=0; v<=nm; v++) {

    j = v % n_1;

    if(j==0 && v!=0 && i<n_0 && v!=nm){i+=1;m[i]=[];}

    m[i][j]=0;

    if (i>1) {m[i][0]=s[0][i-2];}
    if (j>1) {m[0][j]=s[1][j-2];}

    if (i==1 && j>1) {m[i][j]=m[i][j-1] + gap;}
    if (j==1 && i>1) {m[i][j]=m[i-1][j] + gap;}

    if(i>1 && j>1){

        var A = m[i-1][j-1] + f(m[i][0],m[0][j]); // \
        var B = m[i-1][j] + gap;                  // -
        var C = m[i][j-1] + gap;                  // |
        var D = 0;

        m[i][j] = Math.max(A, B, C, D);
    }
}

function f(a1, a2) {
    if(a1 === a2){return Match;} else {return Mismatch;}
}

document.write(SMC(m));

// SHOW MATRIX CONTENT
function SMC(m) {
```

(Continued)

Additional algorithm 3.11 (Continued)

```
    var r = "<table border=1>";
    for(var i=0; i<m.length; i++) {
        r += "<tr>";
        for(var j=0; j<m[i].length; j++){
            r += "<td>"+m[i][j]+"</td>";
        }
        r += "</tr>";
    }
    r += "</table>";

    return r;
}

</script>
```

Output:

0	0	T	G	G	A	T	C	G	A
0	0	-2	-4	-6	-8	-10	-12	-14	-16
T	-2	2	0	0	0	0	0	0	0
G	-4	0	4	2	0	0	0	2	0
A	-6	0	2	3	4	2	0	0	4
A	-8	0	0	1	5	3	1	0	2
T	-10	0	0	0	3	7	5	3	1
T	-12	0	0	0	1	5	6	4	2
C	-14	0	0	0	0	3	7	5	3
A	-16	0	0	0	2	1	5	6	7
G	-18	0	2	2	0	1	3	7	5
T	-20	0	0	1	1	2	1	5	6
T	-22	0	0	0	0	3	1	3	4
A	-24	0	0	0	2	1	2	1	5

Of course, the same implementation can be designed for global alignment by adapting the Additional algorithm 3.8 to a single loop. Nevertheless, although the single-loop implementation is interesting, it does not bring time-related optimizations. The single-loop approach follows the same number of steps as the nested loop, namely $n \times m$ (n_0*n_1). Therefore, to avoid confusions, the single-loop approach will not be used further and it remains only an option. Up to this point, either the score matrix of global alignment (Additional algorithm 3.6) or the score matrix of the local alignment (Additional algorithm 3.8), are ready to be used.

3.5 Traceback

The traceback is the process of deducing and building the best alignment by following the values over the score matrix (Figure 3.2). The traceback is a recursive process that starts from an initial element ($m_{i,j}$) and tends to end in the upper left corner of the matrix ($m_{2,2}$). The elements from which the traceback starts and stops, makes the difference between global and local alignment. In the global alignment, the starting point is represented by the element found in the lower right corner of the matrix ($m_{n,m}$). In local alignment, the starting point is represented by the element which contains the maximum value above the matrix (Table 3.1).

The two methods differ again on the traceback breakpoints. In the case of global alignment, the breakpoint is always the upper left element of the matrix ($m_{2,2}$). In the case of local alignment, the breakpoint is the first element encountered in the path of the traceback that shows a value of zero (Table 3.1). From the starting point of the traceback (whatever it may be), a series of rules simultaneously establish the path taken by the algorithm and the alignment of the two sequences. From the starting point, the path follows three possible moves: diagonally, up, or left (Figure 3.2a,e). The rules for such a navigation are the same with those used in the score matrix initialization stage. An implementation of the traceback stage is shown out of context in Additional algorithm 3.12. The traceback stage receives the matrix initialized by the previous module. First, new variables are declared, such as *AlignmentA* or *AlignmentB* that store the two aligned sequences, and the variables that dictate the starting point for the traceback. A recursive loop starts from the start element (i=n_0;j=n_1;). In the recursive loop, two conditions dictate which characters in the two sequences are added in reverse order inside *AlignmentA* and *AlignmentB* variables.

If the value in the current element (m[i][j]) is equal to the value of the element on the diagonal (m[i-1][j-1]) plus the value returned by the function f (f(Ai, Bj)), then the letter (Bj) to the left (m[0][j]) of the current element will be added in front of the string that is stored in *AlignmentA* variable (Ai+AlignmentA) and the letter (Ai) above (m[i][0]) the current element will be added in front of the string from the *AlignmentB* variable (Additional algorithm 3.12). The value returned by function f determines whether the above condition is true or not. The value returned by function f will be the value from the *Match* variable if the two letters are the same (Ai=Bj), or the value from the *Mismatch* variable will be returned otherwise. If the two letters are not identical, then function f will return the value for *Mismatch* and the condition discussed

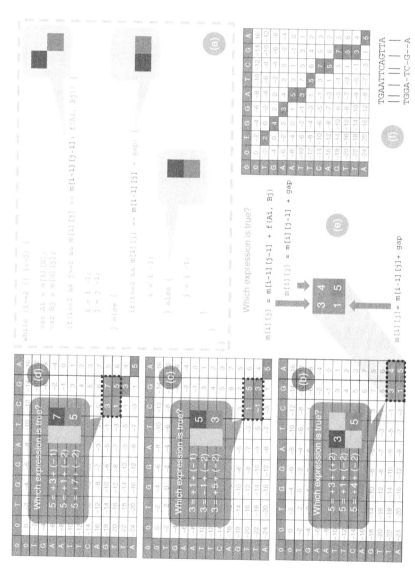

Figure 3.2 Traceback rules. (a) shows the link between the implementation and the relative position of each element. (b–d) Show the first three iterations made by the traceback module in the global alignment case. (e) It shows the positions of the elements against which the equality is being verified. (f) Shows the complete traceback path and the two sequences aligned according to this path.

Table 3.1 Traceback start and stop locations.

	Global	Local
Traceback start location	$m_{n,\,m}$	$Max(m_{ij})$
Traceback stop location	$m_{2,\,2}$	if $m_{ij} = 0$

Beginning and end points of the traceback paths in the case of global and local alignment algorithms.

above will not be true. In that case, a second condition is required, namely if the value from the current element (m[i][j]) is equal to the value from the left element (m[i-1][j]) plus the value of the *gap* variable (gap), then a gap symbol ('-') will be inserted in front of the string from the *AlignmentB* variable and the letter (Ai) above (m[i][0]) the current element will be inserted in front of the string from the *AlignmentA* variable. If none of the conditions are true, a gap symbol ('-') will be inserted in front of the string from the *AlignmentA* variable and the letter (Bj) from the left (m[0][j]) of the current element (m[i][j]) will be inserted in front of the string from *AlignmentB* variable. The meaning of gap symbols: a gap signifies a mutation, namely either a deletion on the sequence on which the gap symbol is present (e.g. ACTG → AC-G) or it can mean an insertion of a letter/nucleotide on the opposite side of the gap symbol (e.g. ACG → ACTG). The position on the alignment that contains a gap symbol on either side, it is called an indel (insertion or deletion).

Additional algorithm 3.12 Note that the source code is out of context and is intended for explanation of the method.

```
var AlignmentA = "";
var AlignmentM = "";
var AlignmentB = "";

var i = n_0;
var j = n_1;

while (i>=2 || j>=2) {

    var Ai = m[i][0];
    var Bj = m[0][j];

    if(i>=2 && j>=2 && m[i][j]==m[i-1][j-1] + f(Ai, Bj)) {
```

(Continued)

Additional algorithm 3.12 (Continued)

```
        AlignmentA = Ai + AlignmentA;
        AlignmentB = Bj + AlignmentB;

        if(Ai==Bj){
            AlignmentM = '|' + AlignmentM;
        } else {
            AlignmentM = ' ' + AlignmentM;
        }

        i = i - 1;
        j = j - 1;

    } else {

        if(i>=2 && m[i][j]==m[i-1][j] + gap) {

            AlignmentA = Ai + AlignmentA;
            AlignmentB = '-' + AlignmentB;
            AlignmentM = ' ' + AlignmentM;

            i = i - 1;

        } else {

            AlignmentA = '-' + AlignmentA;
            AlignmentB = Bj + AlignmentB;
            AlignmentM = ' ' + AlignmentM;

            j = j - 1;
        }
    }
}
```

The conditions described above are checked by the algorithm at each iteration (Additional algorithm 3.12). At the end of the traceback operation, the two strings from *AlignmentA* and *AlignmentB* variables can be placed one below the other for a proper visualization of the alignment. Also, to highlight the correspondence between the two aligned sequences, a third variable was introduced, namely *AlignmentM*. The addition of correspondence characters to this variable is relatively simple. Each time function f returns the value of the *Match* variable, a correspondence symbol (" | ") is added to the left side of the string found in the *AlignmentM* variable. If the f function returns the value of the *Mismatch* variable, then it means that the corresponding letters on the two sequences (*AlignmentA* and *AlignmentB*) are not the same and an HTML space symbol (" ") is added in front of the string from *AlignmentM* variable.

3.6 Global Alignment

The full implementation of the global alignment algorithm is presented here. The use of all the steps discussed so far is mandatory in the functional context of these algorithms. Up to this point, font styles have not been important. However, when constructing the alignment, a monospace font must be declared initially to obtain symbols/letters of equal length. An equal length allows for a visual correspondence between the alignment strings, when positioned one below the other. Therefore, the grand implementation begins with the statement of this font style (Additional algorithm 3.13). Variables from all stages are declared immediately after the font style. Next, the score matrix initialization module is introduced (with both stages included in a single nested loop). The recently discussed traceback module is introduced after the score matrix initialization module and the three variables that store the alignment (*AlignmentA*, *AlignmentM*, and *AlignmentB*) are printed directly in the output of the application for visualization. Finally, at the end of the implementation, the *f* function and the *SMC* function are introduced (Additional algorithm 3.13). Notice that the score matrix initialization module transfers the coordinates of the lower right element ($i=n\_0; j=n\_1;$) to the traceback module.

Additional algorithm 3.13 Note that the source code is in context and works with copy/paste.

```
<style>
    body {
        padding: 1rem;
        font-family: monospace;
        font-size: 18px;
        font-style: normal;
        font-variant: normal;
        line-height: 20px;
    }
</style>

<script>

// variable statement
var s0 = 'TGAATTCAGTTA';
```

(Continued)

Additional algorithm 3.13 (Continued)

```
var s1 = 'TGGATCGA';

var Match = +2;
var Mismatch = -1;
var gap = -2;

var AlignmentA = "";
var AlignmentM = "";
var AlignmentB = "";

var m = [];
var t = [];
var s = [];

// Initialization and completion
s[0] = [] = s0.split('');
s[1] = [] = s1.split('');

var n_0 = s[0].length + 1;
var n_1 = s[1].length + 1;

for(var i=0; i<=n_0; i++) {

    m[i]=[];
    t[i]=[];

    for(var j=0; j<=n_1; j++) {

        m[i][j]=0;
        t[i][j]=0;

        if (i==1 && j>1) {m[i][j]=m[i][j-1]+gap;}
        if (j==1 && i>1) {m[i][j]=m[i-1][j]+gap;}

        if (i>1) {m[i][0]=t[i][0]=s[0][i-2];}
        if (j>1) {m[0][j]=t[0][j]=s[1][j-2];}

        if(i>1 && j>1){

            var A = m[i-1][j-1] + f(m[i][0],m[0][j]);      //'\
            var B = m[i-1][j] + gap;                        //'-
            var C = m[i][j-1] + gap;                        //'|

            m[i][j] = Math.max(A, B, C);
        }
    }
}
```

```
//Traceback
var i = n_0;
var j = n_1;

while (i>=2 || j>=2) {

    var Ai = m[i][0];
    var Bj = m[0][j];

    if(i>=2 && j>=2 && m[i][j]==m[i-1][j-1] + f(Ai, Bj)) {

        t[i][j] = m[i-1][j-1] + f(Ai, Bj);
        AlignmentA = Ai + AlignmentA;
        AlignmentB = Bj + AlignmentB;

        if(Ai==Bj){
            AlignmentM = '|' + AlignmentM;
        } else {
            AlignmentM = ' ' + AlignmentM;
        }

        i = i - 1;
        j = j - 1;

    } else {

        if(i>=2 && m[i][j]==m[i-1][j] + gap) {

            t[i][j] = m[i-1][j] + gap;
            AlignmentA = Ai + AlignmentA;
            AlignmentB = '-' + AlignmentB;
            AlignmentM = ' ' + AlignmentM;
            i = i - 1;

        } else {

            t[i][j] = m[i][j-1] + gap;
            AlignmentA = '-' + AlignmentA;
            AlignmentB = Bj + AlignmentB;
            AlignmentM = ' ' + AlignmentM;
            j = j - 1;
        }
    }

    if(i<=2 && j<=2) {t[i][j] = 0;}
}
```

(Continued)

Additional algorithm 3.13 (Continued)

```
document.write(SMC(m) + '<hr>');
document.write(SMC(t) + '<hr>');

document.write(AlignmentA + '<br>');
document.write(AlignmentM + '<br>');
document.write(AlignmentB + '<br>');

function f(a1, a2) {
    if(a1 === a2){return Match;} else {return Mismatch;}
}

// SHOW MATRIX CONTENT
function SMC(m) {
    var r = "<table border=1>";
    for(var i=0; i<m.length; i++) {
        r += "<tr>";
        for(var j=0; j<m[i].length; j++){
            r += "<td>"+m[i][j]+"</td>";
        }
        r += "</tr>";
    }
    r += "</table>";

    return r;
}
</script>
```

Output:

0	0	T	G	G	A	T	C	G	A
0	0	-2	-4	-6	-8	-10	-12	-14	-16
T	-2	2	0	-2	-4	-6	-8	-10	-12
G	-4	0	4	2	0	-2	-4	-6	-8
A	-6	-2	2	3	4	2	0	-2	-4
A	-8	-4	0	1	5	3	1	-1	0
T	-10	-6	-2	-1	3	7	5	3	1
T	-12	-8	-4	-3	1	5	6	4	2
C	-14	-10	-6	-5	-1	3	7	5	3
A	-16	-12	-8	-7	-3	1	5	6	7
G	-18	-14	-10	-6	-5	-1	3	7	5
T	-20	-16	-12	-8	-7	-3	1	5	6
T	-22	-18	-14	-10	-9	-5	-1	3	4
A	-24	-20	-16	-12	-8	-7	-3	1	5

0	0	T	G	G	A	T	C	G	A
0	0	0	0	0	0	0	0	0	0

T	0	2	0	0	0	0	0	0	0
G	0	0	4	0	0	0	0	0	0
A	0	0	0	3	0	0	0	0	0
A	0	0	0	0	5	0	0	0	0
T	0	0	0	0	3	0	0	0	0
T	0	0	0	0	0	5	0	0	0
C	0	0	0	0	0	0	7	0	0
A	0	0	0	0	0	0	5	0	0
G	0	0	0	0	0	0	0	7	0
T	0	0	0	0	0	0	0	5	0
T	0	0	0	0	0	0	0	3	0
A	0	0	0	0	0	0	0	0	5

```
TGAATTCAGTTA
|| | || |  |
TGGA-TC-G--A
```

These coordinates indicate to the traceback module from where the alignment path should start. This is obvious here, but it becomes more important in the case of local alignment as described later. The global alignment implementation from above shows the score matrix, the traceback matrix, and the main global alignment. The traceback matrix was also introduced in the traceback module. The visualization of the traceback matrix shows the preferred path of the algorithm. Note: Combinations made between different values of the main parameters (*Match*; *Mismatch*; and *gap*) may change the alignment configuration if alternative alignments exist.

3.7 Local Alignment

The full implementation of the local alignment algorithm is presented here (Additional algorithm 3.14). Unlike global alignment, local alignment requires the position of the element containing the maximum value over the matrix. The element showing the maximum value over m will be the starting point for the traceback procedure. Thus, a detection scheme is added to the initialization module as follows:

```
if(m[i][j] > MMax){MMax = m[i][j];x=i;y=j;}
```

where x and y are the new variables that hold the coordinates of the element with the maximum value and the *MMax* variable contains the latest maximum value of an element. The final result of the above scheme establishes the new values for variables i and j, that point out the element from which the traceback starts, as follows:

```
var i = x;
var j = y;
```

Of course, a matrix can show several elements with values close to the maximum value, and therefore those elements represent the alternative traceback start locations of other alignment configurations. However, such alternative traceback start locations require another type of approach called "forced alignment," which is discussed in great detail in the next chapter.

Additional algorithm 3.14 Note that the source code is in context and works with copy/paste.

```
<style>
    body {
        padding: 1rem;
        font-family: monospace;
        font-size: 18px;
        font-style: normal;
        font-variant: normal;
        line-height: 20px;
    }
</style>

<script>

// declaratie variabile
var s0 = 'TGAATTCAGTTA';
var s1 = 'TGGATCGA';

var Match = +2;
var Mismatch = -1;
var gap = -2;

var AlignmentA = "";
var AlignmentM = "";
var AlignmentB = "";

var m = [];
var t = [];
var s = [];

var MMax = 0;
var MMin = 0;

var x = 0;
var y = 0;

// Matrix initialization and completion
s[0] = [] = s0.split('');
s[1] = [] = s1.split('');
```

```
var n_0 = s[0].length+1;
var n_1 = s[1].length+1;

for(var i=0; i<=n_0; i++) {

    m[i]=[];
    t[i]=[];

    for(var j=0; j<=n_1; j++) {

        m[i][j]=0;
        t[i][j]=0;

        if (i>1) {m[i][0]=t[i][0]=s[0][i-2];}
        if (j>1) {m[0][j]=t[0][j]=s[1][j-2];}

        if (i>1) {m[i][0]=s[0][i-2];}
        if (j>1) {m[0][j]=s[1][j-2];}

        if(i>1 && j>1){

            var A = m[i-1][j-1] + f(m[i][0],m[0][j]);    //'\
            var B = m[i-1][j] + gap;                      //'-
            var C = m[i][j-1] + gap;                      //'|
            var D = 0;

            m[i][j] = Math.max(A, B, C, D);

            if(m[i][j] > MMax){MMax = m[i][j];x=i;y=j;}
            if(m[i][j] < MMin){MMin = m[i][j];}
        }
    }
}

//Traceback & text alignment
var i = x;
var j = y;

while (i>=2 || j>=2) {

    var Ai = m[i][0];
    var Bj = m[0][j];

    if(i>=2 && j>=2 && m[i][j]==m[i-1][j-1] + f(Ai, Bj)) {
        t[i][j] = m[i-1][j-1] + f(Ai, Bj);
        AlignmentA = Ai + AlignmentA;
        AlignmentB = Bj + AlignmentB;
```

(Continued)

Additional algorithm 3.14 (Continued)

```
        if(Ai==Bj){
            AlignmentM = '|' + AlignmentM;
        } else {
            AlignmentM = ' ' + AlignmentM;
        }

        i = i - 1;
        j = j - 1;

    } else {

        if(i>=2 && m[i][j]==m[i-1][j] + gap) {
            t[i][j] = m[i-1][j] + gap;
            AlignmentA = Ai + AlignmentA;
            AlignmentB = '-' + AlignmentB;
            AlignmentM = ' ' + AlignmentM;
            i = i - 1;

        } else {
            t[i][j] = m[i][j-1] + gap;
            AlignmentA = '-' + AlignmentA;
            AlignmentB = Bj + AlignmentB;
            AlignmentM = ' ' + AlignmentM;
            j = j - 1;
        }
    }

    if(m[i][j]<=0){break;}
}

// Print matrix m
document.write(SMC(m) + '<hr>');
document.write(SMC(t) + '<hr>');

// Print the alignment
document.write(AlignmentA + '<br>');
document.write(AlignmentM + '<br>');
document.write(AlignmentB + '<br>');

// Matching function
function f(a1, a2) {
    if(a1 === a2){return Match;} else {return Mismatch;}
}

// SHOW MATRIX CONTENT
function SMC(m) {
    var r = "<table border=1>";
```

```
    for(var i=0; i<m.length; i++) {
        r += "<tr>";
        for(var j=0; j<m[i].length; j++){
            r += "<td>"+m[i][j]+"</td>";
        }
        r += "</tr>";
    }
    r += "</table>";
    return r;
}
</script>
```

Output:

0	0	T	G	G	A	T	C	G	A
0	0	0	0	0	0	0	0	0	0
T	0	2	0	0	0	2	0	0	0
G	0	0	4	2	0	0	1	2	0
A	0	0	2	3	4	2	0	0	4
A	0	0	0	1	5	3	1	0	2
T	0	2	0	0	3	7	5	3	1
T	0	2	1	0	1	5	6	4	2
C	0	0	1	0	0	3	7	5	3
A	0	0	0	0	2	1	5	6	7
G	0	0	2	2	0	1	3	7	5
T	0	2	0	1	1	2	1	5	6
T	0	2	1	0	0	3	1	3	4
A	0	0	1	0	2	1	2	1	5

0	0	T	G	G	A	T	C	G	A
0	0	0	0	0	0	0	0	0	0
T	0	2	0	0	0	0	0	0	0
G	0	0	4	0	0	0	0	0	0
A	0	0	0	3	0	0	0	0	0
A	0	0	0	0	5	0	0	0	0
T	0	0	0	0	0	7	0	0	0
T	0	0	0	0	0	0	0	0	0
C	0	0	0	0	0	0	0	0	0
A	0	0	0	0	0	0	0	0	0
G	0	0	0	0	0	0	0	0	0
T	0	0	0	0	0	0	0	0	0
T	0	0	0	0	0	0	0	0	0
A	0	0	0	0	0	0	0	0	0

```
TGAAT
|| ||
TGGAT
```

Note: The t matrix only serves to show the path of the traceback and is not involved in the alignment. As such, this matrix will be removed in the next steps. Also, since the whole focus is on the main alignment, matrix m will not be shown in the following versions. The above result (i.e. the output of Additional algorithm 3.14) can be disappointing, at least up to this point. Notice the difficulty in distinguishing between the above results and those provided by global alignment. Moreover, at this stage, the local alignment results are uncertain, as nothing is easily recognizable about the matched regions from the two sequences.

3.8   Alignment Layout

The local alignment algorithm becomes truly useful and interesting if both sequences (s0 and s1) are shown in their entirety over the alignment (Figure 3.3b). The relative position of one sequence over the other provides a great visual insight. However, for such an approach, one more step is needed. Note that variables x and y retain the position of the element at which the traceback module began the alignment, while $r1$ and $r2$ retain the position of the element at which the traceback module interrupted the alignment. Once the traceback ends, the length of the alignment is equal to the length of the traceback path (see the output of Additional algorithm 3.14).

To observe the two sequences of the alignment in their entirety, the current alignment requires additions, both to the left and the right sides of the alignment strings (i.e. *AlignmentA* and *AlignmentB* variables). Moreover, the correct framing of the alignment can be done by adding an appropriate number of spaces on the left side of the two sequences. In a first phase, the algorithm below uses the four variables (i.e. $r1, r2, x, y$) and the *substr* method to add the missing regions to the old alignment (Additional algorithm 3.15). This is realized by using a copy/paste strategy (Figure 3.3a–c; note that here, x and y can be misleading as they have a different meaning from the classical axis approach). The subsegments that are required to the left side of the string found in the *AlignmentA* variable are taken from the s0 sequence, starting from position 0 up to position $r1$. In contrast, the subsegments that are required to the left side of the string from *AlignmentB* are taken from the s1 sequence, starting from position 0 up to position $r2$. The subsegments that are required on the right side of the string in *AlignmentA* are taken from the sequence s0, starting from position x up to the end of the sequence (n_0-x – the s0 sequence length minus the end of the alignment). Conversely, the subsegments that are required on the right side of the string from *AlignmentB* are taken from the s1 sequence, starting from position y up to the end of the sequence (n_1-y). However, unequal additions ($r1 \neq r2$) made to the left side of the strings, shift the region of the alignment. As a solution, a series of space characters are introduced in order to push the alignment back to the initial position.

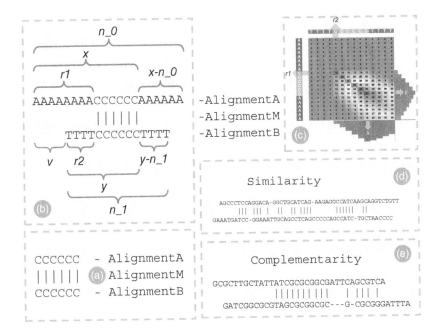

Figure 3.3 Additions and new possibilities for local sequence alignment. (a) It shows the result of classical local alignment. Note that the local alignment algorithm allows only the visualization of the region that constitutes the alignment and it eliminates the adjacent regions from both sequences. (b) Indicates the significance of the variables in context and shows the method of completion and relative placement of the alignment. (c) Presents the score matrix of the classical local alignment from panel a. Here the path followed by the algorithm and the correspondence with the letters from both sequences can be observed. (d) Shows the result of the local alignment for a longer sequence. (e) Shows an alternative method of using local alignment for detection of complementarity between two sequences.

Additional algorithm 3.15 Note that the source code is out of context and is intended for explanation of the method.

```
var r1 = i - 1;
var r2 = j - 1;

// Relative positioning of sequences
var tM='';
var tS='';
```

(Continued)

Additional algorithm 3.15 (Continued)

```
// Add to the end
AlignmentA = AlignmentA + s0.substr(x-1, n_0 - x);
AlignmentB = AlignmentB + s1.substr(y-1, n_1 - y);

// Add to the beginning
AlignmentA = s0.substr(0, r1) + AlignmentA;
AlignmentB = s1.substr(0, r2) + AlignmentB;

if(r1>r2){
    var v = r1 - r2;

    for(var u=1; u<=v; u++) {tS = tS + e;}
    for(var u=1; u<=v+r2; u++) {tM = tM + e;}

    AlignmentB = tS + AlignmentB;
    AlignmentM = tM + AlignmentM;
} else {
    var v = r2 - r1;

    for(var u=1; u<=v; u++) {tS = tS + e;}
    for(var u=1; u<=v+r1; u++) {tM = tM + e;}

    AlignmentA = tS + AlignmentA;
    AlignmentM = tM + AlignmentM;
}
```

Thus, in the next stage, a condition checks which of the two variables ($r1$ or $r2$) contains a higher value. A higher value for $r1$ involves the introduction of space characters to the left side of *AlignmentB* and *AlignmentM*. The number (v) of space characters is established by the result obtained from subtracting $r2$ from $r1$ ($v=r1-r2$). A higher value for $r2$ involves the introduction of space characters to the left side of *AlignmentA* and *AlignmentM*. Consequently, the number (v) of space characters is established by the result obtained from subtracting $r1$ from $r2$ ($v=r2-r1$). The space characters required for the left side of the alignment are progressively added inside two variables (tS and tM). Their temporary content is then pushed into the alignment according to the values shown by $r1$ and $r2$. This concludes the main strategy on the relative positioning of an alignment.

3.9   Local Sequence Alignment – The Final Version

The final implementation begins by declaring the type of font discussed at the beginning of the chapter. Next, the variables used are declared and the appropriate values are set. For a change of scenery, two new sequences may be used (Figure 3.3d):

s0: "AGCCCTCCAGGACAGGCTGCATCAGAAGAGGCCATCAAGCAGGTCTGTT"
s1: "GAAATGATCCGGAAATTGCAGCCTCAGCCCCCAGCCATCTGCTAACCCC"

In a single nested loop, the alignment matrix is initiated and the score values are calculated for each element in the matrix (Additional algorithm 3.16). In a second step, three strings are built for the alignment, namely the string of the first sequence, the intermediate string, which indicates the similarity between the two sequences, and the string of the second sequence. In a third stage, the alignment positions are recalculated. The first sequence is positioned relative to the second sequence by inserting a precise number of space characters where appropriate. In the last step, the three strings are printed directly below each other to display the alignment. Note: the *SMC* function has been removed as the visualization of the score matrix is only optional at this point. Also, the HTML symbol for space is stored in variable *e,* which is used where it is needed throughout the implementation.

Additional algorithm 3.16 Note that the source code is in context and works with copy/paste.

```
<style>
    body {
        padding: 1rem;
        font-family: monospace;
        font-size: 18px;
        font-style: normal;
        font-variant: normal;
        line-height: 20px;
    }
</style>

<script>

// Variable statement
var Match = +2;
var Mismatch = -1;
```

(Continued)

Additional algorithm 3.16 (Continued)

```
var gap = -2;
var s0 = 'AGCCCTCCAGGACAGGCTGCATCAGAAGAGGCCATCAAGCAGGTCTGTT';
var s1 = 'GAAATGATCCGGAAATTGCAGCCTCAGCCCCCAGCCATCTGCTAACCCC';

var AlignmentA = "";
var AlignmentM = "";
var AlignmentB = "";

var e = ' ';

var m = [];
var s = [];

var MMax = 0;
var MMin = 0;

var x = 0;
var y = 0;

// Matrix initialization and completion
s[0] = [] = s0.split('');
s[1] = [] = s1.split('');

var n_0 = s[0].length + 1;
var n_1 = s[1].length + 1;

for(var i=0; i<=n_0; i++) {

    m[i]=[];

    for(var j=0; j<=n_1; j++) {

        m[i][j]=0;

        if (i==1 && j>1) {m[i][j]=m[i][j-1]+gap;}
        if (j==1 && i>1) {m[i][j]=m[i-1][j]+gap;}

        if (i>1) {m[i][0]=s[0][i-2];}
        if (j>1) {m[0][j]=s[1][j-2];}

        if(i>1 && j>1){
            var A = m[i-1][j-1] + f(m[i][0],m[0][j]);   //'\
            var B = m[i-1][j] + gap;                     //'-
            var C = m[i][j-1] + gap;                     //'|
            var D = 0;

            m[i][j] = Math.max(A, B, C, D);

            if(m[i][j] > MMax){MMax = m[i][j];x=i;y=j;}
```

```
            if(m[i][j] < MMin){MMin = m[i][j];}
        }
    }
}

//Traceback & text alignment
var i = x;
var j = y;

while (i>=2 || j>=2) {

    var Ai = m[i][0];
    var Bj = m[0][j];

    A = m[i-1][j-1] + f(Ai, Bj);
    B = m[i-1][j] + gap;
    C = m[i][j-1] + gap;

    if(i>=2 && j>=2 && m[i][j]==A) {

        AlignmentA = Ai + AlignmentA;
        AlignmentB = Bj + AlignmentB;

        if(Ai==Bj){
            AlignmentM = '|' + AlignmentM;
        } else {
            AlignmentM = e + AlignmentM;
        }

        i = i - 1;
        j = j - 1;

    } else {

        if(i>=2 && m[i][j]==B) {
            AlignmentA = Ai + AlignmentA;
            AlignmentB = '-' + AlignmentB;
            AlignmentM = e + AlignmentM;
            i = i - 1;

        } else {
            AlignmentA = '-' + AlignmentA;
            AlignmentB = Bj + AlignmentB;
            AlignmentM = e + AlignmentM;
            j = j - 1;
        }
    }

    var r1 = i - 1;
```

(Continued)

Additional algorithm 3.16 (Continued)

```
      var r2 = j - 1;

      if (m[i][j]<=0) {break;}
}

// LAYOUT
var tM='';
var tS='';

// Check the end
AlignmentA = AlignmentA + s0.substr(x-1, n_0 - x);
AlignmentB = AlignmentB + s1.substr(y-1, n_1 - y);

// Check the beginning
AlignmentA = s0.substr(0, r1) + AlignmentA;
AlignmentB = s1.substr(0, r2) + AlignmentB;

if (r1>r2) {
    var v = r1 - r2;

    for (var u=1; u<=v; u++) {tS = tS + e;}
    for (var u=1; u<=v+r2; u++) {tM = tM + e;}

    AlignmentB = tS + AlignmentB;
    AlignmentM = tM + AlignmentM;
} else {
    var v = r2 - r1;

    for (var u=1; u<=v; u++) {tS = tS + e;}
    for (var u=1; u<=v+r1; u++) {tM = tM + e;}

    AlignmentA = tS + AlignmentA;
    AlignmentM = tM + AlignmentM;
}

// Print the alignment
document.write(AlignmentA + '<br>');
document.write(AlignmentM + '<br>');
document.write(AlignmentB + '<br>');

// Matching function
function f(a1, a2) {
    if (a1 === a2) {return Match;} else {return Mismatch;}
}

</script>
```

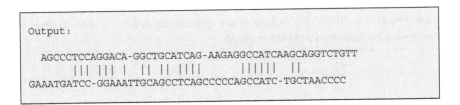

```
Output:

 AGCCCTCCAGGACA-GGCTGCATCAG-AAGAGGCCATCAAGCAGGTCTGTT
     ||| ||| |  || || ||||        |||||| ||
 GAAATGATCC-GGAAATTGCAGCCTCAGCCCCCAGCCATC-TGCTAACCCC
```

The output from Additional algorithm 3.16 shows an intuitive alignment of the two sequences taken into account. It is now perhaps clear that global alignment is less useful than local alignment. Moreover, countless variations and innovations exist on alignment algorithms. However, the local alignment algorithm always shows ideal results. As it will be demonstrated in the next chapter, variations between the values of the parameters, that control the behavior of the local alignment algorithm, allow the detection of all possible alignment configurations. Note: The above implementation is the core of the *BioJupiter* application, which is described in detail in the next chapter (please see the additional materials online).

3.10 Complementarity

Up to this point, the local alignment algorithm has been used as a detection method for the similarity between two sequences. Nevertheless, a creative approach can turn the local alignment algorithm into a complementarity detector (Figure 3.3e). A small addition to Additional algorithm 3.16 can simulate the complementarity between different regions of the same strand or between two separate DNA/RNA strands. For example, this technique may be useful for single-stranded RNA sequences, which tend to form areas of self-complementarity, which may lead further to various biological functions. To repurpose the local alignment algorithm, a function for generating the complement of a sequence must be constructed. The implementation below presents a simple solution for obtaining the complement of a DNA sequence (Additional algorithm 3.17). In the first instance, the function replaces each type of letter in the sequence with a different single-digit integer (ie. 0 to 9). Thus, a sequence of digits is obtained, each digit corresponding to a specific letter from the initial sequence of letters. Then, in order to replace each digit with a complimentary letter, the function takes into account the mapping between digits and letters.

Additional algorithm 3.17 Note that the source code is out of context and is intended for explanation of the method.

```
function complement(c) {

    c = c.replace(/C/g,'1');
    c = c.replace(/G/g,'2');
    c = c.replace(/A/g,'3');
    c = c.replace(/T/g,'4');

    c = c.replace(/1/g,'G');
    c = c.replace(/2/g,'C');
    c = c.replace(/3/g,'T');
    c = c.replace(/4/g,'A');

    return c;
}
```

As perhaps expected, a direct replacement of one letter with its complement is not possible. This concludes the *complement* function. A few simple changes are needed to use the local alignment algorithm for complementarity. Initially, two sequences are considered for complementarity, namely:

s0: "GCGCTTGCTATTATCGCGCGGCGATTCAGCGTCA"
s1: "GATCGGCGCGTAGCGCGGCGCGCGCGGGATTTA"

Next, the *complement* function (Additional algorithm 3.17) is inserted at the end of the algorithm (Additional algorithm 3.16) so that it can be called when needed. As soon as variable *s0* is declared, it is immediately transformed into a complementary sequence by using the *complement* function.

s0: "GCGCTTGCTATTATCGCGCGGCGATTCAGCGTCA" – original version
s0: "CGCGAACGATAATAGCGCGCCGCTAAGTCGCAGT" – complementary sequence

Then, the implementation makes the normal alignment between the complementary version of the *s0* sequence and the unmodified *s1* sequence. Before the alignment is fully displayed, the complementary version of *s0* found in *AlignmentA* is converted back to the original version of *s0* by calling the *complement* function again. In other words, *s0* becomes its complement (s0=complement(s0);), which is ordinarily aligned with sequence *s1*. Before presenting the alignment, the *s0* sequence contained in the *AlignmentA* is converted back (`comple-ment(AlignmentA)`) to the initial *s0* sequence and the complementarity between the two DNA strands is displayed.

Additional algorithm 3.18 Note that the source code is in context and works with copy/paste.

```
<style>
    body {
        padding: 1rem;
        font-family: monospace;
        font-size: 18px;
        font-style: normal;
        font-variant: normal;
        line-height: 20px;
    }
</style>

<script>

// Variable statement
var Match = +2;
var Mismatch = -1;
var gap = -2;

var s0 = 'GCGCTTGCTATTATCGCGCGGCGATTCAGCGTCA';

s0 = complement(s0);

var s1 = 'GATCGGCGCGTAGCGCGGCGCGCGCGGGATTTA';

var AlignmentA = "";
var AlignmentM = "";
var AlignmentB = "";

var e = ' ';

var m = [];
var s = [];

var MMax = 0;
var MMin = 0;

var x = 0;
var y = 0;

// Matrix initialization and completion
```

(Continued)

Additional algorithm 3.18 (Continued)

```
s[0] = [] = s0.split('');
s[1] = [] = s1.split('');

var n_0 = s[0].length + 1;
var n_1 = s[1].length + 1;

for(var i=0; i<=n_0; i++) {

    m[i]=[];

    for(var j=0; j<=n_1; j++) {

        m[i][j]=0;

        if (i==1 && j>1) {m[i][j]=m[i][j-1]+gap;}
        if (j==1 && i>1) {m[i][j]=m[i-1][j]+gap;}

        if (i>1) {m[i][0]=s[0][i-2];}
        if (j>1) {m[0][j]=s[1][j-2];}

        if(i>1 && j>1){

            var A = m[i-1][j-1] + f(m[i][0],m[0][j]); //'\
            var B = m[i-1][j] + gap;                  //'-
            var C = m[i][j-1] + gap;                  //'|
            var D = 0;

            m[i][j] = Math.max(A, B, C, D);

            if(m[i][j] > MMax){MMax = m[i][j];x=i;y=j;}
            if(m[i][j] < MMin){MMin = m[i][j];}
        }
    }
}

//Traceback & text alignment
var i = x;
var j = y;

while (i>=2 || j>=2) {

    var Ai = m[i][0];
    var Bj = m[0][j];

    A = m[i-1][j-1] + f(Ai, Bj);
    B = m[i-1][j] + gap;
    C = m[i][j-1] + gap;
```

```
    if(i>=2 && j>=2 && m[i][j]==A) {

        AlignmentA = Ai + AlignmentA;
        AlignmentB = Bj + AlignmentB;

        if(Ai==Bj){
            AlignmentM = '|' + AlignmentM;
        } else {
            AlignmentM = e + AlignmentM;
        }

        i = i - 1;
        j = j - 1;

    } else {

        if(i>=2 && m[i][j]==B) {
            AlignmentA = Ai + AlignmentA;
            AlignmentB = '-' + AlignmentB;
            AlignmentM = e + AlignmentM;
            i = i - 1;

        } else {
            AlignmentA = '-' + AlignmentA;
            AlignmentB = Bj + AlignmentB;
            AlignmentM = e + AlignmentM;
            j = j - 1;
        }
    }

    var r1 = i-1;
    var r2 = j-1;

    if(m[i][j]<=0){break;}
}

// LAYOUT
var tM='';
var tS='';

// Check the end
AlignmentA = AlignmentA + s0.substr(x-1, n_0 - x);
AlignmentB = AlignmentB + s1.substr(y-1, n_1 - y);

// Check the beginning
AlignmentA = s0.substr(0, r1) + AlignmentA;
AlignmentB = s1.substr(0, r2) + AlignmentB;
```

(Continued)

Additional algorithm 3.18 (Continued)

```
if(r1>r2){
    var v = r1 - r2;

    for(var u=1; u<=v; u++) {tS = tS + e;}
    for(var u=1; u<=v+r2; u++) {tM = tM + e;}

    AlignmentB = tS + AlignmentB;
    AlignmentM = tM + AlignmentM;
} else {
    var v = r2 - r1;

    for(var u=1; u<=v; u++) {tS = tS + e;}
    for(var u=1; u<=v+r1; u++) {tM = tM + e;}

    AlignmentA = tS + AlignmentA;
    AlignmentM = tM + AlignmentM;
}

// Print the alignment
document.write(complement(AlignmentA) + '<br>');
document.write(AlignmentM + '<br>');
document.write(AlignmentB + '<br>');

// Matching function
function f(a1, a2) {
    if(a1 === a2){return Match;} else {return Mismatch;}
}

// Make complementarity
function complement(c) {

    c = c.replace(/C/g,'1');
    c = c.replace(/G/g,'2');
    c = c.replace(/A/g,'3');
    c = c.replace(/T/g,'4');

    c = c.replace(/1/g,'G');
    c = c.replace(/2/g,'C');
    c = c.replace(/3/g,'T');
    c = c.replace(/4/g,'A');

    return c;
}

</script>
```

```
Output:

GCGCTTGCTATTATCGCGCGGCGGATTCAGCGTCA
        ||||||| |||    | ||| |
    GATCGGCGCGTAGCGCGGCGC---G-CGCGGGATTTA
```

In the output from above, the complementarity of the two DNA strands can be observed by following the vertical lines between the two sequences (Additional algorithm 3.18). Notice the beginning of the alignment. Between the two sequences, there is a "C" letter on *s0* (position 8 from left to right) and a "G" letter on *s1* (position 6 from left to right), which are not joined by a vertical line. This is due to the fact that the two letters are not part of the initial result shown by the algorithm (as seen in Additional algorithm 3.14). However, small variations in the parameter values (match: +2; mismatch: −1; and gap: −2) also change the alignment configuration, which, in turn, can increase/decrease the length of the initial alignment. Of course, different initial alignment configurations may include the correspondence symbol (vertical line) between the two letters (for further experiments, please change the parameter values and run Additional algorithm 3.18 to observe the effects discussed so far).

3.11 Conclusions

The technical details for sequence alignment were described by using a series of implementations. The first part of the chapter showed the technical details regarding the initialization of the score matrix. First, the score matrix structure was designed. Secondly, the score values were calculated separately for both the local and global alignment. Next, the principle behind the traceback procedure was described and some useful examples were given. Then, a separate version for global alignment and local alignment was implemented and tested. The output of each version showed the score matrix, the traceback path, and the sequence alignment. The last important step was the description of an algorithm for the alignment layout that was able to rearrange the local alignment in the appropriate context. In addition, a conclusive implementation for the local alignment algorithm was presented. Toward the end of the chapter, a complementarity detector was implemented by using a small addition to the local alignment algorithm.

4

Forced Alignment (II)

4.1 Introduction

This chapter discusses an advanced implementation of the local alignment algorithm and explains novel particularities through a series of experiments. It shows how modern software technologies can help extend and standardize the software development in science and briefly considers the challenges imposed by discretization on data visualization. An introduction to a novel alignment regime is made using a short review of the main similarities and differences between global and local alignment. The local-forced alignment regime allows for new observations and interesting new possibilities. First, the chapter presents different configurations of numerical clusters over the score matrix that call for a reconsideration of the optimal alignment. This is reinforced by simple cases that show multiple optimal alignments with similar levels of significance. Detailed experiments and discussions related to the meaning of randomness prompt the use of a novel mathematical model for estimation of significance in local sequence alignments (please read Chapter 6). Secondly, the chapter presents regime changes on the score matrix that show chaotic behaviors for neighboring traceback start locations. Moreover, it shows how small changes in the alignment parameters led to radical changes in the distribution of numerical clusters over the score matrix, and consequently to abrupt changes in the alignment configuration. Last but not least, the chapter also shows how specially constructed sequences can lead to manipulations of the score matrix. In turn, this controlled process suggests new possibilities for image encoding by reverse sequence alignment. The experiments and figures presented here were made entirely with *BioJupiter*. BioJupiter was specially built for this chapter. Toward the end of the chapter, various textual details are explained about this application. Note: The BioJupiter application is present online in the additional materials of the book.

Algorithms in Bioinformatics: Theory and Implementation, First Edition. Paul A. Gagniuc.
© 2021 John Wiley & Sons, Inc. Published 2021 by John Wiley & Sons, Inc.
Companion website: www.wiley.com/go/gagniuc/algorithmsinbioinformatics

4.2 Global and Local Sequence Alignment

Sequence alignment is usually associated with the field of bioinformatics and computational biology. It is often used for alignments of nucleotide or protein sequences in the hope of finding functional similarities. However, sequence alignment is useful in most scientific fields, from mathematics to comparative linguistics. The numeric sequences in number theory or the relations between words or phrases of spoken languages are good examples.

4.2.1 Short Notes

The genesis of dynamic programming principle behind classical sequence alignment was rooted in the mid-1960s, and it seems to start with a mathematician from the Soviet Union, known today for the Levenshtein distance [240]. Historically, the sequence alignment algorithms are divided into two main categories, namely global alignment and local alignment [241, 242]. In their original form, both methods align two sequences in the hope of finding similarities. A general global alignment technique is the Needleman–Wunsch algorithm [241]. This approach forces an alignment on the entire length of the longest sequence and attempts to align every character/letter in every sequence. By contrast, local alignment, widely known as the Smith–Waterman algorithm, identifies regions of similarity within long sequences [242]. However, only minor differences exist between the two algorithms (Table 4.1). In the global alignment, both positive and negative numbers are considered on the score matrix, whereas in the local alignment, the negative numbers are eliminated by a constant value (zero), which competes within the maximization function. In the global alignment, the traceback procedure starts from the right-bottom element regardless of the values present over the score matrix and ends at the top left element. In local alignment, the traceback starts from the element with the maximum value found on the score matrix and ends at the first element containing zero.

The two algorithms provide perfect solutions for the theoretical problem of sequence alignment. However, the downside was and still is related to their implementation on computers. The number of elements in the score matrix is represented by the length of the first sequence (n) multiplied by the length of the second sequence (m). Thus, such an algorithm requires an $n \times m$ score matrix and proportional computer memory. The consumption of memory and time is directly related to the construction of the score matrix. Consequently, the number of steps that the algorithm must follow increases with the size of the score matrix. Nevertheless, modern solutions use heuristic methods and parallel programming techniques to shrink the time complexity of large-scale implementations [243].

Table 4.1 Global alignment and Local alignment.

Description	Global alignment	Local alignment
Alignment parameters	$\alpha = match\ value$ $\beta = mismatch\ value$ $\gamma = gap\ penalty$	$\alpha = match\ value$ $\beta = mismatch\ value$ $\gamma = gap\ penalty$
Substitution matrix	$G_{ij} = \begin{bmatrix} G_{1,1} & \cdots & G_{1,m} \\ \vdots & \ddots & \vdots \\ G_{n,1} & \cdots & G_{n,m} \end{bmatrix}$	$L_{ij} = \begin{bmatrix} L_{1,1} & \cdots & L_{1,m} \\ \vdots & \ddots & \vdots \\ L_{n,1} & \cdots & L_{n,m} \end{bmatrix}$
The scoring function	$f(x_i, y_j) = \begin{cases} \alpha, & x_i = y_j \\ \beta, & x_i \neq y_j \end{cases}$	$f(x_i, y_j) = \begin{cases} \alpha, & x_i = y_j \\ \beta, & x_i \neq y_j \end{cases}$
Scores	$G_{ij} = max \begin{cases} G_{i-1,\,j-1} + f(x_i, y_j) \\ G_{i-1,\,j} + \gamma \\ G_{i,\,j-1} + \gamma \end{cases}$	$L_{ij} = max \begin{cases} L_{i-1,\,j-1} + f(x_i, y_j) \\ L_{i-1,\,j} + \gamma \\ L_{i,\,j-1} + \gamma \\ 0 \end{cases}$
Traceback start location	$G_{n,\,m}$	$Max(L_{ij})$
Traceback stop location	$G_{1,1}$	$if\ L_{ij} = 0$

The two columns in the table show a side-by-side comparison of each stage in both algorithms. Matrix G represents the global alignment and matrix L represents the local alignment. The two matrices are identical. Note that the differences appear at the maximization function and the traceback start and end locations. The ith character in sequence x is x_i and the jth character of sequence y is y_j. The match (α), mismatch (β), and gap (γ) represent the parameters of the two algorithms.

4.2.2 Understanding the Technology

Internet browsers represent perhaps the most advanced, reliable, and stable software technology. The web pages/applications are represented by three entities that work together: HTML code, Cascading Style Sheets (CSS), and JavaScript (JS) [244]. Over time, syntax preservation and a natural back compatibility made this threesome the most trusted software environment in the world, and allowed the HTML applications a lifetime measured in decades. JS applications made for modern browsers also enjoy the cross-platform compatibility. In this light, the natural course was the implementation of an advanced application of sequence alignment. Based on my personal observations over the years, complex software implementations without maintenance exhibit a mean expiry time of around two years. Classic programming methods have been used to increase the validity period of BioJupiter. These approaches included methods that will be preserved over time in the future versions of Hypertext Markup Language (HTML5). The current approach uses best practices on native JS to avoid deprecation and prolong the lifetime of the application. With the new advent of HTML5 and the immutable objects that represent raw data, such as blob, modern Internet browsers can process large amounts of data at impressive speeds. For big data visualization, modern browsers offer several powerful objects, and probably the most important object of this type is the canvas object. This object it is used to draw real-time graphics by using JS.

4.2.3 Main Objectives

The first primary objective was the creation of an independent and durable application for sequence alignment, built into a single portable HTML file, following the principle: one HTML file – one application. The second primary goal was the interactivity and responsiveness of the application for real-time alignments and the establishment of a link between a heatmap and the main alignment. The third primary goal was the possibility of storage, export or import of experiments, and the implementation of a new file format specific for BioJupiter. The secondary aim was the development of sufficiently advanced graphics to suit modern bioinformatics requirements and expectations regarding manuscript figures or modern lecture presentations.

4.3 Experiments and Discussions

Experimental observations based on the current implementation of the algorithm are described here. The concept of *forced alignment regime* is introduced and a novel mathematical model for estimation of significance in sequence alignments is explored in detail (Figure 4.1g,h and Table 4.2). The forced alignment regime allowed for new observations related to the chaotic behavior of the local alignment algorithm (Figure 4.2d–f). Also, a few notes are added about the idea of reverse alignment and how it could lead to image encoding.

Figure 4.1 Local alignment regime and symbols. (a) A higher resolution of the heatmap, based on an alignment of the first 300b from of the insulin (INS) gene from *Homo sapiens* and *Macaca mulatta*. The heatmap shows that the optimal alignment area does not follow the main diagonal. In this specific case, the optimal alignment can be made by changing the traceback location from the element with the maximum value to one of the elements at the base of the top right spike. (b–e) Indicates the possibility of using any symbols outside the English alphabet. (b) Shows an alignment based on Cyrillic letters. (c) Shows an alignment based on Greek letters. (d) Shows an alignment based on special characters. (e) Shows the alignment of digits. (f) Shows the alignment of two names and demonstrates a semiglobal alignment achieved by the local alignment algorithm. (g, h) It shows the forced sequence alignment regime, which consists of changing the traceback from one initial cell to another arbitrary cell. The alignment results are presented above the heatmaps. The regime change leads to different alignment configurations. Notice that the markers above or below each sequence correspond to the position of the cross marker over the heatmap, and the aligned sequences are positioned relative to each other.

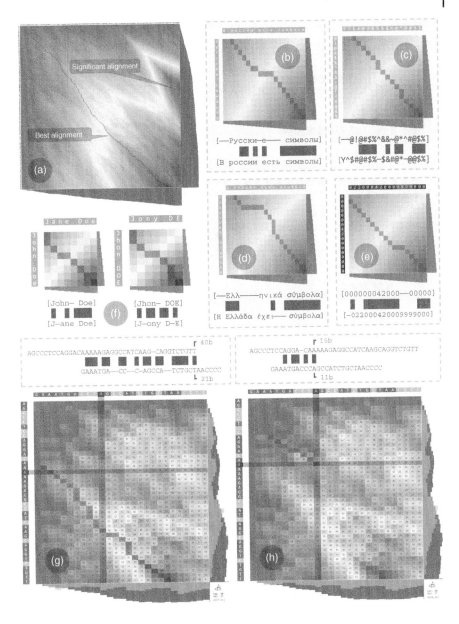

Table 4.2 Examples of alignments and associated scores.

Alignment	$P(S)$	$\sigma(S)$	Score%
5-[AAAAAAAAAAAAAAAAAAAAAAAAAAAAATCTCGAAA]-3 5-[GGGTCTCGGGGAAAAAAAAAAAAAAATTTTTTTTTTTAAAAAAAA]-3	94.11	15.27	78.84
5-[TA—GCCCTATCGGT–CATTTTTTTT]-3 5-[TACGGGCC——CGCTACTTTTTTTTT]-3	88.46	47.55	40.91
5-[AGAGAGAGAGAGAGAGAGAGAGAGAGAGAGAGAGAGAGAGAGAG]-3 5-[ATATATATATATATATATATATATATATATATATATATATATAT]-3	50.94	50.00	0.94
5-[TTTTTAAAAAAAAAAAAAAAGGAAAAAAAAAAAAAAAAAATTTTTT]-3 5-[TTTTTGGGGGGGGGGGGGGGGGGGGGGGGGGGGGGGGGGTTTTTT]-3	100	0.00	100
5-[TAG–CC–CTATCGGTCA]-3 5-[TACGGGCCCGCTA–C]-3	100	43.17	56.83
5-[TAGCCCTATCGGTCA]-3 5-[TAGCCCTATCGGTCA]-3	100	0.00	100
5-[AAAAAAAAAGGGGGGTACGGTGGGGGAGGG]-3 5-[CCCCAAAAAATACGGTAAAAA]-3	100	27.42	72.58
5-[AAAAAAAAAGGGGGGTACGGTGGGGGAGGG]-3 5-[CCCCAAAAAA————TACGGTAAAAA]-3	100	36.88	63.12

The first column shows the alignments. The normal text signifies the segment provided by the local alignment algorithm and the underlined text signifies the completions outside this segment. The second column shows the match + gap percentage in the match sequence. The third column shows the information content of the match sequence and column four shows the significance score.

4.3.1 Alignment Layout

In its raw form, the result of the local alignment algorithm is visually unintuitive and hard to distinguish from the global alignment (Figure 4.1f). An addition to the original local alignment algorithm was made to complete or push the sequences to the appropriate locations to shift the sequences between them when appropriate (Figures 4.1g,h and 4.3a–h). The final result of the alignment is usually constructed from three different sequences stacked over each other. The two sequences that are compared incorporate a "match sequence" made of white spaces and the representative characters for match or gap. The match sequence contains all the information related to the significance of the alignment.

4.3.2 Forced Alignment Regime

The classic local alignment algorithm dictates a traceback starting from the element with the maximum value (Figure 4.1g). However, by clicking the cells with positive integer values on the heatmap, a new regime may be imposed upon the local alignment algorithm (Figure 4.1h). Thus, the traceback can start from any element with a positive integer. By forcing an alignment from the bottom-right corner of the heatmap, the local alignment becomes a semiglobal alignment in some cases (Figures 4.1b–f and 4.3j–l). Moreover, by manipulating the alignment parameters, one can reach a similar result with that from the global alignment (Figure 4.3j–l). The required condition is that the alignment parameters must be set in such a way that a positive integer value is provided for each element in a connecting path between $L_{1,1}$ and the $L_{n,m}$ element of the score matrix (Figure 4.3j–l).

4.3.3 Alignment Scores and Significance

An alignment score can be formulated in several ways [245]. For instance, one version would be a sum over the values from the traceback path. Another option would be a sum of the values given by a score function (δ) over the alignment

---

Figure 4.2 True randomness and chaos. (a) Shows a series of heatmaps from experiments made on random sequences of 300 characters, each built from a different alphabet. The components of the alphabet are embedded in braces and are displayed next to each heatmap. The three dots inside the braces represented all the letters starting from letter "A". Note that the increase of the number of components/letters of an alphabet leads to a visible decrease in the number of matches made by pure chance. (b) An alignment of two random sequences of 300 characters based on a random number generator, (c) and an alignment of two random sequences of 300 characters based on a Markov Chains generator. Note the difference between the two heatmaps: In (b) the image shows a pattern closely related to the observed structure of regular biological sequences, while in (c) the image shows a pattern different from that expected from biological sequences. The top right corner in (b) and (c) shows the contour features of the image from the heatmap. (d) Shows an alignment starting from a specific cell and a regime change (e) started from a neighboring cell. A switch between the two neighboring cells induces completely different results. Thus, (d) and (e) indicate a degree of nonlinearity. (f) It shows a nonlinear behavior in the normal regime of local alignment. Namely, a small change in the parameter values (± 1 in γ) causes a drastic change in the traceback start location.

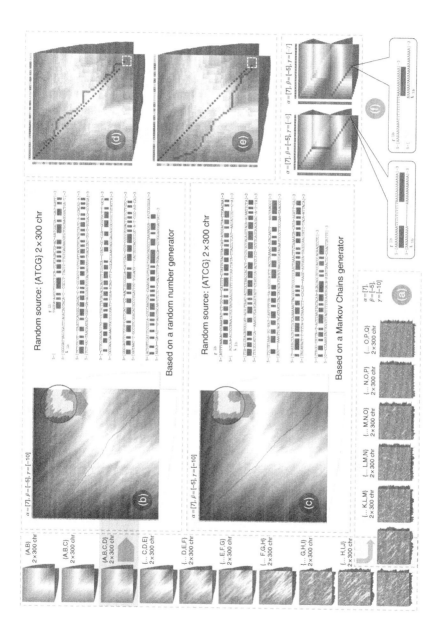

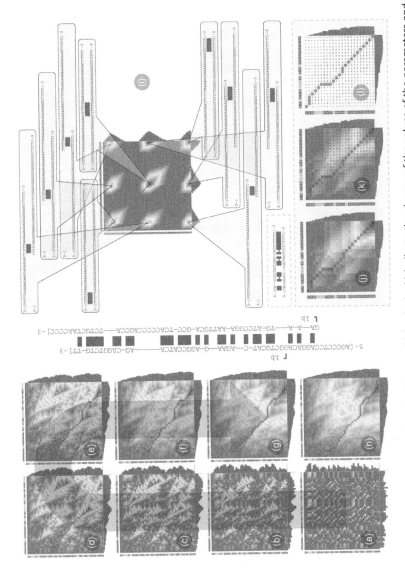

Figure 4.3 Alignment parameters and the significance of islands. (a–h) Indicate the change of the values of the parameters and the real time changes visually perceived on the surface of the heatmap. The alignment configuration is shown to the right of the eight successive heatmaps in an upright position. (i) Shows a direct association between the islands on the score matrix and the alignment configuration. It also suggests the possibility of novel image-encoding strategies (j-l) Indicates the transition from colors to their associated values by maximizing the transparency of the heatmap.

results, for instance:

$$S = \sum_{i=1}^{c} \delta(x_i, y_i), \qquad \delta(x_i, y_i) = \begin{cases} \alpha, & x_i = y_i \\ \beta, & x_i \neq y_i \end{cases}$$

where S represents the alignment score and c is the length of the alignment. The i index represents the position of the characters in the alignment. Also, α represents the value for a match and β represents the value for a mismatch. The maximum score possible is represented by a multiplication between the length of the alignment (c) and the match value (α). Thus, c multiplied by α represents the score for a perfect alignment between two sequences. Consequently, the alignment score may also be expressed as a percentage:

$$S\% = \frac{100}{(c \times \alpha)} \times \sum_{i=1}^{c} \delta(x_i, y_i)$$

Although alignment scores are useful units of measure, best alignment scores are often confused with the significance of optimal alignments. The local alignment ensures the best alignment considering the starting position over the score matrix. However, local alignment does not guarantee a significant optimal alignment. A simple test on the insulin gene from *Homo sapiens* and *Macaca mulatta* indicates that the local alignment algorithm will not guarantee a significant optimal alignment (Figure 4.1a). The alignment behavior in this case is closer to what an alignment of two random sequences will provide (Figure 4.2b,c). For example, the alignment of two random DNA sequences can result in a large number of matches; however, this would not mean that the alignment is optimal or meaningful. For a more significant optimal alignment, a forced alignment regime can be set to enable a different configuration of the alignment by changing the traceback start location (Figure 4.1a,g,f).

In practice, different parameters lead to dynamic values inside the score matrix, which make the notion of optimal alignment even more subjective (Figure 4.3a–h).

4.3.4 Optimal Alignments

The heatmap is a direct mirror of the score matrix where the high–low values in the matrix are transformed into a three-color-gradient for an intuitive visualization (Figure 4.1g,h). An overview on all types of alignments was given by the observations that can be made on the heatmap. The total number of possible alignments is equal to the number of matrix elements containing positive integers (Figure 4.1g,h). Thus, the standard local alignment allows for all possible alignment configurations. The optimal alignment topic has been discussed about over the years. In certain types of experiments, alignment islands and/or bridges can be observed (Figures 4.3i and 4.4a–d). These formations represent irregularly shaped clusters of high numerical values and specific local maxima that indicate possible alignments (Figures 4.3i and 4.4a–k). Based on experiments performed with this application, the optimal alignment does not exist in complex cases. For

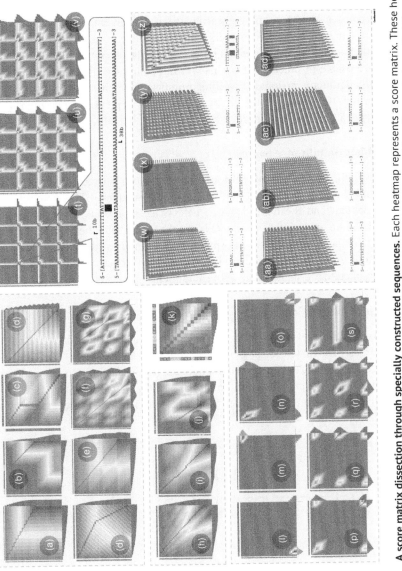

Figure 4.4 A score matrix dissection through specially constructed sequences. Each heatmap represents a score matrix. These heatmaps represent different sequence alignment configurations based on specially constructed sequences. The alignment shows that the optimal alignment area does not follow the main diagonal. (a–k) Show different patterns that can appear on the score matrix if special sequences are used. (k) Shows the symmetry present on the heatmap in the case of a perfect alignment between two sequences of the same size. (l–s) Show islands of high value numbers on the score matrix and presents the meaning of different areas on the score matrix. (t–v) Shows a patterned grid. (w-ad) It shows a variety of possibilities for clean granular patterns.

example, Figure 4.4h shows two optimal alignments. Moreover, another naive case explains the correspondence between alignment and the areas on the heatmap (Figures 4.3i and 4.4l–s). There, the panel shows nine equidistant optimal alignments, all with the same level of significance (Figure 4.3i). In forced alignment, the traceback can generate an alignment starting from any element with a positive integer (Figure 4.1g,h). The positive integers found over the score matrix represent the total number of possible alignments (Figure 4.1g,h). However, not all possible alignments are usable in practice since this will not guarantee the existence of matches in all alignment versions. Even if the number of matches is high for an alignment variant, the information content of the match sequence can be high, which makes the alignment less significant.

4.3.5 The Main Significance Scores

In this proposal, the optimal alignment represents a balance between the match percentage and the information content of the match sequence. The smallest information content and the highest match percentage represents the optimal alignment. The main analysis was conducted over the match sequence present in between the two aligned sequences. Information is an emergent property. Thus, the number and the order of characters in the match sequence contain all the information related to the significance of the alignment.

4.3.6 The Information Content

A three-character alphabet defines the match sequence: a white space for mismatch, a character for match (e.g. "$\blacksquare$"), and a character for gap (e.g. "-"). The model below computes the information content of the match sequence. It is based on a successive comparison between different areas of the match sequence. The equation of the information content is defined below (please see this model in the next chapter):

$$s = \{x_1, \ldots, x_{|s|}\}$$

$$\sigma(s) = 100 - \left(\frac{\sum_{u=1}^{|s|-1} \left(\frac{\sum_{i=1}^{|s|-u} f(x_i, x_{u+i})}{(|s|-u) \times 100} \right)}{(|s|-1)} \right)$$

$$f(x_i, x_{u+i}) = \begin{cases} +1, & x_i = x_{u+i} \\ 0, & x_i \neq x_{u+i} \end{cases}$$

where s is the match sequence composed of any type of symbol x from the alphabet mentioned above and f is a match function. Also, $|s|$ represents the total number of symbols in the match sequence (e.g. $s = $ "$\blacksquare$ -$\blacksquare$-$\blacksquare$"; $|s| = 7$). The ith character in s is compared to the $u + i$th character in s. If the two characters match, then function f will return 1 and if the two characters do not match then function f will return 0. Thus, the $\sigma(s)$ function will provide a number between 0 and 100. The information content was represented on a growing scale, where the maximum information content was represented by 100 and the lack of information was represented by zero.

4.3.7 The Match Percentage

To obtain a value independent of the alignment length, the number of matches was translated to percentages. The percentage of matches is extracted from the match sequence as follows:

$$P(s) = \frac{100}{|s|} \times (\text{matches} + \text{gaps})$$

where $P(s)$ represents the percentage of matches and gaps and $|s|$ is the total length of the match sequence. The final significance score is obtained from the information content value and the match percentage. Thus, the information content is subtracted from the match percentage:

$$Score\% = P(s) - \sigma(s)$$

where $Score\%$ represents the level of significance for an alignment, $P(s)$ is the percentage of match and gap, and $\sigma(s)$ is the information content function. Note that function $\sigma(s)$ provides a value of zero under two conditions: whether the match sequence is composed exclusively of white characters or it is composed exclusively of characters representative for match. A maximum value of $P(s)$ indicates a sequence composed of match (i.e. "▐") and gap (i.e. "-") characters. A high value of $P(s)$ and a small value of $\sigma(s)$ indicates that the alignment tends toward the highest significance. A complementarity between $P(s)$ and $\sigma(s)$ can be observed. For instance, when the value of $P(s)$ is equal to 100, the information content is automatically zero, but the main score is 100 ($Score\% = 100 - 0 = 100$). However, when $P(s)$ is equal to zero and the information content is zero, the main score is zero ($Score\% = 0 - 0 = 0$). In contrast, a high information content demands a uniform distribution of the match characters over the match sequence, regardless of $P(s)$. Thus, an increase of the information content leads to a natural decrease in the match + gap percentage ($P(s)$). Another approach to extract the alignment significance can be made locally by using sliding windows on the match sequence. In this manner, a series of signals can be obtained along the match sequence. The above can be tested in an intuitive manner by using the *BioJupiter* HTML file from the online supplementary materials. Note that here a novel score was described. This score is relative to the length of the aligned segment provided by the local alignment algorithm (Table 4.2). Absolute scores can be developed by taking into account the length and the offset between the two sequences; however, these matters are outside the scope of those discussed here.

4.3.8 Significance vs. Chance

A reference system for evaluating the alignment significance is particularly important [246]. This reference system was extracted by experiment from observations made on repeated alignments of random sequences. In random sequences, the main interest is related to the number of matches that can be formed by pure chance and the information content provided by the order of the match characters. The normal local alignment regime was used, in which the alignment was made automatically from the maximum value found over the score matrix.

4.3.9 The Importance of Randomness

The importance of the quality of random sequences is crucial, especially when these are used for setting a significance threshold or for extracting different constants [247, 248]. A series of tests were made using random DNA sequences from two sources (Figure 4.2a). The first source was a pseudo random number generator (PRNG) and the second source was a Markov Chains generator (MCG) based on a true random number generator (TRNG). PRNG provides random numbers based on seeds (one initial numeric value). The process starts from the seed and makes the final result seemingly random and yet fully predictable. On the other hand, TRNG provides a random result which is, at least in theory, unpredictable.

A probability distribution can also indicate the difference between the two methods. For instance, a random generation of n integers between 1 and 10, yield an equal probability of occurrence by using TRNG ($p(1...n) = 1/10$, $\sum_{i=1}^{n} p(i) = 1$) and different probabilities of occurrence by using PRNG ($p(1) \neq p(2) \neq \cdots \neq p(n)$, $\sum_{i=1}^{n} p(i) = 1$). In the past, I have implemented and explained the theory behind a true MCG and shown that MCGs are able to generate true random sequences even when PRNGs are used as the main source for random numbers [249]. Certain programming environments use PRNG. Nevertheless, PRNG is a reference system and is crucial in all scientific fields because it allows a comparison between what is and what is not truly random. Thus, a PRNG is intentionally biased precisely for this reason. Random character sequences generated on the basis of PRNG (without MCG), can accurately mimic the expectations seen in biological sequences (Figure 4.2b). The experiments from Figure 4.2b,c point out that some alignment results can be misleading if overlooked. Figure 4.2b,c suggests that successive matches of up to 6 letters or more can occur by pure chance (with high frequency) in the alignment of two random DNA/RNA sequences. However, in recent years the results of different PRNG methods are placed on a spectrum and can be closer to or further from TRNG. Thus, special attention should be paid to these two methods in Bioinformatics and Computational Biology, as these approaches can establish the difference between objective and subjective results.

4.3.10 Sequence Quality and the Score Matrix

TRNG are responsible for providing a uniform distribution of probabilities for a set of numbers. In the case of DNA, a TRNG can be used to provide random numbers with equal frequency, which in turn are associated with the letters from a set (e.g. 1 means A, 2 means C, 3 is T, 4 is G). Thus, the resulting sequence of random numbers is replaced with the corresponding letters from the set and a new random DNA sequence can be obtained (e.g. "1,4,2,2,1,2,4,3" to "AGC-CACGT"). The quality of a TRNG consists in providing a uniform probability distribution across all numbers of a set. An uneven probability distribution in random sequences leads to a false reference system and an overestimation or underestimation in the statistical significance of the alignment (please see the next subchapter). For special experiments in bioinformatics, random sequences are not necessarily represented by uniform probability distributions. In such cases the probability distribution of a TRNG can be changed arbitrarily by using a MCG. For instance, MCGs can generate a sequence of letters using the following probability distribution: $p(A) = 0.3$, $p(T) = 0.2$, $p(C) = 0.1$, $p(G) = 0.4$. MCGs and their implementation has been widely described in [249]. However, in the current experiment, true random sequences have been obtained using an MCG, regardless of the TRNG quality (Figure 4.2c). The characteristic footprints of random sequences generated using a TRNG or a PRNG can be observed on the main heatmap of the implementation (Figure 4.2b). The alignment of a pair of random sequences generated using PRNG closely resembles the domain shuffling characteristics observed in biological sequences.

A raw example of DNA shuffling is shown in Figure 4.4b, where an artificially constructed sequence shows a characteristic pattern. More complex biological sequences indicate a high frequency of domain shuffling on the evolutionary timeline and seem to suggest that such rearrangements are an important part of the mutation process (Figure 4.1a). Visually, domain shuffling is represented by numerous spikes inclined with the tip toward the diagonal of the heatmap (Figure 4.1a).

4.3.11 The Significance Threshold

Observations on different alignments made between random sequences can help in determining a significance threshold. Sequence pairs of 300 random characters were generated using two methods: TRNG and MCG (i.e. $p(A) = p(T) = p(C) = p(G) = 0.25$ for DNA) [249]. These pairs were aligned in $t = 100$ trials for each method. Normal local alignment was used, in which the traceback started from the maximum value found on the score matrix. The match sequence from each alignment provided a significance score (*Score%*) based on the information content $\sigma(s)$ and the match percentage ($P(s)$). An average between *Score%* values was made over t trials, yielding a final threshold score:

$$RScore\% = \frac{\sum_{i=1}^{t}[P(s_i) - \sigma(s_i)]}{t}$$

where *RScore%* was calculated based on random sequences generated with MCG or TRNG (Table 4.3). The significance threshold for TRNG was assessed at *RScore%* = 22.8 ± 2.81 and the significance threshold based on MCG was *RScore%* = 32.47 ± 3.36 (Table 4.3). The percentage of match + gap in random sequences was high and relatively constant, around 81.42% for TNRG and 85.64% for MCG. Interestingly, successive matches from 6 up to 8 characters are common in alignments made between random sequences, for both TRNG and MCG (Figure 4.2b,c). The information content in random sequences is also constant, at 58.58 ± 0.96 for TNRG and 53.17 ± 1.33 for MCG (Table 4.3).

The *RScore%* values represent a reference system for alignment significance. MCGs generate true random sequences. Thus, a *RScore%* (32.47 ± 3.36) based on MCGs should be used for such thresholds. A comparison with *RScore%* shows whether an alignment is significant or not. Alignments with *Score%* ≤ *RScore%* are not considered significant because such values can also be obtained by pure chance. On the other hand, an alignment shows a degree of significance only if the value of *Score%* exceeds the *RScore%* value.

4.3.12 Optimal Alignments by Numbers

The number of optimal alignments over the score matrix is related to the number of islands and the threshold for their local maxima. The number of significant alignments could be detected directly on the score matrix using unsupervised learning techniques such as K-Means clustering. Thus, the islands present on the score matrix can be detected automatically. The number of clusters detected would be equal to the number of best alignments. Local maxima of each cluster/island

Table 4.3 The significance thresholds based on TRNG and MCG.

	TRNG (100 trials)			MCG (100 trials)		
	$P(s)$	$\sigma(s)$	RScore%	$P(s)$	$\sigma(s)$	RScore%
AV	81.42	58.58	22.83	85.64	53.17	32.47
SD	±2.11	±0.96	±2.81	±2.35	±1.33	±3.36

The table shows the average RScore% values for 100 TRNG trials and 100 MCG trials. In each trial, an alignment was made between two random sequences. The match sequence resulting from the alignment was analyzed and the significance score (Score%) was determined. The mean value between significance scores represented the reference system, namely the RScore%.

on the score matrix could be the element from which a regime change can be made automatically. Note: In this version of the implementation, such an automatic detection was not the main focus.

4.3.13 Chaos Theory on Sequence Alignment

One observation worth mentioning was the connection between the complexity of the score matrix and nonlinearity (*chaos theory*). The complexity of the score matrix can easily lead the local alignment algorithm into a nonlinear behavior, where regime changes in close proximity over the score matrix can lead to radically different alignments (Figure 4.2d,e). For example, a regime change near the border of two nearby islands can induce such a behavior.

Also, in some situations, small changes in parameter values lead to radical changes in the dynamics of the values over the score matrix (Figure 4.2f). These observations can be verified by experimentation (please use the case from Figure 4.3h and the online supplementary materials). Chaos theory may open a new research path for Bioinformatics in regard to the sequence alignment techniques.

4.3.14 Image-Encoding Possibilities

Special alignments were built to show the density areas in the form of islands or bridges (Figure 4.3). The experiments suggest the possibility of a new image-encoding strategy that can use specially constructed sequences for an arbitrary manipulation of the values in the score matrix (Figure 4.4a–d). For instance, images showing granular patterns on the heatmap may be constructed through specially designed sequences (Figure 4.4w-ad). In turn, the rules behind these artificial sequences may lead to new principles for archiving images using an $O(n + m)$ encoding. As far as I know, these rules are unknown and unexplored in the past. In the future, a mathematical model for reverse alignment can be developed. For this endeavor, a new model published in the journal *Chaos* could be useful for predictions that involve special matrix operations and 3D tensors [250]. This model is called *Spectral forecast* [250]. Note: Reverse alignment is the process of transforming any score matrix back into two separate sequences.

4.4 Advanced Features and Methods

This chapter describes a complex open source implementation of sequence alignment as an offline web application (please see the supplementary materials online). The application is designed in JS/CSS/HTML and is executed by Internet browsers. It is fully embedded in a single HTML file entirely independent of external resources and can be downloaded from the online supplementary materials. The graphical user interface (GUI) of the application is divided into four main areas: menu, heatmap, alignment of sequences in text format, and the data storage module.

The main alignment parameters can be changed in real time and are represented by match(α), mismatch(β), and gap (γ), each ranging from -100 to 100. Manipulation of alignment parameters allows three types of behaviors that can be identified with global alignment, local alignment, and a new behavior called forced alignment. The traceback in the classical local alignment algorithm starts from the element with the largest positive integer. A type of alignment called forced alignment is introduced to force new alignments from arbitrary positions (Table 4.4).

4.4.1 Sequence Detector

The application was designed for the analysis of biological sequences; however, its purpose is wider in use. A sequence-type detector allows the application to distinguish between DNA, RNA, PROTEIN sequences, number sequences, or plain text. Depending on the type of sequence, the application will add specific information around the alignment and will use different sets of color codes [251].

4.4.2 Parameters

The alignment parameters are represented by the values of match (α), mismatch (β), and gap (γ), and can take values between -100 and 100. Parameter changes are made using an HTML5 slider object. For short sequences below 100 characters, the parameter change is followed by a real-time instant alignment. This type of rapid modification of the parameters leads to a fast experimentation procedure and to an intuition related to the results of the main algorithm.

4.4.3 Heatmap

The information display technique includes a heatmap and the text of the aligned sequence (Figure 4.5a–c). The heatmap supports full customization for colors, transparency, and size. The canvas object was divided into five areas of interest: the actual heatmap, the two sequences from the top and the left of the heatmap, the two density graphs, and the information window in the lower right corner (Figure 4.5a).

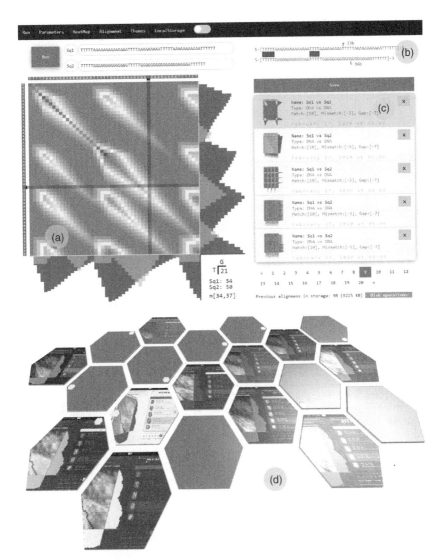

Figure 4.5   BioJupiter interface and functionalities. (a–c) The application interface. (a) main heatmap, (b) the main alignment result, (c) the local storage browser and disk operations, and (d) graphic themes.

Table 4.4 Forced alignment and local alignment.

Description	Forced alignment	Local alignment
Alignment parameters	$\alpha = match\ value$ $\beta = mismatch\ value$ $\gamma = gap\ penalty$	$\alpha = match\ value$ $\beta = mismatch\ value$ $\gamma = gap\ penalty$
Score matrix	$M_{ij} = \begin{bmatrix} M_{1,1} & \cdots & M_{1,m} \\ \vdots & \ddots & \vdots \\ M_{n,1} & \cdots & M_{n,m} \end{bmatrix}$	$L_{ij} = \begin{bmatrix} L_{1,1} & \cdots & L_{1,m} \\ \vdots & \ddots & \vdots \\ L_{n,1} & \cdots & L_{n,m} \end{bmatrix}$
Penalty/reword function	$f(x_i, y_j) = \begin{cases} \alpha, & x_i = y_j \\ \beta, & x_i \neq y_j \end{cases}$	$f(x_i, y_j) = \begin{cases} \alpha, & x_i = y_j \\ \beta, & x_i \neq y_j \end{cases}$
Scores	$M_{ij} = max \begin{cases} M_{i-1,\,j-1} + f(x_i, y_j) \\ M_{i-1,\,j} + \gamma \\ M_{i,\,j-1} + \gamma \\ 0 \end{cases}$	$L_{ij} = max \begin{cases} L_{i-1,\,j-1} + f(x_i, y_j) \\ L_{i-1,\,j} + \gamma \\ L_{i,\,j-1} + \gamma \\ 0 \end{cases}$
Traceback start location	$M_{ij} > 0$	$Max(L_{ij})$
Traceback stop location	if $M_{ij} = 0$	if $L_{ij} = 0$

The two columns in the table show a side-by-side comparison of the local alignment algorithm in two different regimes. To differentiate between the two regimes, M and L represent different notations for the same matrix. The regime change is based on the traceback start location. In the forced alignment regime, the alignment can start from any element in the score matrix that contains a positive integer. In the Local alignment regime, the alignment starts from the element with the maximum value found on the score matrix. In the experimental implementation, the match (α), mismatch (β), and gap (γ) parameters can take values between -100 and 100.

Heatmap Area

The heatmap is a direct mirror of the score matrix, where the high–low values in the matrix are transformed into a three-color-gradient for an intuitive visualization (Figure 4.3a–h). The heatmap colors, representative of the values from the score matrix, are customizable. By default, color blue was used inside the heatmap cells for score values close to zero, yellow for intermediate values, and red for score values close to the maximum. A cross marker functionally linked with the mouse was constructed over the heatmap to underline the corresponding characters from the top and left of the heatmap at arbitrary cell positions. The application uses an adaptive method for scaling the fronts of integers, which allowed an increase or decrease of the font size depending on the resolution of the heatmap. The height and width of the heatmap cells were considered when calculating the size and position of monospaced fonts with respect to the dimensions of the canvas object. Transparency was supported for the entire heatmap area, for the traceback path, and for the fonts inside the cells. The customization of the heatmap width and height was also supported for presentations or paper figures that require high resolutions. The maximum size of the heatmap was limited at 2000 pixels on width and height.

Fractional Parts and Discretization

Naturally, the first impulse in setting the cell size inside a heatmap is the division between the width (hW) of the heatmap and the length of the first sequence (m) and the division between the height of the heatmap (hH) and the length of the second sequence (n):

$$cH = \frac{hH}{n}, \quad cW = \frac{hW}{m}$$

where cH is the height and cW is the width of a cell. However, from a computer graphics point of view, this simplistic approach raises a problem related to discretization. For instance, the division of hW or hH by any positive integer (n or m) may result in fractional parts and the division will rarely provide an integer. The unit of measure for the height (hH) or width (hW) of the heatmap is the pixel, which can be represented only by using whole numbers. A direct division can result in fractional dimensions, otherwise incompatible with pixel mapping. Thus, to represent a constant cell size in a discretized environment, two integers were mandatory for cH and cW. The rounding process was an acceptable solution for discretization. In this manner, cells of equal size have been translated to pixels. However, rounding leads to changes in the cell size and consequently in changes of the cell grid size over the heatmap, potentially exceeding hH and/or hW. Nevertheless, the rounding step was followed by a recalculation and reset of the heatmap area prior to the drawing process:

$$hH = [cH] \times n$$

$$hW = [cW] \times m$$

where square brackets denote the standard rounding function. To determine and set the new size of the heatmap, $[cH]$ was multiplied by the length of the first sequence (m) and $[cW]$ was multiplied by the length of the second sequence (n). Thus, the new integer values for hH and hW were used to establish the relative height and width of the heatmap on the canvas object. In this manner, equal cell dimensions have been obtained and the cell grid perfectly matched the recalculated size of the heatmap.

Top-Left Sequences

Special background colors were used to distinguish between the characters shown in the top and left of the heatmap when the heatmap cells become too small for displaying fonts. The background colors of the characters on the top and on the left of the heatmap are particular to each type of sequence. The application supports separate color coding for each type of sequence and the representing colors can be changed manually. For instance, a separate set of colors will be displayed for DNA sequences and another set of colors will be displayed for protein sequences or digit sequences. The font color for all top-left characters can be changed manually and by default the color white is chosen according to the compatibility with the background color set.

Heatmap Charts

The heatmap is accompanied by two charts meant to show the density of the values on the lines and columns of the score matrix. The chat on the right of the heatmap shows the average of all positive integers from the lines of the score matrix, whereas the chart below the heatmap shows the average of all positive integers on the columns. The X and Y coordinates of the mouse over the heatmap leads to a secondary set of information presented in real time on the same two charts. The values on the row corresponding to the Y coordinate of the mouse represent the secondary information shown in the chart below the heatmap. The Y axis represents 100% and the X axis represents the length of the sequence. Consequently, the values on the column corresponding to the X coordinate of the mouse represent the secondary information shown in the chart on the right. The Y axis represents the length of the sequence and the X axis represents 100%, where 100% in both charts represents the maximum value found on the score matrix. To plot the bars in the charts, a vector was associated with each chart. Vector R was associated with the right chart and vector B was associated with the chart below the heatmap. For the chart on the right of the heatmap the following equation applies:

$$R_i = \left(\frac{w}{Max(M_{ij})} \right) \times \left(\frac{\sum_{j=1}^{m} M_{ij}}{m} \right)$$

where M_{ij} represents the score matrix and R_i represents the vector of the chart to the right of the heatmap which holds the average values on the lines of the score matrix. Variable w represents 100% and is the total width of the bars in the right chart. In the first part of the equation, w is divided by the maximum value found on the score matrix. The result is multiplied by the second part of the equation, namely by the sum of the values from line i, divided by the total number of columns (m) in the score matrix. Each value from component i of vector R was plotted with respect to the origin as a horizontal line in the chart to the right of the heatmap.

For the chart at the bottom of the heatmap, a similar equation applies:

$$B_j = \left(\frac{h}{Max(M_{ij})} \right) \times \left(\frac{\sum_{i=1}^{n} M_{ij}}{n} \right)$$

where B_j represents the vector of the chart below the heatmap, which holds the average values on the columns of the score matrix. Variable h represents 100% and is the total height of the bars in the bottom chart. As before, in the first part of the equation, h is divided by the maximum value found on the score matrix. The result is multiplied in the second part of the equation, namely by the sum of the values from columns j, divided by the total number of lines (n) in the score matrix. Each value from component j of vector B was plotted with respect to the origin as a vertical line in the chart at the bottom of the heatmap. The second set of bars on the two charts is activated on mouse position and shows the percentages represented by raw values of each cell from the lines or columns of the heatmap. Thus, the secondary bars of the charts are linked to the vertical and horizontal cells of the cross marker. Empirically, the secondary bars of the chart at the right of the heatmap are calculated as:

$$RC_i = \frac{w}{Max(M_{ij})} \times M_{i,x}$$

where RC_i represents each cell on the vertical line of the cross marker, whereas x and y represent the intersection cell between the vertical and the horizontal lines of the cross marker. Consequently, the secondary bars of the chart at the bottom of the heatmap are calculated as:

$$BC_j = \frac{h}{Max(M_{ij})} \times M_{y,j}$$

where BC_j represents each cell on the horizontal line of the cross marker. According to their relative position, the two charts automatically adopt the width or the height of the heatmap. The customization of the bottom chart height or the right chart width is limited for a maximum size of 1000 pixels. Transparency was supported for both charts. The mean values on columns/rows and the raw values on current column / current row have separate transparency filters and separate color codes. For even more refined customization, the colors of the axes on the two charts can also be changed.

Information Window

Longer sequences lead to higher resolutions of the heatmap (Figure 4.1a). Thus, the resolution limit is dependent on the size of the heatmap. The maximum resolution can shrink the size of a heatmap cell to the size of a single pixel, namely one score matrix value – one pixel. At such resolutions, the visual identification of the background colors of the top and left cells of the heatmap may be difficult. In this eventuality, the information window is functionally linked to the cross marker and shows the cell value and the top and left characters/letters corresponding to the intersection. The information window also shows the length of each sequence and the size of the score matrix. Also, the font color of the information window is customizable.

4.4.4 Text Visualization

A character represents a symbol with meaning: a letter, a digit, a diacritic, or an Egyptian hieroglyph; are classic examples. The implementation can align sequences containing any type of symbol and is not restricted only to letters or numbers, which can be useful for many scientific fields, especially for linguistics where diacritics are the key in researching the relationships between dialects (Figure 4.1b–e) [252, 253]. The alignment construct was positioned to the right of the heatmap in rows, so that aligned characters appear in successive columns (Figure 4.5b). Monospace fonts were used to ensure the same amount of horizontal space for each character/letter in the representation of the alignment. Similar characters were indicated with a system of symbols. Symbols from the 8-bit Unicode Transformation Format (UTF-8) were used for alignment construction. Thus, text alignments with custom graphics can be obtained through different UTF-8-character combinations. Different options have been made available for the text alignment such as *hide/show DNA notation, append names, detect sequence type,* or *sequence match completeness.* The *detect sequence type* option turns on or off the sequence type detector module. When this module is switched on, the application detects if the sequence introduced by the user is of type DNA, RNA, protein, number, or ASCII. Depending on the verdict, the application treats these types of sequences differently. For instance, the *hide/show DNA notation* option is dependent on the sequence-type detector and will add the 5′ and 3′ ends to the alignment, if DNA/RNA is present. For larger portions of the alignment, where matches are not present, the visual field ceases to correctly provide an intuitive location of the first and second sequence. Thus, the *append names* option allows the additions of names after each sequence, for instance the name of the species from which each sequence originates. The initial result of the local alignment algorithm was composed of segments from the two sequences that participate to the alignment, eliminating the areas of the sequence that do not participate in the alignment. Thus, the application automatically completes the sequences in front and behind these segments to shift the sequences to their correct positions for an intuitive visual representation (Figure 4.1g,h). By pure chance, the sequences added to the initial result can also contain matches. Thus, *sequence match completeness* represents the option of showing the matches that exist outside of the main results provided by local alignment.

4.4.5 Graphics for Manuscript Figures and Didactic Presentations

In the case of the alignment representation, customization can be made either by changing the colors or by changing the representative characters for match, mismatch, or gap. All UTF-8 symbols representative of match, mismatch, and gap, can be changed by the user from the optional set of characters. Other options consist of customization of the number of letters per alignment line, the font scale for the alignment (zoom in text) or by modifying the spacing between the alignment lines. The alignment has been specially constructed in text format to be used in other editing media, such as Excel or Word. The image drawn on the canvas object can be saved in the default PNG format and used in manuscript figures and scientific or didactic presentations. A figure with the maximum resolution of 3000×3000 pixels was in line with modern requirements. Moreover, for more special and artistic figures, the application also allows for a 3D perspective of the heatmap.

4.4.6 Dynamics

In BioJupiter, option changes have an immediate effect. For instance, changes in the parameters, match, mismatch, and gap, result in real-time instant changes in the alignment text, the heatmap matrix, and the heatmap charts. Total customization of graphics is supported. A set of predefined graphics themes can be chosen according to the visual preferences of the user (Figure 4.5d). A cross marker indicates real-time cell position over the heatmap and underlines the corresponding characters in the two sequences. The fonts of the characters adapt to the size of the cells in the heatmap. The markers on the text alignment are linked by the events over the heatmap. The text markers were equipped with the ability to skip over the gaps areas to maintain the correct count of characters in the alignment.

4.4.7 Independence

The main strategy to develop an application independent of external files was based on the Base64 encoding. All the links that should contain paths to external files were replaced with the file content encoded in Base64. Inside the application, the w3.css file was embedded using Base64 encoding to pay a tribute to the great W3Schools organization.

4.4.8 Limits

The size of the canvas object is inversely proportional to the speed of response of the application. It seems that a large width and height of the canvas object affects the performance with which JS draws in real time. In other words, a resolution of the canvas object of 3000 × 3000 pixels will visibly affect the responsiveness of the application. Moreover, the canvas object is used only if the length of the sequences in the alignment does not exceed the resolution capacity of the heatmap.

The time consumption is related to the calculation of the score matrix. Thus, the speed of the application is inversely proportional to the length of the aligned sequences. However, sequences below 100–200 characters allow for real-time changes in the alignment. These observations were made on modest computers and the assessment of the limitations may be subjective. Most likely, the performance of modern computers will greatly increase the responsiveness of the application.

4.4.9   Local Storage

A detailed description of the local storage system and of a strategy for developing multivalue records may be of great importance in future complex applications that use Internet browsers for bioinformatics and computational biology (Figure 4.5c).

Web Storage API

The Web Storage API has been added to HTML5 as a more efficient method for storing client-side data and it is supported by all modern browsers. It is probably one of the most useful and requested properties for Internet browsers. The emergence of the Web Storage API module came as an extension of browser cookies. The Web Storage API provides mechanisms by which browsers can store values associated with unique keys. Local Storage has a capacity of around 10Mb and stores data with no expiration date. Local Storage data persists even when the browser is closed and reopened and can be manipulated or deleted through JS (or manually).

Format and Limitations

Within Web Storage, the Local Storage mechanism was used to save the alignment experiments. For each experiment, the record contains the alignment parameters, the sequence names, the sequence types, and the date/time they were saved; as well as an image capture of the heatmap (Figure 4.4c). The strategy for storing the heatmap in the local storage was based on Base64 encoding, which allowed the conversion of the image information to a string variable. Moreover, this strategy eliminated a simple and classic method of storing a record as an object. Thus, JSON format in not used, as Base64 encoding generates ASCII characters that may interfere with the JS objects. To protect the application from exceeding the allowed storage space, a limitation was imposed for a maximum of 150 records/experiments.

Record Navigation

To browse the records from the local storage, a navigation system was developed and positioned to the right of the heatmap (Figure 4.5c). By default, the navigation window displays five records at a time in the form of rows. Each row displays information related to sequence names, sequence types, the alignment parameters, and the date and time at which the experiment was saved to local storage. Each row also contains a button for record deletion. A saved experiment can be loaded in the application by row selection.

Local Storage Expansion Capabilities

In local storage, a unique key is representative for one value. Thus, records that hold multiple values are not possible by default. To create a multivalue record associated with a single key, a strategy based on two string components was developed: A master key (MK) and a set of specific keys (SKs). A record identification key (RIK) is a composed signature and it was built from a MK followed by a SK from a set (Table 2.1). The MK consists of a unique identifier string and represents the record signature for an experiment. The unique identifier string representing MK is the number of milliseconds between midnight of January 1, 1970 and the current date (i.e. MK = "1583021473940"). SK consists of preset identifier strings. The set of identifier strings from SK (i.e. SK = "param, s0, s1, s0name, s1name") is equal with the number of values, which are part of a record. Thus, the values that belong to a record will be accompanied by the MK in the first part of the RIK and then by the name of the value in the second part of the RIK, namely a SK from the set (e.g. RIK = "1583021473940param"). In other words, MK is the main key for identifying a record and SK is the secondary key for identifying a particular value in the record (Table 4.5).

A special record (SR) holds all MKs associated with the experiments. The application shows the previous alignments in storage by counting the MKs present in the SR. The record deletion process also uses the MK from the SR to find all RIK associated with an experiment.

Import and Export

To make a distinction between experiments saved by the user on the local computer and the imported experiments, a duality of the MKs was intentionally chosen (Table 2.1). Thus, MKs are produced in two ways. New records saved by the user are represented by MKs from the internal clock of the computer (i.e. MK = "1583021473940"). The imported records have another basis for generating MKs. When importing a set of experiments from file, a function is responsible for generating a new MK for each record. This MK is constructed of 8 random characters (i.e. MK = "A7OcnaNu"). The approach helps the application to attach proper labels and distinguish between records made by the user or those imported

Table 4.5 RIK examples for one multivalue record from local storage.

RIKs on local computer	RIKs for imported files
1583021473940	A7OcnaNu
1583021473940param	A7OcnaNuparam
1583021473940s0	A7OcnaNus0
1583021473940s1	A7OcnaNus1
1583021473940s0name	A7OcnaNus0name
1583021473940s1name	A7OcnaNus1name

RIK is a unique key for a value. The RIK components are shown in detail: MK as normal text and SK as underlined text. For instance, MK without the presence of SK is the root of the record and stores the heatmap image in Base64 format. The SK named "param" holds the values of the alignment parameters in the form of a string, in which the values are delimited by a unique character (i.e. "|", "10|-5|7|...") and are then decoded when needed based on the same unique character. In this case, for simplicity, the creation of a RIK for each parameter was avoided. The SK "s0" and "s1" hold the two sequences that participate in the alignment. The SK "s0name" and "s1name" hold the names representing the two sequences.

from other computers. In this manner, collisions are avoided between imported MKs and those produced on the local computer.

Disk Operations

The application calculates and displays the space occupied in local storage under the navigation window. The *disk operations* option is positioned in the same location and is represented by the import/export module and contains three types of operations: (i) *file to local storage* – imports a file with experiments into local storage. The application calculates the number of records before import. It sums the number of records already existing in the local storage and the number of records from the file to verify that the total number of records will not exceed 150 after the import process. If this number is exceeded, the application cancels the import from the file until the necessary space is released by the user. (ii) *Local storage to file* – the operation saves all the records from the local storage in a text file. Exported files contain the all the records from local storage separated by a unique string. (iii) *Delete local storage* – the operation permanently deletes all records from local storage.

4.5 Conclusions

Here, a novel alignment approach was introduced, namely the *forced alignment* regime. The *BioJupiter* implementation uses the forced alignment regime to provide a simple framework for both research and teaching activities in bioinformatics and genetics. This fully independent tool serves as a demonstration of the applicability of native JS in computational molecular biology as well as a useful tool to quickly gain an intuitive visual overview of DNA, RNA, proteins, or other types of sequences (i.e. digit sequences). A dissection of the results was presented using several alignment cases. The main examples involved different types of sequences with multiple arrangements. These cases were presented from simple to complex to provide an intuitive understanding of the mechanisms of the algorithm. Moreover, a novel method for assessing the significance of an alignment was described and tested. The use of UTF symbols provides a wide applicability outside the field of bioinformatics. Thus, *BioJupiter* is mainly addressed to geneticists and bioinformaticians, but also to mathematicians in the field of number theory and to scientists in the field of behavioral analysis and linguistics.

5

Self-Sequence Alignment (I)

5.1   Introduction

Self-sequence alignment represents a novel mathematical model for measuring the information content in biological sequences, such as DNA, RNA, or proteins. Nevertheless, the use of this model is not limited to biological sequences. Thus, any symbol inside a sequence can be considered. The information content is calculated using a dedicated function $\sigma(s)$, which provides values expressed in percentages, where 100% represents the highest information content and zero the lowest. Two principles are explored: A global principle whereby a single value is obtained for a sequence of any length and a local approach in which a series of values are used to construct signals along the sequence. To argue the rationale behind the model, the issue of information is debated through associations with natural phenomena encountered in other fields. By following an experimental approach, the distribution of $\sigma(s)$ values is interpreted and the meaning of the maximum and minimum values is explained in the context of information. The laws behind information dynamics in biology represent a future research direction in bioinformatics and computational biology [254, 255]. To rationalize the inner workings and the utility of this model, a discussion is made on the meaning of true randomness in relation with the limitations shown by all electronic devices and compression algorithms. This chapter further shows how these natural constraints are integral in different biological phenomena and how their exploitation could advance our understanding of biological information and associated functions. Toward the end of the chapter, the Self-Sequence Alignment model is implemented in two versions and tested. Note: The model is described and implemented in this chapter for the first time and the information is a primary source.

Algorithms in Bioinformatics: Theory and Implementation, First Edition. Paul A. Gagniuc.
© 2021 John Wiley & Sons, Inc. Published 2021 by John Wiley & Sons, Inc.
Companion website: www.wiley.com/go/gagniuc/algorithmsinbioinformatics

5.2 True Randomness

Our human construct over the meaning of information is often related to significance. However, the issue of information content is closely related to the meaning of randomness, namely: what constitutes information? There are many sources of randomness and some of them are more familiar then others. For example, the white noise in TV sets translates to images, which contain uniformly distributed white and black pixels, whereas the white noise in radio receivers is translated to white sound (static). In these cases, both the individual pixels and discretized sounds are considered independent random variables with a uniform probability distribution. Thus, in ideal environments, the white noise images and static sounds are considered sources of true randomness [256, 257]. Thermal noise produced by the inner electronics and electromagnetic signals prompted by cosmic microwave background radiation are the main sources for white noise [258, 259]. On computers, a random source of information can take many numerical forms, from a sequence of random numbers to sequences of random characters/ symbols.

5.3 Information and Compression Algorithms

In computer science, classical compression methods can easily suggest what information represents. Lossless compression algorithms encode the source information into a format smaller in size than the original, which can in turn be decoded back to the original format without loss. The aim of such algorithms is shrinking and preserving the information for potential reversibility. Compression algorithms base their operation on progressive elimination of redundancies and reach their limit when the information content lacks any pattern [260, 261]. Such a limiting case can be seen in sources that contain white noise data, where the data do not contain significant repetitions for a possible reduction in size. For example, a simple test can be done for a compression attempt of any BMP/PNG image that contains white noise in a two-dimensional regime. The compression algorithm will be unable to exploit any significant statistical redundancy in the BMP/PNG image. The original BMP/PNG file and the compressed file will show the same size. Thus, true randomness shows the highest information content and represents pure information. This example can also be extrapolated for character sequences, which have a one-dimensional regime. Character sequences with repetitions contain less information compared with those without repetitions. This observation underlines the method described and discussed here.

5.4   White Noise and Biological Sequences

> Note: Biological noise can only be discussed in the evolutionary context, which includes events spanning on tens of thousands to hundreds of thousands of years.

The phenomena responsible for sensitivity limitations in electronic devices have also affected the evolutionary process that stretched over millions of years [258, 259, 262–266]. The evolutionary process uses a diverse set of forces that act upon mutations responsible for DNA information [262–266]. At a fundamental level, the *thermal agitation* phenomenon makes the affinity between distant molecules possible and is also one of the factors responsible for mutations in DNA [262–266]. Thus, thermal agitation brings a constant added randomization in the cycle of biochemical reactions. Based on the above, the consideration is that genetic mutations tend to maintain an imbalance in order to decrease the information content. Following this principle, genetic mutations such as simple sequence repeats (SSRs) or short tandem repeats (STRs) break the symmetry of pure information (noise) [31]. To draw a parallel between information technology and genetics, consider the following: Images that contain white noise show a uniformity in any direction over the 2D plane and for this reason are considered symmetrical. The insertion of a white spot over the 2D plane breaks the symmetry of the image. The same is true with one-dimensional information of DNA, RNA, or proteins. Thus, repeat expansion mutations are crucial molecular mechanisms of evolution and their role is to decrease the information content in one of the directions indicated by selection. As suggested above, thermal agitation and other factors lead to point mutations (biological white noise), as the biological equilibrium dictates a constant lift of the information content in DNA in a natural manner. In other words, genetic mutations with active molecular mechanisms destroy DNA information in favor of a set of phenotypic traits. Nevertheless, here my aim was the creation of a unit of measure under the assumption that repetitive sequences contain less information than random sequences.

5.5   The Mathematical Model

The proposed model defines the information content of a single sequence. It is based on a successive comparison between different areas of the same sequence to determine the number of matches. A sum of the matches over these successive alignments indicates the symmetry breaks inside the sequence, and consequently the information content as understood and discussed above. The information content (σ) equation is defined below as:

$$s = \{x_1, \ldots, x_{|s|}\}$$

$$\sigma(s) = 100 - \left(\frac{\sum_{u=1}^{|s|-1} \left(\frac{\sum_{i=1}^{|s|-u} f(x_i, x_{u+i})}{(|s|-u) \times 100} \right)}{(|s|-1)} \right)$$

$$f(x_i, x_{u+i}) = \begin{cases} +1, & x_i = x_{u+i} \\ 0, & x_i \neq x_{u+i} \end{cases}$$

where f is a match function, s is the sequence composed of any type of symbol x, and $|s|$ represents the total number of symbols in the sequence s (e.g. $s =$ "TACGTATC"; $|s| = 8$). The ith character in s is compared to the $u+i^{th}$ character in s. If the two characters match, then function f will return 1 and if the two characters do not match then function f will return 0. Thus, the $\sigma(s)$ function will provide a number between 0 and 100. Note: Normally, the summation result divided by $|s| - 1$ shows values on a reversed scale, where the maximum information content is represented by zero and the lack of information is represented by 100. For a more intuitive representation, a complementary result was obtained. Namely, the information content was represented on a growing scale by subtracting the final result from 100.

5.5.1 A Concrete Example

A straightforward example may decrease the apparent complexity of the above equation. Please consider the following sequence:

$$s = \{ATAAG\}$$

where s is a five-character sequence. Different parts of this equation will be considered for detailed exemplifications of the principle behind the determination of the information content. The two summations from $\sigma(s)$ are the core of the equation.

$$\sigma(s) = 100 - \left(\frac{\sum_{u=1}^{|s|-1} \left(\frac{\sum_{i=1}^{|s|-u} f(x_i, x_{u+i})}{(|s|-u) \times 100} \right)}{(|s|-1)} \right)$$

$$f(x_i, x_{u+i}) = \begin{cases} +1, & x_i = x_{u+i} \\ 0, & x_i \neq x_{u+i} \end{cases}$$

The first two summations are calculated considering the example sequence (s):

$$\sum_{u=1}^{|s|-1} \left(\frac{\sum_{i=1}^{|s|-u} f(x_i, x_{u+i})}{(|s| - u) \times 100} \right)$$

$$= \left(\frac{(f(A,T) + f(T,A) + f(A,A) + f(A,G))}{(5-1) \times 100} \right) +$$

$$+ \left(\frac{(f(A,A) + f(T,A) + f(A,G))}{(5-2) \times 100} \right) + \left(\frac{(f(A,A) + f(T,G))}{(5-3) \times 100} \right)$$

$$+ \left(\frac{(f(A,G))}{(5-4) \times 100} \right) = \left(\frac{(0+0+1+0)}{(5-1) \times 100} \right) + \left(\frac{(1+0+0)}{(5-2) \times 100} \right)$$

$$+ \left(\frac{(1+0)}{(5-3) \times 100} \right) + \left(\frac{(0)}{(5-4) \times 100} \right)$$

$$= 25 + 33 + 50 + 0 = 108$$

By replacing the summation result from the above in the final information content formula, the following can be obtained:

$$\sigma(s) = 100 - \left| \frac{\sum_{u=1}^{|s|-1} \left(\frac{\sum_{i=1}^{|s|-u} f(x_i, x_{u+i})}{(|s|-u) \times 100} \right)}{(|s| - 1)} \right| = 100 - \left(\frac{108}{(5-1)} \right)$$

$$= 100 - \left(\frac{108}{4} \right) = 100 - 27 = 73\%$$

Thus, the information content σ for sequence "ATAAG" shows a value of 73%.

5.5.2 Model Dissection

The relationship between the above summations covers two regions of the same sequence. The two regions shrink in different directions: one region remains with a fixed point to the first character in the main sequence (s) and the second region remains with a fixed point at the last character in the main sequence (s). To map the internal organization of the main sequence, successive alignments between the two areas follow the number of matches. Thus, a large number of matches indicate low levels of information and a small number of matches indicate high levels of information (Table 5.1).

The calculation method for a short sequence (i.e. "ATAAG") is shown below. Each column in the table represents a step in the calculation method. The first column shows parts of the equation in which variables are replaced with actual values. The second column shows the match between different sections of the same sequence. On the third column, the location of these sections is shown in comparison with the original sequence s. The last column shows the result

Table 5.1 An intuitive step-by-step calculation example.

Total steps of *u*: 4	Self-alignment	Location	Match per step
Step u=1:	X[i] = **ATAA** X[u+i] = **TAAG**	**ATAA** [i] ATAAG **TAAG** [u+i]	f(**A**, **T**) = 0 | u=1 | i=1 f(**T**, **A**) = 0 | u=1 | i=2 f(**A**, **A**) = 1 | u=1 | i=3 f(**A**, **G**) = 0 | u=1 | i=4
$\displaystyle\sum_{u=1} \frac{\sum_{i=1}^{4} f(x_i, x_{u+i})}{(4\times100)}$	ATAA - - ■ - TAAG		$\displaystyle\sum_{i=1}^{4} f(x_i, x_{u+i}) = 1$
$\|s\| = 5$ $\|s\| - u = 5 - 1 = \mathbf{4}$			$\displaystyle\sum_{u=1} \frac{1}{4\times100} = 25$
Step u=2:	X[i] = **ATA** X[u+i] = **AAG**	**ATA** [i] ATAAG **AAG** [u+i]	f(**A**, **A**) = 1 | u=2 | i=1 f(**T**, **A**) = 0 | u=2 | i=2 f(**A**, **G**) = 0 | u=2 | i=3
$\displaystyle\sum_{u=2} \frac{\sum_{i=1}^{3} f(x_i, x_{u+i})}{(3\times100)}$	ATA - - AAG		$\displaystyle\sum_{i=1}^{3} f(x_i, x_{u+i}) = 1$
$\|s\| = 5$ $\|s\| - u = 5 - 2 = \mathbf{3}$			$\displaystyle\sum_{u=2} \frac{1}{3\times100} = \sim 33$

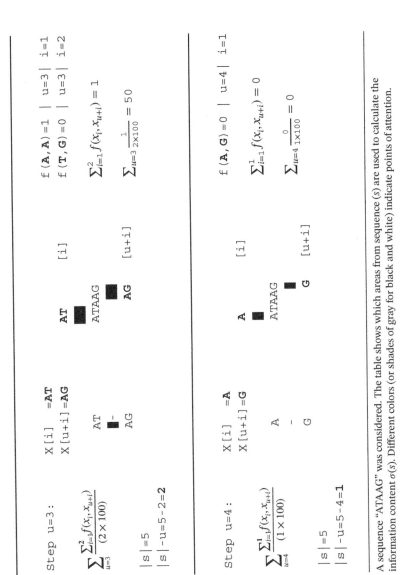

Step u=3:

X[i] = **AT**
X[u+i] = **AG**

$$\sum_{u=3} \frac{\sum_{i=1}^{2} f(x_i, x_{u+i})}{(2\times100)}$$

|s| = 5
|s| - u = 5 - 2 = **2**

	AT	**[i]**	f(**A**, **A**) = 1	u=3	i=1
AT		ATAAG	f(**T**, **G**) = 0	u=3	i=2
- ▇			$\sum_{i=1}^{2} f(x_i, x_{u+i}) = 1$		
AG	**AG**	[u+i]	$\sum_{u=3} \frac{1}{2\times100} = 50$		

Step u=4:

X[i] = **A**
X[u+i] = **G**

$$\sum_{u=4} \frac{\sum_{i=1}^{1} f(x_i, x_{u+i})}{(1\times100)}$$

|s| = 5
|s| - u = 5 - 4 = **1**

	A	**[i]**	f(**A**, **G**) = 0	u=4	i=1
A ▇		ATAAG	$\sum_{i=1}^{1} f(x_i, x_{u+i}) = 0$		
-			$\sum_{u=4} \frac{0}{1\times100} = 0$		
G	**G**	[u+i]			

A sequence "ATAAG" was considered. The table shows which areas from sequence (s) are used to calculate the information content $\sigma(s)$. Different colors (or shades of gray for black and white) indicate points of attention.

returned by function f for each match. The result of the sum over the values returned by function f in each step is further added together in a final value. The final value is divided by the number of characters in sequence s ($|s| - 1$) and the new result represents the complementary value of the information content shown by the "ATAAG" sequence. At this point, a sequence with low information shows a high value and vice versa. To reverse the scale, the result is subtracted from 100.

5.5.3 Conditions for Maxima and Minima

Maximum values ($\sigma(s) = 100$) can be obtained only if all the characters in the sequence are different. For example, in the case of DNA sequences, maximum values can be obtained by permutations of the four types of characters in sequences with a length of 2, 3, and 4 characters (e.g. ATCG TACG, ..., TCGA). Thus, the number of permutations between the characters of an alphabet will dictate the total number of possible sequences that will show an information content of 100. A general formulation for the number of possible permutations from a given group of n characters, taken r at a time is:

$$^nP_r = \frac{n!}{(n-r)!}, \quad r \le n$$

where ^nP_r is the total number of sequences that will provide 100% information content for certain values of r and n. The above formulation can be further generalized for an interval of r. The total number of sequences with maximum information content for an interval of r between two and n is:

$$\delta = \sum_{\substack{r=2 \\ n \ge r}}^{n} {^nP_r} = \sum_{\substack{r=2 \\ n \ge r}}^{n} \frac{n!}{(n-r)!}$$

where δ represents the total number of sequences that can provide maximum information content with respect to the number of characters in the alphabet (n) and r represents the sequence length. For instance, δ applied to DNA or RNA can show a total of 60 unique sequences with a maximum information content ($\sigma(s) = 100$):

$$\delta = \sum_{\substack{r=2 \\ n \ge r}}^{n} \frac{n!}{(n-r)!}$$

$$= \left(\frac{4!}{(4-2)!} \right) + \left(\frac{4!}{(4-3)!} \right) + \left(\frac{4!}{(4-4)!} \right)$$

$$= (12) + (24) + (24) = 60$$

On the other hand, sequences of any size that contain only one type of character show minimum scores ($\sigma(s) = 0$) and indicate a lack of information content (e.g. "AAAA..." or "TTTT..."). Thus, maxima lowers with an increase in the length of the sequence ($r > n$) and the minima remains zero regardless of length (r).

5.6 Noise vs. Redundancy

In vivo DNA exhibits a natural tendency toward noise; a probability distribution with zero mean and finite variance. To avoid a zero mean probability distribution, the DNA replication apparatus uses repeat expansions and other molecular mechanisms to induce constant and guided asymmetries [31]. In computer systems, a compression algorithm eliminates redundancy. Much like the limitations imposed by pure information on compression algorithms, DNA too does not allow for compression/tight packing of nucleotide areas with high information content [31, 267]. In contrast, DNA redundancies allow nucleosomes to self-organize into fractal-like macromolecular structures, which can be highlighted by the heterochromatin organization. One can easily observe a universal law in which high information content does not allow for self-organization, while low information content can lead to self-organization. Thus, fractal-like structures could not exist in the absence of redundancies. In the past, this idea prompted the *DNA recycle hypothesis*, which proposes heterochromatin and euchromatin as the main canvas for evolutionary forces [31].

5.7 Global and Local Information Content

Above, the model has been applied to sequences of any size. However, on biological sequences the model can provide signals based on localized values. Thus, one of the possibilities would be to find the local information content in a long DNA sequence by using sliding windows ($\sigma(x_i \ldots x_{i+w-1} \mid x \in s)$). For example, the content of each sliding window can be passed to function σ and the results can be stored by using a vector. The components i of a vector v can hold the individual results of function σ:

$$v_i = \sigma(x_i \ldots x_{i+w-1} \mid x \in s), \quad 1 \geq i \leq (|s| - w + 1)$$

where $v[i]$ will contain the values for each sliding window i of length w. Both global and local regimes use the same function σ to determine the information content. The difference between the two approaches can be elucidated by using the following example:

$$s = \text{"TATATTCGGC"}$$

Table 5.2 A signal extraction based on local information content.

$x_i...x_{i+w-1} \mid x \in s$	$v_i = \sigma(x_i...x_{i+w-1} \mid x \in s)$
s=TATATTCGGC	$i = 1...(\mid s \mid - w + 1) = 1...(10 - 5 + 1)$
$\mid\mid\mid\mid\mid\mid\mid\mid\mid\mid$	
TATAT$\mid\mid\mid\mid\mid$	$v[1] = \sigma(\text{TATAT}) = 50.00$
ATATT$\mid\mid\mid\mid$	$v[2] = \sigma(\text{ATATT}) = 64.58$
TATTC$\mid\mid\mid$	$v[3] = \sigma(\text{TATTC}) = 72.92$
ATTCG$\mid\mid$	$v[4] = \sigma(\text{ATTCG}) = 93.75$
TTCGG$\mid$	$v[5] = \sigma(\text{TTCGG}) = 87.50$
TCGGC	$v[6] = \sigma(\text{TCGGC}) = 81.25$

A 10-character sequence is considered as an example, namely: "TATATTCGGC." The first column in the table 5.2 shows the positions of the sliding windows over the main sequence (s), while the second column shows the information content obtained from each.

Note: Note that the maximum value for i is the total length of the sequence ($\mid s \mid$) minus the length of the sliding window (w) plus 1.

Globally, the calculation of information content is made for the entire sequence as discussed earlier:

$$\sigma(s) = \sigma(\text{TATATTCGGC}) = 86.12$$

Nevertheless, the local information content can be particularly important for long sequences (i.e. whole genomes). In the local regime, a sliding window (e.g. $w = 5$) and a vector $v[i]$ are introduced, where ($\mid s \mid - w + 1$) denotes the number of components i in vector v (Table 5.2).

Thus, to visualize a signal, each value from components i of vector v can be plotted on a chart. The y-axis can represent the information content and the x-axis can represent the number of sliding windows ($\mid s \mid - w + 1$).

5.8 Signal Sensitivity

As a concept, sensitivity applies only to the local approach and is represented by the amount of variation between the values provided by σ. The sliding window length (w) shows a direct impact on sensitivity and may increase or decrease the sensitivity of these signals. Small sliding windows tend to exhibit extreme variations, while large ones tend to show small variations. As the length of a sliding window (w) tends more toward the length ($\mid s \mid$) of the main sequence (s), the results

Table 5.3 A convergence test of the local regime for sliding windows of different sizes.

	$w = 5$	$w = 6$	$w = 7$	$w = 8$	$w = 9$	$w = 10$
$i = 6$	50					
$i = 5$	64.58	85.33				
$i = 4$	72.92	92.00	90.28			
$i = 3$	93.75	84.33	86.94	87.82		
$i = 2$	87.50	79.33	86.39	88.30	89.14	
$i = 1$	81.25	44.33	69.17	79.63	83.81	86.12
AV	**75.00**	**77.06**	**83.19**	**85.25**	**86.47**	**86.12**
SD	±16.03	±18.85	±9.50	±4.87	±3.77	±0.00

The sliding window dimensions (w) are represented on the columns of the table 5.3, while their position (i) in the main sequence (s) is represented on the lines. The last two rows show the average and standard deviation per each experiment. The bold values in this table 5.3 represent averages (noted AV).

get closer to the global result provided by $\sigma(s)$:

$$\lim_{w \to |s|} \sigma(x_i \ldots x_{i+w-1} \mid x \in s) = \sigma(s)$$

where the variations between the results provided by function σ also decrease. To test the above limit, the same 10-character sequence is considered from the above example. Also, different sliding window lengths are used, between $w = 5$ and $w = 10$, where $w = 10 = |s|$ (Table 5.3). Thus, Table 5.3 shows a nonlinear convergence toward $\sigma(s)$. Note that nonlinearity is information dependent. Thus, sequences with different structures can lead to characteristic variations for a certain interval of w.

A high sensitivity leads to a time series problem, while a low sensitivity leads to loss of important features in the information content. Thus, both extremes lead to a loss of information. Optimal signal sensitivity is still an open debate in bioinformatics and computational biology for all models that use sliding windows and remains a future research topic. The mathematical model of Self-Sequence Alignment is described and tested here for the first time. The main efforts were focused strictly on the technical part of the current model. The correlation and interpretation of biological signals against known biological functions is one of the most stringent endeavors. Further developments will likely include sophisticated implementations that may allow genome-wide analysis based on the current perspective of information content.

5.9 Implementation

The implementation phase includes a global and a local approach, both discussed above. In a first phase, the main equation is implemented inside a function called *Sigma*. The *Sigma* function is implemented in two versions. The first version uses the *substr* function (Additional algorithm 5.1) and the second version the *split* function (Additional algorithm 5.2). Later, the second version of the *Sigma* function is included into a scanner that reads the local information content of a sequence z.

5.9.1 Global Self-Sequence Alignment

As a reminder, the global approach is directly represented by the main equation and may consider sequences of any length. Sequence "TATATTCGGC" has been previously discussed and will be the representative input for Additional algorithm 5.1. Therefore, a strict JavaScript implementation of the $\sigma(s)$ function can be found below (Additional algorithm 5.1):

Additional algorithm 5.1 Note that the source code is in context and works with copy/paste.

```
// SELF-SEQUENCE ALIGNMENT

<script>

document.write(Sigma("TATATTCGGC"));

function Sigma(s)
{
    var t = 0;
    var m = 0;

    for (var u=1; u<=(s.length - 1); u++)
    {
        for (var i=0; i<=(s.length-u); i++)
        {
            m += f(s.substr(i,1), s.substr(u+i,1));
        }
    }
```

```
                t += (m / (s.length-u) * 100);
                m = 0;
         }
    }

    return (100 - (t / (s.length - 1))).toFixed(2);
}


function f(x,y){
     if (x == y) {
           return 1;
     } else {
           return 0;
     }
}

</script>
```

```
Output:
86.12
```

The correspondence between the main equation and the implementation is shown below (Figure 5.1). Notice that inside Additional algorithm 5.1, the $x[i]$ variable is represented by "s.substr(i,1)" and the $x[u+i]$ variable is represented by "s.substr(u+i,1)". For the implementation of function f, variables $x[i]$ and $x[u+i]$ are renamed x and y for ease:

$$f(x_i, x_{u+i}) = \begin{cases} +1, & x_i = x_{u+i} \\ 0, & x_i \neq x_{u+i} \end{cases}$$

$$f(x,y) = \begin{cases} +1, & x = y \\ 0, & x \neq y \end{cases}$$

$$f(x_i, x_{u+i}) \equiv f(x,y)$$

The self-alignment implementation depends more on the programming style. Thus, the model can also be implemented with the help of an array and the *split* function which is native to JavaScript (Additional algorithm 5.2). The implementation from Additional algorithm 5.2 is more direct and more similar to the main equation of the model (Figure 5.1).

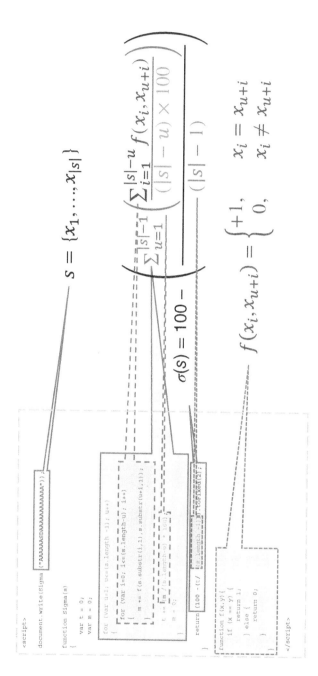

$$s = \{x_1, \ldots, x_{|s|}\}$$

$$\sigma(s) = 100 - \dfrac{\sum_{u=1}^{|s|-1}\left(\dfrac{\sum_{i=1}^{|s|-u} f(x_i, x_{u+i})}{(|s|-u) \times 100}\right)}{(|s|-1)}$$

$$f(x_i, x_{u+i}) = \begin{cases} +1, & x_i = x_{u+i} \\ 0, & x_i \neq x_{u+i} \end{cases}$$

Figure 5.1 Self-sequence alignment – implementation vs. model. The figure shows the connection between the equation and the JavaScript implementation. Note that each part of the equation shows a modular correspondence in the main implementation.

Additional algorithm 5.2 Note that the source code is in context and works with copy/paste.

```
<script>

// SELF-SEQUENCE ALIGNMENT

document.write(Sigma("TATATTCGGC"));

function Sigma(s)
{
    var t = 0;
    var m = 0;
    var x = [];

    x = s.split('');

    for (var u=1; u<=(s.length - 1); u++)
    {
        for (var i=0; i<=(s.length-u); i++)
        {
            m += f(x[i], x[u+i]);
        }

        t += (m / (s.length-u) * 100);
        m = 0;
    }

    return (100 - (t / (s.length - 1))).toFixed(2);
}

function f(x1, x2){
    if (x1 == x2) {
        return 1;
    } else {
        return 0;
    }
}
</script>
```

(Continued)

Additional algorithm 5.2 (Continued)

```
Output:
86.12
```

In the implementation above, each variable retains its significance, exactly as seen in the main equation. Individual symbols from sequence s are stored in variable x in the same order. This conversion is done using the *split* function ($x=s.split('')$). Then these symbols are traversed above x by using the index of the array. Different parts of x are then compared by using an index value based on the interplay between the values of i and u, as dictated by the mathematical formula of the model (Figure 5.1).

These discussions conclude the global analysis based on Self-Sequence Alignment. Since the *Sigma* function processes sequences of any size, this function can also analyze segments of constant length from a larger sequence (i.e. a large gene or even an entire genome). Thus, based on this function, a series of adaptations can be made that lead to the determination of the local information content along a sequence.

5.9.2 Local Self-Sequence Alignment

On biological sequences, the model can provide localized values in the form of signals. Sliding windows are used to find the local information content in a long DNA sequence, denoted by a variable z. For a change of scenery in the experimentation, the following sequence z will be used:

$$z = \text{"AAAAAACAGGTGAGTAAAAAAAA"}$$

Also, a sliding window with a length of nine positions is chosen ($wl=9$). The content of each sliding window over z ($z.substr(l,wl)$) can be processed by function σ ($Sigma(w)$) and the results can be stored by using a vector. The components l of a vector v hold the individual results of function σ as follows:

Additional algorithm 5.3 Note that the source code is in context and works with copy/paste.

```
<script>

// SELF-SEQUENCE ALIGNMENT

function Sigma(s)
```

```
{
    var t = 0;
    var m = 0;
    var x = [];

    x = s.split('');

    for (var u=1; u<=(s.length - 1); u++)
    {
        for (var i=0; i<=(s.length-u); i++)
        {
            m += f(x[i], x[u+i]);
        }

        t += (m / (s.length-u) * 100);
        m = 0;
    }

    return (100 - (t / (s.length - 1))).toFixed(2);
}

function f(x1, x2){
    if (x1 == x2) {
        return 1;
    } else {
        return 0;
    }
}

// THE INFORMATION CONTENT SCANNER

var z = "AAAAAACAGGTGAGTAAAAAAAA";

var wl = 9; // SLIDING WINDOW LENGTH
var w = ''; // SLIDING WINDOW CONTENT
var v = ''; // VECTOR v

var u = z.length - wl + 1;
```

(Continued)

Additional algorithm 5.3 (Continued)

```
for(var l=0; l<u; l++) {

    w = z.substr(l, wl);
    v = Sigma(w);

    document.write('sw('+(l+1)+') = "'+w+'"<br>v['+l+']
    ='+v+'<hr>');
}

</script>
```

Output:
```
sw(1) = "AAAAAACAG"
v[0]=50.76

sw(2) = "AAAAACAGG"
v[1]=66.50

sw(3) = "AAAACAGGT"
v[2]=78.60

sw(4) = "AAACAGGTG"
v[3]=83.29

sw(5) = "AACAGGTGA"
v[4]=67.26

sw(6) = "ACAGGTGAG"
v[5]=76.00

sw(7) = "CAGGTGAGT"
v[6]=81.53

sw(8) = "AGGTGAGTA"
v[7]=66.95

sw(9) = "GGTGAGTAA"
```

```
v[8]=78.51

sw(10)  =  "GTGAGTAAA"
v[9]=78.81

sw(11)  =  "TGAGTAAAA"
v[10]=73.50

sw(12)  =  "GAGTAAAAA"
v[11]=61.82

sw(13)  =  "AGTAAAAAA"
v[12]=40.04

sw(14)  =  "GTAAAAAAA"
v[13]=55.45

sw(15)  =  "TAAAAAAAA"
v[14]=33.97
```

The output above shows the content of each sliding window and the value returned by function *Sigma* for that specific content. Note that the index of vector v starts from 0 and the index of the first sliding window starts from 1. As discussed in the previous subchapters, redundancies (i.e. "TAAAAAAAA") lead to small σ values, and vice versa (i.e. "AAACAGGTG" or "CAGGTGAGT"). Note that here, the sequence "TATATTCGGC" was no longer used because the results were presented in detail in Table 5.2. Nevertheless, the results from Table 5.2 can be verified by using Additional algorithm 5.3, the "TATATTCGGC" sequence and a sliding window of five positions (wl=5).

5.10 A Complete Scanner for Information Content

A few deletions from the previous implementation lead to a scanner, which produces a clean signal (Additional algorithm 5.4). The components of the signal indicate the information content on each sliding window above the z-sequence.

Additional algorithm 5.4 Note that the source code is in context and works with copy/paste.

```
<script>

// THE INFORMATION CONTENT SCANNER

var z = "AAAAAACAGGTGAGTAAAAAAAA";

var signal = '';
var wl = 12;
var w = '';

var u = z.length - wl + 1;

for(var l=0; l<u; l++) {
    w = z.substr(l, wl);
    signal += Sigma(w) + ',';
}

document.write(signal);

// SELF-SEQUENCE ALIGNMENT

function Sigma(s)
{
    var t = 0;
    var m = 0;
    var x = [];

    x = s.split('');

    for (var u=1; u<=(s.length - 1); u++)
    {
        for (var i=0; i<=(s.length-u); i++)
        {m += f(x[i], x[u+i]);}

        t += (m / (s.length-u) * 100);
        m = 0;
```

```
    }

    return (100 - (t / (s.length - 1))).toFixed(1);
}

function f(x1, x2){
    if (x1 == x2) {return 1;} else {return 0;}
}

</script>
```

```
Output:
75.2,60.1,71.3,78.7,67.7,66.8,75.6,62.3,76.8,72.5,
64.8,52.9,
```

The scanner section was moved at the beginning of the implementation whereas the *sigma* function and the *f* function were moved to the end. Compared to the previous implementation, the sliding window has been enlarged to 12 symbols only to try other parameters. Note: In the *Sigma* function the values have been reduced to one decimal place to fit in the output window.

5.11 Conclusions

The current chapter described a new mathematical model for measuring information in biological sequences. Self-sequence alignment was developed based on the premise that point mutations represent the biological noise. Depending on their importance, older information structures sink and disappear into the biological noise at different evolutionary speeds. Based on guidance from selection pressures, new mutations build over biological noise by strengthening old information structures or by eliminating them entirely. Thus, DNA is continually degenerated and revitalized over time. In turn, this process reverberates over protein structures and their spatiotemporal production. Self-sequence alignment detects the footprints of these evolutionary mechanisms and may be a promising approach for unveiling the interactions between different biological functions. Thus, self-sequence alignment may be used as a novel computational signal for genomics and proteomics.

6

Frequencies and Percentages (II)

6.1 Introduction

In multicellular eukaryotes, genomic landscapes are crucial for transitions between different cellular states (cell types) and their characteristic patterns of gene expression. The order and frequency of different groups of nucleotides along the DNA molecules dictate specific stochastic distributions (natural preferences) for torsion, bending, and the interactions of DNA with different proteins in the nuclear space. The same stochastic distributions are valid for RNA molecules and proteins. Thus, the distribution of nucleotides is prime for understanding life. In research, the frequency presented by different combinations of nucleotides along the genome of some organisms, allows for relaxed correlations, without immediate practical implications. Over the years, the innumerable possibilities of interpretation led to fewer and fewer relevant connections between the frequency of different nucleotide groups and the mechanisms of life. Thus, although the frequencies shown by different groups of nucleotides (di-nucleotides, tri-dinucleotides, tetra-nucleotides, and so on) along the DNA molecule involves the most common computational methods, their significance, and interpretation in the global context is still elusive and to be elucidated in the future. Nevertheless, this chapter presents the computational methods by which these frequencies can be automatically extracted from a DNA/RNA or a protein sequence. In a first stage, a general function for frequency detection is briefly described. Both the frequencies of nucleotides and the frequencies of groups of nucleotides are taken into consideration. Two scanners are described in detail in the second stage of the chapter; both sharing the capability of generating a discrete signal from a sequence of symbols/letters/characters. The first scanner computes the frequencies over different sequences without any discrimination on the main

Algorithms in Bioinformatics: Theory and Implementation, First Edition. Paul A. Gagniuc.
© 2021 John Wiley & Sons, Inc. Published 2021 by John Wiley & Sons, Inc.
Companion website: www.wiley.com/go/gagniuc/algorithmsinbioinformatics

signal, while the second scanner uses a threshold value to filter the uninteresting regions from the main signal. The methods described here are further used in the following chapters in conjunction with other methods to reveal new sequence features that may indicate the relationship between different functional parts of the DNA molecule.

6.2 Base Composition

The nucleotide content is the very first approach in characterizing a nucleic acid sequence. Usually, frequencies refer to counts (absolute frequency). However, one way of looking at the frequency of values is through the use of percentages (relative frequency expressed as a percentage). A percentage reflects the proportion of counts of a particular symbol or set of symbols inside a sequence. For instance, the term "GC% content" or "$(G + C)\%$" or "$(C + G)\%$" represents the percentage of guanine (G) plus the percentage of cytosine (C) bases in a DNA or RNA fragment (a region, a gene, an entire genome). Please consider the sequence below:

ATA<u>CC</u>GGT<u>C</u>A<u>CGCGCGGCGCGCGCAC</u>

The total length of the above sequence is 26 b and the number of cytosines plus the number of guanines adds up to 20 b. To find out the $G + C$ percentage, the length of the sequence divides the number of $G + C$ bases found in the sequence:

$$(G + C)\% = \frac{(G + C)}{sequence\ length} = \frac{20}{26} = 0.77 = 77\%$$

Like many other formulas, the above expression can be rewritten in several ways. For example, an equivalent formula can directly show the result in percentages, as follows:

$$(G + C)\% = \frac{100}{sequence\ length} \times (C + G) = \frac{100}{26} \times 20 = 77\%$$

Thus, for the above sequence, the percentage of cytosines and guanines shows a value of 77%. From this result, one can automatically deduct the value for adenine and thymine (i.e. $(A + T)\% = 100\% - (G + C)\% = 100\% - 77\% = 33\%$) or their percentage can be calculated in the same way by using the above formula.

6.3 Percentage of Nucleotide Combinations

On the other hand, the CpG% indicates the percentage of "CG" dinucleotides (cytosine fallowed by guanine in the linear sequence of bases) in a DNA or

RNA fragment. Letter "p" signifies the phosphate group that connects the two nucleotides together. For instance, consider the sequence below:

ATAC<u>CG</u>GTCA<u>CG</u>C<u>CG</u>G<u>CG</u>C<u>CG</u>CAC

As before, the total length of the sequence is 26 b; however, the number of dinucleotides (groups of cytosines fallowed by guanines) adds up to 7. To find out the CpG percentage, the number of nucleotides in the group (two nucleotides) is multiplied by the number of groups found (seven dinucleotides). The result is then divided by the total length of the sequence:

$$CpG\% = \frac{(members \times groups\ found)}{sequence\ length} = \frac{(2 \times 7)}{26} = \frac{14}{26} = 0.54 = 54\%$$

The CpG% value indicates that "CG" occupies 54% of the above sequence. Note that terms such as: "CpG sites" or "CG sites" or "CpG dinucleotides" or "CG dinucleotides" have the same meaning. Moreover, any combination of nucleotides can be referred to in the same manner. For instance, consider ApA instead of CpG. The terms "ApA sites" or "AA sites" or "ApA dinucleotides" or "AA dinucleotides" indicate an adenine molecule followed by an adenine molecule in the linear sequence of bases along the 5′–3′ direction. According to the same reasoning, the percentage can be obtained for trinucleotides or tetranucleotide combinations (e.g. ApTpC%; GpGpApT%), and so on. However, the notations are malleable. For instance, combinations above three nucleotides make the notation hard to follow (e.g. GpGpApT%) and it can be written as GGAT%. For instance, please consider the "GGAT" group of four members/letters and the sequence from below:

ATACGTCA<u>GGAT</u>CGGCG<u>GGAT</u>CAC

The above sequence shows two occurrences for "GGAT." The same formula from above applies:

$$GGAT\% = \frac{(members \times groups\ found)}{sequence\ length} = \frac{(4 \times 2)}{24} = \frac{8}{24} = 0.33 = 33\%$$

Thus, the GGAT% value indicates that the group of letters "GGAT" occupies 33% of the sequence. Note that the above formula is also valid for groups that contain repetitions (e.g. "GCGC," or "AAAA," and so on).

6.4 Implementation

A general function can be designed to calculate both the individual percentage of letters and the percentage of different groups of letters/nucleotides (Additional algorithm 6.1). The implementation below incorporates what has been discussed so far:

Additional algorithm 6.1 Note that the source code is in context and works with copy/paste.

```
<script>

// GF - Group Frequencies

var c = 0;
var g = 0;

c = Number(GF('ATACCGGTCACGCGCGGCGCGCGCAC','G'));
g = Number(GF('ATACCGGTCACGCGCGGCGCGCGCAC','C'));

document.write('GC%=' + (c+g) + '<br>');

document.write('CpG%=' + GF('ATACCGGTCACGCGCGGCGCGCGCAC','CG')+'<br>');
document.write('GGAT%=' + GF('ATACGTCAGGATCGGCGGGATCAC','GGAT'));


function GF(s, m)
{
    var p = 0;
    s = s.toLowerCase();
    m = m.toLowerCase();

    p = s.split(m).join("").length;
    p = (s.length - p)/s.length;

    return (p * 100).toFixed(2);
}

</script>
```

```
Output:

GC% = 76.93
CpG% = 53.85
GGAT% = 33.33
```

For simplicity, a *GF* (group frequencies) function uses the "split" and "join" method to remove all instances of a group of letters/nucleotides (m) from a given DNA/RNA sequence (s). The length of the remaining sequence (p) is subtracted from the length of the initial sequence (s.length) to obtain the total number of letters/nucleotides belonging to the group (s.length - p). The result is then divided by the length of the initial sequence (s.length) to deduce the proportion occupied by group *m* inside sequence *s*. This last result is multiplied by 100 for percentage conversion. To evaluate the percentage of cytosine and guanine (C+G)% above sequence *s*, the *GF* function is called independently for each letter and the results are summed. The *GF* function is a universal function. For instance, both the percentage of single nucleotides and the percentage of different groups of nucleotides can be found without additional changes to the function. Moreover, the use of the function for proteins (or any other type of sequence) does not require any modifications (it can be used as it is). Another step-by-step version for determining the (C+G)% is shown below. However, this version does not process groups of letters as the *GF* function does (Additional algorithm 6.2).

Additional algorithm 6.2 Note that the source code is in context and works with copy/paste.

```
<script>

// a step-by-step version for (C+G)%

document.write(CG_content('ATACCGGTCACGCGCGGCGCGCGCAC'));

function CG_content(s)
{
    s = s.toLowerCase();

    var a = 0;
    var t = 0;
    var c = 0;
    var g = 0;

    for (var u=0; u<=s.length; u ++)
    {
        var n = s.substr(u,1);
        if (n == "a") {a = a + 1;}
        if (n == "t") {t = t + 1;}
        if (n == "g") {g = g + 1;}
        if (n == "c") {c = c + 1;}
    }

    return (((c + g)/s.length) * 100).toFixed(2);
}

</script>
```

Output:

```
GC% = 76.92
```

The above implementation shows a rudimentary function called "*CG_content,*" which handles each nucleotide/letter independently. It uses a counter variable for each letter type in a DNA sequence. The function traverses the sequence from the first to the last letter. Each time it encounters a letter, the associated integer variable is incremented by 1. At the end of this computation, the sum of the counts from the associated variables (a+t+g+c) is equal to the length of the sequence (s.length). Thus, for instance, the cytosine and guanine counts are divided by the length of the sequence to determine the proportion of these two nucleotides/letters. The result is multiplied by 100 for a conversion to percentages. As expected, the adaptation of the function for protein sequences involves the addition of one variable to each amino acid (one variable – one type of amino acid).

6.5   A Frequency Scanner

The computation of CpG% for entire sequences (e.g. a long gene or a genome) provides a global value that is less useful for interpretation. However, the computation of CpG% (or any other combination of nucleotides) on successive shorter regions over a DNA/RNA or protein sequence may indicate different structural properties (details also discussed in other chapters). But how is the distribution detected? To answer this question, the implementation from below (Additional algorithm 6.3) shows a CpG% scanner whose *GF* function generates a discrete signal from a sequence *s*.

Additional algorithm 6.3 Note that the source code is in context and works with copy/paste.

```
<script>

// A frequency scanner

var s = 'ATACCGGTCACGCGCGGCGCGCGCAC';

var w = 9;
var t = ";
var n = ";

for (var u=0; u<=s.length-w; u ++)
```

```
{
    n = s.substr(u,w);
    x = GF(n,'CG');
    //x = Number(GF(n,'C')) + Number(GF(n,'G'));
    t += x + ',';

    document.write('['+n+']; CpG% = ' + x + '<br>');
}

document.write('Signal:<br>' + t);


function GF(s, m)
{
    var p = 0;
    s = s.toLowerCase();
    m = m.toLowerCase();

    p = s.split(m).join("").length;
    p = (s.length - p)/s.length;

    return (p * 100).toFixed(0);
}

</script>
```

Output:

```
[ATACCGGTC] CpG% = 22
[TACCGGTCA] CpG% = 22
[ACCGGTCAC] CpG% = 22
[CCGGTCACG] CpG% = 44
[CGGTCACGC] CpG% = 44
[GGTCACGCG] CpG% = 44
[GTCACGCGC] CpG% = 44
[TCACGCGCG] CpG% = 67
[CACGCGCGG] CpG% = 67
[ACGCGCGGC] CpG% = 67
```

(Continued)

Additional algorithm 6.3 (Continued)

```
[CGCGCGGCG] CpG% = 89
[GCGCGGCGC] CpG% = 67
[CGCGGCGCG] CpG% = 89
[GCGGCGCGC] CpG% = 67
[CGGCGCGCG] CpG% = 89
[GGCGCGCGC] CpG% = 67
[GCGCGCGCA] CpG% = 67
[CGCGCGCAC] CpG% = 67

Signal:
22,22,22,44,44,44,44,67,67,67,89,67,89,67,89,67,67,67,
```

The scanner uses a sliding window to extract short regions from the original sequence (s). The current example uses a sliding window of 9 positions ($w=9$) for ease. The content of each region (n) is then injected into the *GF* function ($x = GF(n,'CG')$). The successive results returned by the *GF* function are added to a discrete signal stored into a t variable ($t+=x+','$). The successive values coming from these regions (sliding windows) can show a series of particularities, which indicate different properties of the sequence under consideration.

6.6 Examples of Known Significance

CpG dinucleotides are the most studied group of nucleotides. It is known that CpG sites occur with high frequency on short genomic regions. For instance, in the human genome, these areas are about 0.5–2 Kb in length and their number reaches ~30 000 [268]. These short genomic regions are known as "CpG islands" or "CG islands" or "CGI" or "clusters of CpG dinucleotides," and point to a series of high CpG% values across a discrete signal similar to that shown in the output of Additional algorithm 6.3. Many CpG islands are usually detected in or near gene promoters (Figure 6.1) [269]. CpG islands are attractive for a subset of transcription factors (TFs) [269]. In active genes, the CpG sites remain unmethylated throughout a CpG island (e.g. promoters of housekeeping genes – always active for basic functions of cells) [269]. Methylation of CpG dinucleotides in these regions prevents the binding of TFs, which leads to gene silencing (no expression) [270]. Of course, this is one of the mechanisms of gene silencing (or activation) and is preset during cell division upon transitions made from one cell state to another [270]. In the overall picture, the DNA methylation is critical for cell differentiation

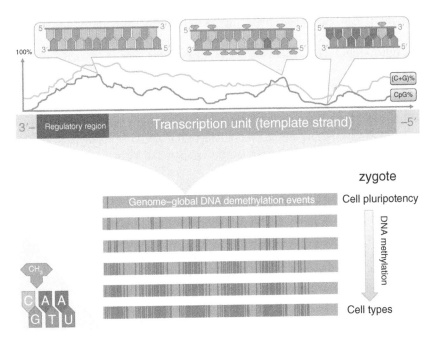

Figure 6.1   CpG% and the epigenetic programming. The top of the figure shows a hypothetical distribution of (C+G)% and CpG% above a gene. It further indicates that intergenic CpGs are methylated and CpGs in the regulatory regions are generally unmethylated. For an intuitive overview, the methylation pattern is represented on the two strands above the chart. The lower part of the figure shows the importance of epigenetic reprogramming that leads, among other mechanisms, to cell types in some eukaryotic organisms. The patterned lines depict an interpretation of DNA methylation across cell generations. The significance of the geometric shapes used for illustration is shown in the lower left corner.

and embryonic development in some organisms [271, 272]. For instance, during mammalian development, two global demethylation events (a partial reset of DNA methylation) are followed by waves of de novo DNA methylation [273]. The first global demethylation event erases the parental imprints from the germline (sperm and egg). The second demethylation event starts in the zygote and continues during the two- to eight-cell stages [274]. Cell pluripotency (e.g. stem cells) is progressively narrowed at each cell division during development, leading to the final states, namely to different cell types. Loss of pluripotency is made (among other mechanisms) through selective DNA methylation of CpGs and other dinucleotide combinations. Thus, a DNA methylation pattern dictates for some of the genes, which of them will continue to be expressed in future cell generations (Figure 6.1). In this way, the initial cellular potency that led to the formation of an organism is progressively diminished and scattered in different steady states (cell types).

But why is CpG so important in the end? DNA methylation occurs at the cytosine bases. In double-stranded DNA, cytosine on one strand binds to guanine on the other strand. Thus, there are two methylated cytosine residues positioned diagonally to each other on the opposing strands (i.e. CpG on one strand and a GpC on the other stand) (Figure 6.1). This symmetry and alternation of methyl groups on the outside of the two strands seems to be important in decreasing the probability of TF binding [275]. In contrast, for example, a CpA will correspond to a GpT on the opposing strand and only cytosine would be methylated. Thus, such an arrangement would be asymmetric compared to a CpG.

6.7   Observation vs. Expectation

The four types of nucleotides, which constitute the DNA molecule, may form a total of 16 dinucleotide combinations (i.e. four possibilities on the first position multiplied by four other possibilities on the second position; $4 \times 4 = 16$). Thus, in the null model, the expected frequency for any of the 16 dinucleotides would be $1/16 = 0.0625$ (i.e. ~6.2%). Consequently, the expected frequency for CpG dinucleotides is also 6.2%. Some organisms use DNA methylation as an epigenetic mechanism and other organisms do not [270]. For instance, vertebrates use DNA methylation mechanisms. During evolution, DNA methylation leads to a conversion of cytosine to thymine and consequently to a CpG deficiency (i.e. CpG leads to TpG). As a consequence, CpG dinucleotides are observed at a frequency below of that expected by chance (6.25%). On the other hand, invertebrates rarely methylate their DNA. Consequently, the observed frequency of CpG sites in these organisms is closer to the expected frequency (around the 6.25% value). Such examples can be found in the genomes of biological models such as *Drosophila melanogaster* (vinegar fly) or *Caenorhabditis elegans* (nematode worm) [270]. Methylation mechanisms are also reported in other kingdoms of life. For instance, DNA methylation mechanisms are mandatory during plant growth and development [276]. Moreover, viral genomes are directly impacted by the nucleotide composition found in the genome of their hosts (e.g. CpG sites are underrepresented in RNA viruses) [277]. In conclusion, CpG sites are underrepresented in organisms that rely heavily on methylation mechanisms. The CpG detection over different GC-rich regions at a frequency much higher than expected may indicate important biological functions. Nevertheless, CpG is just one of the 16 combinations of dinucleotides surrounding the genome organization. DNA methylation is also found at sites other than CpG sequences (i.e. CpT, CpA, CpC, and so on) [278, 279]. Moreover, combinations of three (trinucleotides – 64 combinations) or more than three nucleotides may indicate other types of functional relationships, completing the picture of life even further, in the distant future.

6.8   A Frequency Scanner with a Threshold

Above different signals, only certain values may be of interest. For example, different evolutions along the signal (structures) can be highlighted when the values of the components are above a specific threshold. The implementation from Additional algorithm 6.4 uses a specially constructed sequence (s) to test the threshold approach:

CATTAA<u>CG</u>GAG<u>CGCGCGCG</u>GTCTCTCT

In Additional algorithm 6.4, variable b represents the threshold. The value of variable b can be calculated or declared. For instance, an already-known threshold value can be imposed on Additional algorithm 6.4, either from the literature (e.g. a global CpG% for an entire genome, which may be below or above the expected value) or from some objective research angle. Note: A threshold value of zero will show no difference between the output of Additional algorithm 6.4 and the output of Additional algorithm 6.3. A second possibility (used here) is the detection of a threshold value based on the nucleotide content of the entire sequence (s).

Additional algorithm 6.4 Note that the source code is in context and works with copy/paste.

```
<script>

// A frequency scanner with a threshold filter

var s = 'CATTAACGGAGCGCGCGCGGTCTCTCT';

var w = 9;
var d = 'CG';

var t = '';
var n = '';
var b = GF(s,d);

document.write('Global CpG% = ' + b);

for (var u=0; u<=s.length-w; u++)
{
    n = s.substr(u,w);
```

(Continued)

Additional algorithm 6.4 (Continued)

```
    x = GF(n,d);
    if(x < b){x='-';}
    t += x + ',';
}

document.write('<hr>Signal:<br>' + t);

function GF(s, m)
{
    var p = 0;
    s = s.toLowerCase();
    m = m.toLowerCase();

    p = s.split(m).join("").length;
    p = (s.length - p)/s.length;

    return (p * 100).toFixed(0);
}

</script>
```

```
Output:

Global CpG% = 37
Signal:
-,-,-,-,44,44,67,44,67,67,89,89,67,67,44,44,-,-,-,
```

Note that *s* contains the main sequence and variable *n* (the sliding window of length *w*) temporarily stores successive parts from *s*. The content of each sliding window (*n*) is processed by the *GF* function and the returned value (*x*) is compared to the threshold value (*b*). The returned value (*x*) may be added to the signal (*t*) only if it exceeds the threshold value (*b*), or a line symbol ("-") can be added otherwise. Thus, the scanner above only shows the values that are above the overall CpG% calculated from sequence *s*. Additional experiments are shown in Table 6.1.

The contents of variable *d* can be changed to detect the localized frequency for other combinations of nucleotides (e.g. d='ATC';). For instance, Table 6.1 shows some experiments performed with Additional algorithm 6.4 on different

Table 6.1 Additional experiments on frequencies.

Sequence 1	CCCCCCCCCCGCGCGCGCCCCCCCCCCCCGCGCGCGGCGCG
Signal (CpG%)	-,-,-,-,-,-,67,67,89,89,67,67,-,-,-,-,-,-,-,-,-,-,-,-,67,67,89,89,67,89,67,89,
Sequence 1	CCCCCCCCCCGCGCGCGCCCCCCCCCCCCGCGCGCGGCGCG
Signal (CpC%)	89,89,89,67,67,-,-,-,-,-,-,-,-,-,67,67,89,89,89,89,89,67,67,-,-,-,-,-,-,-,-,-,
Sequence 1	CCCCCCCCCCGCGCGCGCCCCCCCCCCCCGCGCGCGGCGCG
Signal (CpCpC%)	-,-,67,67,67,-,-,-,-,-,-,-,-,67,67,67,-,-,-,67,67,67,-,-,-,-,-,-,-,-,-,
Sequence 2	AAAAAAAAAAAAAAATCATCAAAAAAAAAAAAAAAAAAAA
Signal (ApTpC%)	-,-,-,-,-,-,-,33,33,33,67,67,67,67,33,33,33,-,-,-,-,-,-,-,-,-,-,-,
Sequence 3	@#@#@@#@@##@##@@@@@@##@##@#@#@#@#########@
Signal (@@%)	-,-,-,-,44,-,44,44,67,67,67,67,44,44,-,-,-,-,-,-,-,-,-,-,-,-,-,

A series of sequences and their signals are presented here. Dinucleotide and trinucleotide elements are tested on a series of specially constructed sequences. In each experiment, the overall percentage of the elements (dinucleotides or trinucleotides in this case) determines the threshold value. Moreover, a sequence of symbols and the corresponding signal are shown at the bottom of the table (Sequence 3). Sequence 3 points out that such methods are adaptable to any one dimensional sequence of symbols.

sequences and combinations of nucleotides. The table also points out the universality of these methods by using sequences of different symbols instead of letters (e.g. the "@" symbol and the "#" symbol). Note that all signals are presented in their raw form. However, each signal can be easily plotted in a graphical format for an intuitive view of the evolution of values above sequence s. Moreover, the interplay between several signals can also be observed (please see the online implementations).

6.9 Conclusions

The relationship between different nucleotide combinations may reveal the genomic landscapes of various organisms. The computational simplicity of frequency detection and the interpretation of frequency in the biological context are two stages with a complementary difficulty. The functional significance of the fabric of nucleotide combinations is largely unknown. The interweaving of nucleotide frequencies reflects over coding and noncoding sequences and impacts the genetic code (codon usage), the variations in DNA-binding sites (small regions where proteins prefer to bind to the DNA), the variations in RNA binding sites, the cellular development and differentiation, and so on [271]. The biological context of these frequencies is one of the main concerns in bioinformatics, as it may lead to a clear picture on the fundamental laws by which life works. Of the 16 possible combinations, the CpG dinucleotides have been studied extensively over time. To point out the methods used here, the CpG dinucleotides were taken as an example and their function was discussed in the context of biology. A few methods have been described by which different nucleotide combinations can be extracted from a DNA sequence. First, some technical details were discussed by which frequencies can be calculated globally (on a DNA sequence of any length). Secondly, two scanners were presented. A first scanner allowed the possibility to extract the frequencies without any imposed discrimination. In the implementation of the second scanner, threshold values were used as a filter.

7

Objective Digital Stains (III)

7.1 Introduction

The objective digital stain (ODS) is represented by a distribution of points on a two-dimensional (2D) image, which reflects the information structure inside a DNA/RNA sequence. In turn, the shape of the distribution pushes the association of the method with a kind of "digital stain," thus safely establishing the name of the method. In the past, ODSs were simply called "DNA patterns." On ODSs, the information content (IC) is represented vertically on the y-axis and the frequency of different letters is represented horizontally on the x-axis. The overall idea surrounding this method is that similar distributions of two or more DNA/RNA sequences may show similar functions. Interestingly, dissimilar DNA/RNA sequences may show similar ODSs. This intriguing proprietary has been demonstrated repeatedly over time. The first observation was made in 2012 in the journal *BMC Genomics* [269]. There, different ODS patterns have been shown for eukaryotic gene promoters that indicate several generic classes of promoters (more than 10 classes of promoters in eukaryotes). Moreover, in 2013 in the same journal, a correlation was made between chromosomal territories and ODS patterns shown by gene promoters [31]. More recently, in 2015, a publication in the journal *PLoS ONE* showed the connection between ODSs of promoter sequences and the genes associated with type 1 and 2 diabetes [280]. This latest publication showed more clearly that ODSs are able to link the DNA sequence characteristics to different biological functions. ODSs are not limited to gene promoter sequences and can be used for any DNA or RNA region. Thus, this chapter describes in detail how an ODS can be built and what is its significance. The ODS patterns of several genes, viral genomes, and organellar genomes are also shown. Note that these ODS distributions are called "objective" because they reflect the content of a DNA/RNA sequence.

Algorithms in Bioinformatics: Theory and Implementation, First Edition. Paul A. Gagniuc.
© 2021 John Wiley & Sons, Inc. Published 2021 by John Wiley & Sons, Inc.
Companion website: www.wiley.com/go/gagniuc/algorithmsinbioinformatics

7.2   Information and Frequency

The previous chapters presented two preparatory methods. A novel method of measuring information along a sequence was described in Chapter 5. Next, Chapter 6 presented the methods by which frequencies of different nucleotides or combinations of nucleotides can be computed along a DNA/RNA sequence. However, what could the two methods indicate when they are found in a common context? How could they be linked? In the first instance, the two methods are merged here in a single scanner. For an initial implementation, the following sequence is used:

$$z = \text{"ATACCGGTCACGCGCGGCGCGCGCAC"}$$

The scanners from Additional algorithms 5.4 and 6.3 are combined and adapted here to generate two separate signals (Additional algorithm 7.1). The scanner is built in three parts. The third part is represented by the self-sequence alignment method, which indicates the IC by using the *IC* function (known previously as the *Sigma* function) and the adjacent function f (Additional algorithm 5.1). In practical terms, the *IC* function uses the equation presented below, which has been described in detail in Chapter 6:

$$s = \left\{ x_1, \ldots, x_{|s|} \right\}$$

$$IC(s) = 100 - \left(\frac{\sum_{u=1}^{|s|-1} \left(\frac{\sum_{i=1}^{|s|-u} f\left(x_i, x_{u+i}\right)}{(|s| - u) \times 100} \right)}{(|s| - 1)} \right)$$

$$f\left(x_i, x_{u+i}\right) = \begin{cases} +1, & x_i = x_{u+i} \\ 0, & x_i \neq x_{u+i} \end{cases}$$

where s is the contents of a sliding window. The second part of Additional algorithm 7.1 includes a function that computes the $(C + G)$ content of the same sliding window (such as Additional algorithms 6.1 or 6.2). Both the *CG* (cytosine and guanine – Additional algorithm 6.2) and the *GF* function (group frequencies – Additional algorithm 6.1) can be used to compute the frequency of single letters; however, only the *GF* function can be used for computation of single-letter frequency and frequency of combinations of letters. The *CG* function is time consuming but easy to understand, while the *GF* function contains an approach that leads to a particularly high processing speed. Although faster and more advanced, the *CF* function must be called independently for each letter, whereas the more primitive *CG* function is called only once. The *CG* function may use any of the two equations presented below, which are described in detail in

Chapter 6. For instance, the content represented by "C" and "G" over the sliding window can be found by applying the following equation, which provides values between 0 and 1:

$$(G + C)\% = \frac{(G + C)}{sw}$$

where *sw* is the length of the sliding window. The result can be converted to percentages by a multiplication with 100. Alternatively, the result can be obtained directly as a percentage by using an equivalent formula:

$$(G + C)\% = \frac{100}{sw} \times (C + G)$$

For single letters, but also for combinations of letters, the following equation can be used by the *GF* function (adapted to the context used inside the function):

$$N\% = \frac{(\text{members} \times \text{groups found})}{sw} \times 100$$

where *N*% can represent the percentage of one letter (members = 1) or the percentage represented by a specific combination of several letters (members > 1; please see Chapter 6). Nevertheless, the first part of Additional algorithm 7.1 involves the main scanner, which calls the two functions (*IC* and *GF*) at each step performed by a sliding window. The values returned by the two functions are stored successively in two strings of numbers, called signals.

Additional algorithm 7.1 Note that the source code is in context and works with copy/paste.

```
<script>

// INFORMATION CONTENT VS FREQUENCY

var z = "ATACCGGTCACGCGCGGCGCGCGCAC";

var signal = ";
var wl = 9;
var w = ";

var x = ";
var y = ";

var u = z.length - wl + 1;

for(var l=0; l<u; l++) {
    w = z.substr(l, wl);

    x += (Number(GF(w,'C')) + Number(GF(w,'G'))) + ',';
    y += IC(w) + ',';
}
```

(Continued)

Additional algorithm 7.1 (Continued)

```
document.write('CG [x]=' + x + '<br>IC [y]=' + y);

function GF(s, m)
{
    var p = 0;
    s = s.toLowerCase();
    m = m.toLowerCase();

    p = s.split(m).join(").length;
    p = (s.length - p)/s.length;

    return (p * 100).toFixed(0);
}

// SELF-SEQUENCE ALIGNMENT

function IC(s)
{
    var t = 0;
    var m = 0;
    var x = [];

    x = s.split(");

    for (var u=1; u<=(s.length - 1); u++)
    {
        for (var i=0; i<=(s.length-u); i++)
        {m += f(x[i], x[u+i]);}

        t += (m / (s.length-u) * 100);
        m = 0;
    }

    return (100 - (t / (s.length - 1))).toFixed(1);
}

function f(x1, x2){
    if (x1 == x2) {return 1;} else {return 0;}
}

</script>


Output:

CG [x]=55,55,66,77,77,77,77,77,88,88,100,100,100,100,100,100,88,89,
IC [y]=85.3,80.8,72.8,71.8,65.9,64.5,74.3,79.4,72.6,69.6,56.4,55.8,
       55.8,56.4,54.9,57.9,71.0,58.5,
```

Up to this point, two signals have been produced based on the same z-sequence. The first signal shows the local IC along the z-sequence, while the second signal shows the frequency, namely the CG% content along the same z-sequence. The two signals contain the same number of components, each component representing a sliding window. To observe the relationship between the two signals, the values from the homologous components of the two vectors can be plotted. This can be done in two ways. The first possibility involves the graphical representation of signals in parallel on an x-axis that represents the z-sequence length and a y-axis that represents the signal level. In a second version, the two signals can be plotted against each other as points on a 2D surface, where the x-axis of a point represents the frequency and the y-axis of the same point represents the IC. The second approach is directly related to the method discussed in this chapter. It involves the use of an HTML object that allows the algorithmic drawing of a distribution. This object is called "canvas" and has been used in previous chapters for real-time drawing of heat maps.

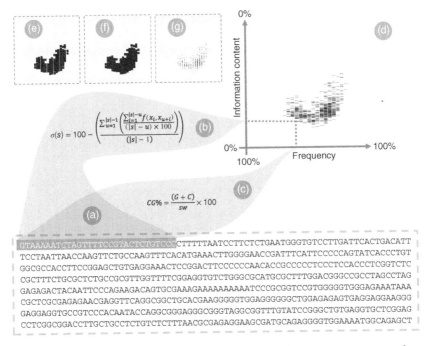

Figure 7.1 The objective digital stain method. (a) Shows the promoter sequence of a gene and the contents of the first sliding window. (b) Indicates the computation method for the information content on the current sliding window. (c) Indicates the CG% content on the current sliding window. (d) Shows an objective digital stain (ODS) – a chart on which dots/lines are plotted using the values from (b) and (c). For sufficiently long sequences, the set of dots/lines form a characteristic pattern/shape, which is particular to the sequence taken into consideration. (e–g) Show different embodiments of the ODS, resulting from various graphical representations of a "point" on the image. Note that there is no special position from which the first dot/line can start. It all depends on the content of the sliding windows.

7.3   The Objective Digital Stain

Two values are calculated from the contents of a sliding window (Figure 7.1a). A value that represents the IC and a value that represents the CG% content (Figure 7.1b,c). The two values represent the coordinates of a single point (pixel) on a graph (Figure 7.1d). Consequently, the set of sliding windows will be represented in a 2D space by a distribution of points. This distribution is called here an ODS or a "pattern." The set of points on this graph shows a series of well-defined structures that may reflect the function of the sequence. To make this graph more visible, a point is represented as a short line. Of course, the geometric shapes that represent this point lead to several visual possibilities (Figure 7.1e–g). Regardless of the graphic scheme, the content of a *z*-sequence produces a characteristic image (Figure 7.1d). The implementation below produces such an image and it includes all that has been described so far (Additional algorithm 7.2).

Additional algorithm 7.2 Note that the source code is in context and works with copy/paste.

```
<canvas id="bio" height="500" width="500"></canvas>

<script>

var z = 'GTAAAAATCTAGTTTTCCGTACTCTGTCCCCTTTTTAATCCTTCTCTGAATGGGTGTCC' +
        'TTGATTCACTGACATTTCCTAATTAACCAAGTTCTGCCAAGTTTCACATGAAACTTGGG' +
        'GAACCGATTTCATTCCCCCAGTATCACCCTGTGGCGCCACCTTCCGGAGCTGTGAGGAA' +
        'ACTCCGGACTTCCCCCCAACACCGCCCCCTCCCTCCACCCTCGGTCTCCGCTTTCTGCG' +
        'CTCTGCCGCGTTGGTTTTCGGAGGTGTCTGGGCGCATGCGCTTTGGACGGGCCGCCTAG' +
        'CCTAGGAGAGACTACAATTCCCAGAAGACAGTGCGAAAGAAAAAAAAAATCCCGCGGTC' +
        'CGTGGGGGTGGGAGAAATAAACGCTCGCGAGAGAACGAGGTTCAGGCGGCTGCACGAAG' +
        'GGGGTGGAGGGGGCTGGAGAGAGTGAGGAGGAAGGGGAGGAGGTGCCGTCCCACAATA' +
        'CCAGGCGGGAGGGCGGGTAGGCGGTTTGTATCCGGGCTGTGAGGTGCTCGGAGCCTCGG' +
        'CGGACCTTGCTGCCTCTGTCTCTTTAACGCGAGAGGAAGCGATGCAGAGGGGTGGAAAA' +
        'TGGCAGAGCT';

Pattern(z);

function Pattern(s) {

    var n = 30;
    var sp = 1;
    var sw;
```

```
    var x;
    var y;

    var canvas = document.getElementById('bio');

    var w = canvas.width;
    var h = canvas.height;

    if (canvas.getContext) {

        var ctx = canvas.getContext('2d');

        ctx.clearRect(0, 0, canvas.width, canvas.height);
        ctx.fillStyle = 'black';

        for (var u=0; u<=s.length - n; u += sp)
        {
            sw = s.substr(u,n);

            x = (w/100) * CG(sw);
            y = (h/100) * IC(sw);

            ctx.fillRect(x, y, (w/sw.length)-2, 1);
        }
    }
}

// Information content
function IC(s)
{
    var t = 0;
    var m = 0;
    var x = [];

    x = s.split(");

    for (var u=1; u<=(s.length - 1); u++)
    {
        for (var i=0; i<=(s.length-u); i++)
        {m += f(x[i], x[u+i]);}

        t += (m / (s.length-u) * 100);
        m = 0;
    }

    return (100 - (t / (s.length - 1))).toFixed(2);
}
```

(Continued)

Additional algorithm 7.2 (Continued)

```
function f(x1, x2){
    if (x1 == x2) {return 1;} else {return 0;}
}

// CG% content
function CG(s)
{
    s = s.toLowerCase();

    var a = 0;
    var t = 0;
    var c = 0;
    var g = 0;

    for (var u=0; u<=s.length; u ++)
    {
        var n = s.substr(u,1);

        if (n == "a") {a = a + 1;}
        if (n == "t") {t = t + 1;}
        if (n == "g") {g = g + 1;}
        if (n == "c") {c = c + 1;}
    }

    return ((100 / (c + g + t + a)) * (c + g)).toFixed(2);
}

</script>


Output:
```

The implementation above consists of four functions and uses a longer z-sequence from the promoter region of a gene (ABI1 – *Homo sapiens* abl interactor 1; found on chromosome 10). The *Pattern* function successively reads the z-sequence with the help of a sliding window and calls two other functions at each step, namely the *IC* and the *CG* function. Note that the *CG* function has the same role as the *GF* function used earlier. Both the *IC* and the *CG* function have been discussed in detail in the previous chapters. The *IC* function (known as the *Sigma* function earlier) returns the IC of a sliding window and the *CG* function returns the CG% value of a sliding window. Function f is also present in the main implementation as a component part of the *IC* method. Notice that the *CG* function used in Additional algorithm 7.2 is the raw version discussed in Additional algorithm 6.2. The values returned by the two functions are then used by the main function (i.e. the *Pattern* function) to represent the points on a canvas object (i.e. id="bio"), thus forming a characteristic image of the z-sequence (Additional algorithm 7.2).

The x-axis represents the frequency (CG%) and the y-axis represents the *IC*. Notice that in this case the two vectors from Additional algorithm 7.1 are removed and the computed values are directly used to plot the representation of the points. Notice that the number of points (represented using small lines), which constitutes an image, is equal to the length of the z-sequence (s.length) minus the length of the sliding window (n). Also, the total number of columns that make up an image is equal to the length of the sliding window (n). Thus, the space in between the points (the length of the small line) is computed by using the width of the canvas object divided by the length of the sliding window (w/sw.length). To plot the two values returned by the *IC* and the *CG* functions, the width and height of the canvas object are each divided by 100. The appropriate result is multiplied by the value returned by the *IC* function and the value returned by the *CG* function, respectively, to establish the x and y coordinates of the point representation. Of course, Additional algorithm 7.2 represents a point by using a small line to allow for a visual distinction above the main distribution. Thus, the result shows the shape and density of a 2D distribution of points, which is particular to each DNA/RNA sequence.

7.3.1 A 3D Representation Over a 2D Plane

However, many points on the image overlap as the content of a sliding window may repeat, especially for small-sized sliding windows. Such overlaps can be important when density is important inside the main distribution. To observe the overlaps, it is possible to use the representation of a third dimension in a 2D plane by gradually changing the color intensity with which the ODS is drawn. Moreover, a gradual transition from one color to another can be an alternative for detecting the overlapping counts. This idea can be implemented by reading the color of the current pixel and by imposing a condition according to which any color other than white (the background) is transitioned to another color by gradual additions. Thus, the overlapping counts will correlate with the intensity of the second color to which the transition is made. A new implementation that makes use of this idea is presented in Additional algorithm 7.3. In the implementation of Additional algorithm 7.3, the colors used are black (first color) and red (second color), and the intensities of red represent the frequency of superpositions at a certain point. To be able to visualize the superpositions, a longer sequence is used, namely the genome of the human immunodeficiency virus 1 (HIV-1; GenBank: AF033819.3).

Additional algorithm 7.3 Note that the source code is in context and works with copy/paste.

```
<canvas id="bio" height="500" width="500"></canvas>

<script>

// AF033819.3 HIV-1, complete genome
var z = 'GGTCTCTCTGGTTAGACCAGATCTGAGCCT ...............';

Pattern(z);

function Pattern(s) {

    var n = 30;
    var sp = 1;
    var e = 10;
    var sw;
    var r;

    var x;
    var y;

    var canvas = document.getElementById('bio');
```

```
    var w = canvas.width;
    var h = canvas.height;

    if (canvas.getContext) {

        var ctx = canvas.getContext('2d');

        ctx.clearRect(0, 0, canvas.width, canvas.height);

        for (var u=0; u<=s.length - n; u += sp)
        {
            sw = s.substr(u,n);
            r = Number(GF(sw,'C')) + Number(GF(sw,'G'));

            x = (w/100) * r.toFixed(2);
            y = (h/100) * IC(sw);

            var o = ctx.getImageData(x, y, 1, 1);
            var d = o.data;

            ctx.fillStyle = 'black';

            if(d[3] !== 0 && (d[0] + e) < 255) {
                ctx.fillStyle = color(d, e);
            }

            x = Math.floor(x);
            y = Math.floor(y);

            ctx.fillRect(x, y, Math.ceil(w/sw.length), 1);
        }
    }
}

function color(d, e){
    var rgba = 'rgba('+(d[0]+e)+', '+d[1]+', '+d[2]+', '+d[3]+')';
    return rgba;
}

// Information content
function IC(s)
{
    var t = 0;
    var m = 0;
    var x = [];

    x = s.split("");

    for (var u=1; u<=(s.length - 1); u++)
```

(Continued)

Additional algorithm 7.3 (Continued)

```
    {
        for (var i=0; i<=(s.length-u); i++)
        {m += f(x[i], x[u+i]);}

        t += (m / (s.length-u) * 100);
        m = 0;
    }

    return (100 - (t / (s.length - 1))).toFixed(2);
}

function f(x1, x2){
    if (x1 == x2) {return 1;} else {return 0;}
}

// X% content
function GF(s, m)
{
    var p = 0;
    s = s.toLowerCase();
    m = m.toLowerCase();

    p = s.split(m).join(").length;
    p = (s.length - p)/(s.length);

    return (p * 100).toFixed(2);
}

</script>

Output:
```

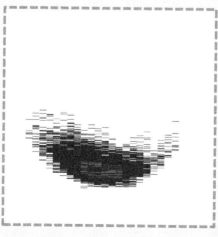

The ODS from the output of Additional algorithm 7.1 shows a nonuniform distribution of superpositions that exhibits interesting features, which can be further exploited for interpretation. Note that the HIV-1 genome (GenBank: AF033819.3) must be inserted in the above implementation before the algorithm is used. For reasons of space, only the first few nucleotides from AF033819.3 (out of a total of ~9.1 Kbp) are present in the z-sequence above. Nevertheless, some additions and changes can be seen in Additional algorithm 7.3. First, the previous frequency function (Additional algorithm 6.2) was replaced with a more professional function named *GF* (Additional algorithm 6.1). Notice that the new *GF* function can return the frequencies of different letter combinations (please see Chapter 6). However, for single-letter frequencies, the *CF* function is called separately for "C" and "G" and the results of the two calls are summed:

```
r = Number(GF(sw,'C')) + Number(GF(sw,'G'));
```

Secondly, a new function called *color* has been added. Before drawing the point representation (the small horizontal line which gradually builds the ODS), a condition in the *pattern* function calls the *color* function only if the pixel color at the x and y positions is anything but pure white (background). If the *rgb*(red, green, blue) value of the pixel is different from pure white, then the *color* function receives the *rgb* code. Once received, the *color* function increments the red value of the *rgb* format and then it returns the modification in the same format. The returned *rgb* format becomes the new color of the point representation. The color of the point representation is changed again each time the content of a sliding window generates the same x and y values for that point. In other words, the *rgb* value for the red color is increased at each overlap up to a maximum of 255 (*rgb*(255, 0, 0); pure red) by using a constant value (*e*). Moreover, the space between the columns of the ODS has been removed for a better visualization of the overlapping details. Also, the Javascript *Math.floor* function is used to round the x and y values (the point coordinates), leading to a much clearer image when compared to the previous implementation.

7.3.2 ODSs Relative to the Background

The ODS method described above (Additional algorithms 7.2 or 7.3) remains the standard approach from which various additions can start. Thus, another option for ODSs can account for the global CG% content of the z-sequence. In Additional algorithm 7.2, the CG% content of a sliding window was calculated according to the formula below, where 100 represented the maximum CG%. Please note that CG% and (C+G)% have the same meaning.

$$CG\% = \frac{100}{(A + T + C + G)} \times (C + G)$$

where A, T, C, or G represent the individual counts of each type of letter in the sliding window and their sum represents the sliding window length. Thus, the above formula is the same as:

$$CG\% = \frac{100}{sw} \times (C + G)$$

where *sw* is the length of the sliding window. However, the reference system can be set to another value, other than 100, namely to the global CG% content shown by the *z*-sequence. The global CG% content of the *z*-sequence can be computed as:

$$b\% = \frac{100}{L} \times (C + G)$$

where *L* represents the *z*-sequence length and *b%* represents the global CG% of sequence *z*. However, when calculating the CG% of the sliding window, the value of *b%* replaces 100. This would change the reference of the ODS as follows:

$$CG\% = \frac{b\%}{sw} \times (C + G)$$

The implementation from Additional algorithm 7.4 uses all of the above. In a first instance, the calculation of the CG% content is made for the whole sequence *z*. The calculated value is stored in variable *b*. Previously, the length of the sliding window divided the unit, namely the value 100. In this approach, the length of the sliding window divides the value in the *b* variable, which represents the overall CG% content of the sequence *z*. Next, the result of the division is multiplied by the counts for "C" and "G" found in the sliding window, as it was done in the previous implementations. In other words, the background CG% value of the *z*-sequence becomes the new reference for the ODS method.

Additional algorithm 7.4 Note that the source code is in context and works with copy/paste.

```
<canvas id="bio" height="500" width="500"></canvas>

<script>

// AF033819.3 HIV-1, complete genome
var z = 'GGTCTCTCTGGTTAGACCAGAT .......';

Pattern(z);

function Pattern(s) {

    var n = 30;
    var sp = 1;
    var e = 10;
    var sw;
    var r;

    var x;
```

```
    var y;

    var canvas = document.getElementById('bio');

    var w = canvas.width;
    var h = canvas.height;

    var b = (Number(GF(s,'C',100)) + Number(GF(s,'G',100))).toFixed(2);;

    if (canvas.getContext) {

        var ctx = canvas.getContext('2d');

        ctx.clearRect(0, 0, canvas.width, canvas.height);

        for (var u=0; u<=s.length - n; u += sp)
        {
            sw = s.substr(u,n);
            r = Number(GF(sw,'C',b)) + Number(GF(sw,'G',b));

            x = (w/100) * r.toFixed(2);
            y = (h/100) * IC(sw);

            var o = ctx.getImageData(x, y, 1, 1);
            var d = o.data;

            ctx.fillStyle = 'black';

            if(d[3] !== 0 && (d[0] + e) < 255) {
                ctx.fillStyle = color(d, e);
            }

            x = Math.floor(x);
            y = Math.floor(y);

            ctx.fillRect(x, y, Math.ceil(w/(sw.length/b*100)), 1);
        }
    }
}

function color(d, e){
    var rgba = 'rgba('+(d[0]+e)+', '+d[1]+', '+d[2]+', '+d[3]+')';
    return rgba;
}

// Information content
function IC(s)
{
    var t = 0;
    var m = 0;
    var x = [];

    x = s.split("");
```

(Continued)

Additional algorithm 7.4 (Continued)

```
    for (var u=1; u<=(s.length - 1); u++)
    {
        for (var i=0; i<=(s.length-u); i++)
        {m += f(x[i], x[u+i]);}

        t += (m / (s.length-u) * 100);
        m = 0;
    }

    return (100 - (t / (s.length - 1))).toFixed(2);
}

function f(x1, x2){
    if (x1 == x2) {return 1;} else {return 0;}
}

// CG% content
function GF(s, m, b)
{
    var p = 0;
    s = s.toLowerCase();
    m = m.toLowerCase();

    p = s.split(m).join(").length;
    p = (b/s.length)*(s.length - p);

    return p.toFixed(2);
}

</script>


Output:
```

The background CG% value will be less than or equal to 100 in some cases. Dividing a value less than 100 by the length of the sliding window will narrow the columns from which the ODS is built. Note that to obtain the x coordinate, a division of the width of the canvas object by 100 was first made, namely:

$$x = \frac{w}{100} \times r$$

where x represents the coordinate value for a point on the x-axis, and r represents the CG% content from the sliding window, which is based on the value of the b variable.

Thus, the proportionality of the lines is recalculated to match the new width of the columns. The length of the sliding window is divided by the global CG content (b) and the result is multiplied by 100 (`sw.length/b*100`). Next, this last result further divides the width of the canvas object to obtain the width of the lines which constitute an ODS (`w/(sw.length/b*100)`):

$$lw = \frac{w}{\left(\frac{sw}{b} \times 100 \right)}$$

where lw is the length of the line that represents a point on the ODS, w is the width of the canvas object, sw represents the length of the sliding window, and b represents the overall CG% content of sequence z. Here, the b value is dynamic and was calculated from the z-sequence. However, at the genome level, known global CG% values could be used for the b variable without a special requirement for expensive computations. When compared to the standard ODS, the optional ODS approach allows for a small advantage over the details. To observe the main differences between the standard version and the optional version presented here, some examples of ODSs are presented in Figure 7.2. More precisely, Figure 7.2 shows both versions of ODSs for a total of four randomly selected genes and the ODS of the mitochondrial genome from *H. sapiens*.

7.4 Interpretation of ODSs

The manner in which the image of a distribution/ODS can be interpreted is debatable and can be done in several ways. In the past, a number of parameters have been approached for the interpretation of these images. But what can be analyzed in such an ODS? An option that does not include sophisticated classification algorithms would be the average of the numbers for each axis, which reduces the entire ODS to the x and y coordinates of a single global point. For example, in Additional algorithm 7.1, two vectors are presented, one for the IC and one for the (C+G)%. The average of the values from the components of the vector that represents the IC would result in the y value for the global point.

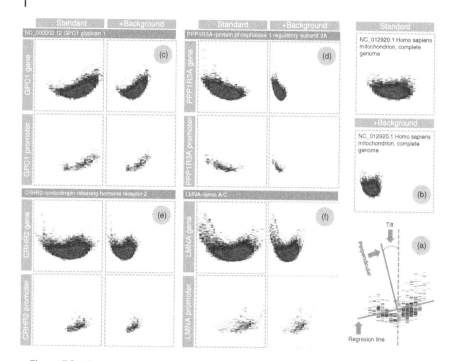

Figure 7.2 Examples of ODSs. (a) Shows an approach to the ODS analysis using a regression line and a line perpendicular to the regression line which indicates the angle to the vertical position. (b) Shows both the standard ODS (above) and the optional ODS (relative to the global CG% content) for the mitochondrial genome of *Homo sapiens*. (c–f) Each of the four panels shows the ODSs of four genes and the ODSs of their promoters (DNA sequences from position −499 to 100 b, relative to the transcription start site – TSS). In each panel, the left side shows the standard ODS of the gene (top) and the standard ODS of the promoter of the respective gene (bottom). Optional ODSs that are relative to the overall CG% content are shown on the right side of each panel for the same genes and promoters. Note that the information content is shown in descending orientation from top to bottom on the vertical axis of each ODS and the ascending frequency is shown from left to right on the horizontal axis.

The average of the values from the components of the vector that represents the relative frequency would represent the x value for the global point. Thus, such a global point could be seen as the center of weight of a distribution/ODS. In large-scale experiments, global points based on multiple ODSs/images can form interpretable distributions of their own. Of course, this approach loses important information, as different distributions of points in several ODSs can show, by chance, similar averages for the x-axis and the y-axis. Thus, multiple distributions that show different shapes and positions on the main image may weight in the same x–y global point. Another option would be the computation of a regression line and an accompanying line perpendicular to the regression line to indicate the inclination/tilt of the distribution (Figure 7.2a).

For such an approach, the coordinates of the ODS points must be reused. Thus, the use of vectors as described in Additional algorithm 7.1 is required both for tracing the ODS distribution and for calculating the regression line. Another option would be the use of image classifiers, such as neural networks for optical character recognition (OCR). Competent algorithms for image classification can differentiate between density, position, and the shape of the distribution. This approach has been used in the past to classify eukaryotic gene promoters [269].

7.5 The Significance of the Areas in the ODS

The distribution of points on the image can be interpreted in several ways. The denser parts of the distribution may be seen as the background of the z-sequence, while the parts that are outside the dense areas can be seen as regions above the background. On the vertical axis, the bottom points of a distribution come from sliding windows whose content shows a high entropy (100% – high IC – low repetitions) and the points that tend to 0% come from sliding windows whose content has a low entropy (low IC – high repetitions). The frequency is represented on the horizontal axis up to a maximum value of 100% (Figure 7.3a). For a change in the reference system, the y-axis of the ODS can be optionally reversed. The inversion can indicate the same maximum value of 100% on both axes but the ODS discussed up to this point would, of course, be mirrored vertically (Figure 7.3b). Points on the left indicate content based more on "A" and "T," culminating in points situated on the far left for single-letter repetitions.

For example, repetitions of 30 letters (the length of the sliding window in this case) consisting exclusively of the letter "A" will lead to a point in the upper left corner of the image. Moreover, a single repetition of the letter "T" is plotted in the same position (top left). In contrast, the right side of the image has the same properties related to the letters "G" and "C." The length of the sliding window impacts the details of the ODS. In the above cases, a sliding window of 30 letters/symbols in length has been chosen for an optimal representation of details. For observing the ODS details in different conditions, a series of experiments show the same ODS of the HIV-1 virus genome with the help of sliding windows of different sizes (Figure 7.3c,d). Figure 7.3c shows the standard ODSs with sliding windows with a length between 10 characters ($n = 10$) and 80 characters ($n = 80$), while Figure 7.3d shows the optional ODS with the same sliding windows parameters. Note that the IC of a one-dimensional representation (the DNA sequence) cannot reach 100% (bottom) in the case of long sliding windows. On the other hand, minimum sliding windows can produce IC values of 100%. Note that the

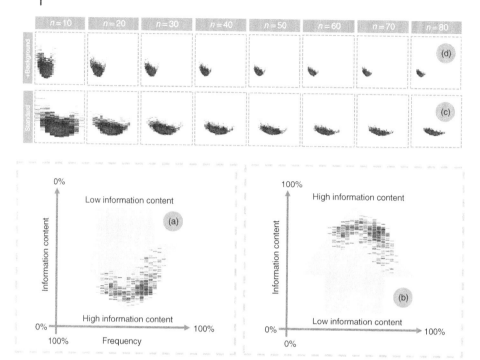

Figure 7.3   The ODS reference system. (a) Shows the standard reference system discussed so far, where the information content is shown in descending orientation from top to bottom on the vertical axis of each ODS, and the ascending frequency is shown from left to right on the horizontal axis. (b) Shows an alternative reference system in which the axis of the information content is inverted for an ascending synchronization with the frequency (both axes leading to 100%). (c) It shows the HIV-1 genome in eight versions by using standard ODSs. Each ODS is arranged in a line, where the first ODS originates from a small sliding window (n = 10 letters) and the last ODS that originates from a long sliding window (n = 80 letters). (d) It shows the optional ODSs for the HIV-1 genome arranged in the same order and the same sliding window parameters.

minimum sliding window is represented by the total letters in the alphabet of the sequence (four in the case of DNA/RNA sequences). To try this experiment, the length of the sliding window in Additional algorithm 7.3 can be set to four positions (n=4). However, only four positions in the sliding window lead to an ODS of four columns, and a gross loss of important details.

7.6   Discussions

Up to this point, two strings of numbers have been produced based on the same DNA sequence. The first string of numbers presented the values of the IC in each sliding window, while the second string of numbers presented the contents of two nucleotides, namely "C" and "G." Furthermore, the two strings were used to generate an ODS, which shows the characteristic details of the main sequence. Moreover, two versions of ODS were described: (i) a standard version in which the CG% content was represented up to a maximum value of 100%, and (ii) an optional version in which the CG% content of a sliding window was shown up to a maximum limit imposed by the CG% content over the entire sequence z. Nevertheless, the two versions represented one type of ODS, namely the IC on the *y*-axis and the CG% content on the *x*-axis. But how many types of distributions/ODSs can be obtained for DNA alone? The IC can be plotted against (C + G)% as discussed along this chapter, but it can also be plotted against dinucleotides (e.g. CpG%) or trinucleotides (not shown here) and so on. Take, for instance, the implementation from Additional algorithms 7.3 or 7.4. In these implementations, the *r* variable contains the sum of relative frequencies for "C" and "G," as follows:

```
r = Number(GF(sw,'C')) + Number(GF(sw,'G'));
```

To obtain a new type of ODS, the above line can be replaced with:

```
r = Number(GF(sw,'CG'));
```

The new type of ODS will show the IC values against the corresponding CpG% values. Thus, since the *GF* function allows the determination of relative frequencies for nucleotide combinations, these ODS representations can take a variety of forms. Moreover, the introduction of the *GF* frequency function allows the method to be used for both DNA/RNA or proteins. In the case of proteins, however, the approach would get quite complicated and the method may consider only a subset of the types of amino acids on the *x*-axis. Of the 20 amino acids, the frequency of a subset can be considered (e.g. "A" – alanine and "L" – leucine and "P" – proline and "V" – valine; (A + L + P + V)%), whereas the IC is only restricted by the types of letters found in the alphabet (usually 20 letters for proteins) of the main sequence (please see Chapter 6). For example, the content of individual letters: "A," "L," "P," and "V" can be determined as follows:

```
r += Number(GF(sw,'A'));
r += Number(GF(sw,'L'));
r += Number(GF(sw,'P'));
r += Number(GF(sw,'V'));
```

where variable *r* from above represents the (A + L + P + V)% content of a sliding window. On the other hand, the relative frequency of a combination of amino acids can be determined as discussed in the case of DNA or RNA sequences. For instance, the ALPV% content of a sliding window can be determined as follows:

```
r = Number(GF(sw,'ALPV'));
```

Of course, the minimum requirement for the above example is a sliding window length (n), which exceeds the length of the combination itself. The IC provided by the self-sequence alignment algorithm can be plotted against other types of signals, not only against the relative frequency. Nevertheless, whether it is the standard type or another type of ODS, the observations so far show a series of correlations between the ODS distribution and biological function [31, 269, 280].

7.6.1 A Similarity Between Dissimilar Sequences

As in the sequence alignment approach, ODSs describe the functional similarity of different DNA/RNA regions of an organism or between organisms. However, unlike sequence alignment that requires two sequences to verify similarity, ODS uses a single DNA/RNA sequence. Moreover, different sequences may lead to similar ODS distributions, which further emphasize the value of the method for finding functional relationships between different DNA/RNA regions. In the past, it has been shown that an ODS distribution can be radically changed if point mutations occur in critical areas of a sequence (replacing or deleting a letter). These critical areas were detected by manual experiments with repeated generation of the ODS distribution. These experiments can be found in the additional materials of the first main publication of the ODS method [269]. However, the rules behind these critical areas are still unknown and under research for the future. To view these ODS changes, please try these investigations using any of the algorithms presented in this chapter and various short sequences that allow ease of handling.

7.7 Conclusions

This chapter was focused on the technical aspects of a particular type of distribution called an ODS. Two separate methods have been merged to form an ODS. One method was the self-sequence alignment, which calculates the IC. The second approach was the relative frequency method. The IC values and the relative frequency values from a sliding window were plotted against each other as points on a 2D surface to obtain the ODS. In a first phase, an implementation was described for the standard ODSs. In the second phase, different options were added to the ODS model. For example, an option was described in which the points that overlap on the ODS distribution can be represented in 3D. Such an approach was based on the intensity of some colors that represented a z-axis (depth or elevation). Next, an implementation was presented that generates ODSs taking into account the global relative frequency of an analyzed sequence (ODSs in relation to the background frequency). Toward the end, the chapter approached several methods concerning the interpretation of ODSs and their significance was discussed in detail.

8

Detection of Motifs (I)

8.1 Introduction

Detection of evolutionary conserved DNA elements with functional relevance is a quest as old as the field of bioinformatics [281]. The location of such elements among the genomes of different species can help us in the future to understand the molecular feedback interactions that allow life to maintain itself above of the background. This chapter explains the role of DNA elements known in the literature as "DNA motifs" and points out the notion of degeneracy often mentioned in biology. Some objective examples of motifs are discussed. The implications of motif elements for DNA-binding proteins and functional RNA are also discussed. A concrete example follows the role that DNA motifs can have for RNA splicing. Moreover, one of the motifs involved in RNA splicing is chosen as a focus for exemplification across several chapters. Note: The description of motifs and their detection extends over five chapters (from I to V).

8.2 DNA Motifs

A motif sequence (or simply "a motif") is a nucleotide or amino acid sequence pattern with a biological function. Motifs are incredibly common inside a genome because they underlie the interface between DNA information and other biological processes in the cell or outside the cell. However, only a minority proportion of motifs are truly common inside a genome. Other, more rare motifs can be present across the genome of a species with a low frequency, but their importance can be vital for many functions. Such motifs are still in research today and to be discovered in the future.

Algorithms in Bioinformatics: Theory and Implementation, First Edition. Paul A. Gagniuc.
© 2021 John Wiley & Sons, Inc. Published 2021 by John Wiley & Sons, Inc.
Companion website: www.wiley.com/go/gagniuc/algorithmsinbioinformatics

8.2.1 DNA-binding Proteins vs. Motifs and Degeneracy

To orchestrate all biological processes of an organism, DNA sequences contain small areas of interaction with different protein molecules known as "DNA-binding proteins." Thus, DNA motifs usually represent conserved or partially conserved interaction areas along the DNA sequence. Nevertheless, all DNA sequences undergo mutations over time and DNA motifs make no exception. Mutations in DNA motifs may increase or decrease the affinity of DNA-binding proteins. Motif conservation was maintained over time by an interdependence (complementarity) between specific amino acid areas (DNA-binding domains [DBD] – a protein-binding motif) from proteins and the order of nucleotides on small areas of DNA (DNA motifs). An overview of the structures of protein–DNA complexes can be found here [282]. Without conservation, the protein–DNA interaction/affinity is diminished or disappears altogether. Furthermore, the lack of interaction between a DNA-binding protein and a DNA motif may interrupt one or more vital biological functions. For instance, the inactivation of such biological functions may have important repercussions over the development of an organism and therefore is usually incompatible with life. Nevertheless, mutations that change one or more noncritical bases in a DNA motif are accepted most of the time by DNA-binding proteins. Thus, such mutations are unlikely to drastically alter the protein–motif interaction. The acceptance of noncritical mutations by DNA-binding proteins allows for the emergence of degeneracy in DNA motifs (functional motifs that look slightly different from case to case). This property makes the genetic code more fault tolerant to point mutations. However, critical point mutations that do alter the protein-motif interaction are removed by selection.

8.2.2 Concrete Examples of DNA Motifs

But how do such motif sequences look like? Proteins involved in the regulation of gene expression are called transcription factors and contain DBD. The DBD of transcription factors exhibit a maximum affinity for specific DNA motifs. Below, six truncated examples of DNA motifs are presented, which objectively show the variability of degeneracy (Table 8.1) [283]:

Alignment of multiple DNA sequences (or protein sequences) from closely related species often leads to the discovery of conserved locations of various lengths. Such locations are candidates for detection of specific DNA-binding sites and consequently lead to the discovery of new functional DNA or protein motifs. The rationale behind this approach is that conserved locations in DNA (or in proteins) are critical for life and therefore must be biologically functional. This simple assumption has led to advances in bioinformatics over the years. Thus,

Table 8.1 Examples of DNA motifs.

SKN7	PHD1	RPN4	FKH1	FKH2	CBF1
GCCCGCCC	AGGCGC	GGGTGGCAAAC	CCTAGAAAACAA	GGAAAGTGTAAAACAT	GTCACGTG
CCTCGGCC	AGGAAC	CGGTGGCAAAG	AAGTGTAAACAA	AAAAAACGTAAACAA	ATCACGTG
TCCGGACC	AGGAAC	GAGTGGCAAAA	ATTCGCAAACAA	GCATGAGGAAAACAA	ATCACGTG
TCTGTGCC	AGGCAC	CAGTGGCAAAT	TGAAGTAAACAT	GGTATATATAAACAA	GGCACGTG
GCTGAGCC	AGGGAC	GAGTAGCAAAC	GAGGGTAAATAA	AAATGAGGTAAACAA	TTCACGTG
GCCAGACC	AGGAAC	CGGCGGCAAAA	AAACGTAAACAA	TGCAAATGTAAATAC	ATCACGTG
GCCGTGCG	AGGAAC	AGGTGGCAAAA	TGAGGAAAACAA	AAAAAAGTAAACAA	GTCACGTG
TCGGGGCC	AGGCAA	AGGTGGCGAAA	AAAAATAAACAA	AGTAAATGTAAACAA	GTCACGTG
GCCGTGCC	AGGCAA	TGGTGGCAAAA	TGAGGTAAACAA	AAAAATGTGAAACAA	TTCACGTG
CCCCGGCC	AGGAAC	CGGTGGCAAAA	GTATGTAAATAT	GAATTAGGAAAACAA	GTCACATG
TCTCGGCC	AGGCAC	TGGTGGCAAAT	AAATGTAAACAA	TGAAGAAGAAAACAA	ATCACGTG
TCTATACC	AGGAAA	AGGTGGCAAAA	AATGGTAAACAA	TGCCACAGTAAACAA	GTCACGAG
TCCGAGCG	AGGCAC	GAGTGGCAATA	ACATGTACACAA	GAACGGAGTAAAAAA	AGCACGTG
TCTGGGCG	AGGCAA	CGGTGGCAAAA	CACAGTAAACAA	GAAAAACATAAACAA	ATCACGTG
GCCGGCCG	AGGAAC	AGGTGGCTAGG	AACGATAAATAT	GGATGAGGTAAACAA	ATCACGTG
TCGCGGCC	GGGCAC	GGGTGGCAAAA	AAAGGTAAACAA	GAAGGAAAAAAACAA	AGCACGTG
CCGGGCCC	AGACAC	CGGTGGCGAAA	AATGGGCAACAG	TTAAAAAATAAACAA	GTCACGTG

(continued)

Table 8.1 (Continued)

SKN7	PHD1	RPN4	FKH1	FKH2	CBF1
ACCGGGCC	AGGAAA	CGGTGGCAAAA	ACAAATAAACAA	GAAGGATGAAAACAA	AGCACGTG
TCGGGACG	AGGCAG	GGATGGCGACA	GATTTTAAACAA	AGAATATGTAAATAA	ATAACGTG
CCCCTGCC	AGGCAC	TGATGGCAAAA	AGAGGTAAACAA	CAAAATCAAAAACAA	AGCACGTG
GCGGGACC	AGGCAC	AAGTGGCCTAC	GAAAGTAAAAAA	AGAGAAGGAAAACAA	ATCACGTG
GCGAGGCC	AGGAAA	TAGTGGAAAAT	CAGTATAAACAA	GGAAAAAAAAACAA	ATCACGTG
GCTGTGCC	GGGCAC	TGGTGGCGAAA	ATTAATAAACAA	GATATCAGTAAACCA	GTCACGTG
GCCGCGCC	AGGCAA	AAGTGGCGTCT	ATTAGTAAACAT	ATAATTAGTAAACAT	ATCACGTG
TCTGCGCC	AGGCAA	GGGTGGCAAAC	AAAAGAAAACAA	TTAAAAGGTAAACAA	ATCACGTG
CCCCGCCC	AGGCAC	GGGTGGCAAAA	GGGCGTAAATAA	CTCAAATGTAAACAA	GTCACGTG

The table shows six types of DNA motifs for DNA-binding domains of different transcription factors (proteins): SKN7, PHD1, RPN4, FKH1, FKH2, and CBF1. Each type of motif is represented by 26 variations. A visual inspection on the columns of the table shows a similarity between different variations of motifs.

motifs that are found among multiple DNA or protein sequences often correlate with important biological functions. The above motif variations (of SKN7, PHD1, RPN4, FKH1, FKH2, and CBF1) represent sections taken from multiple sequence alignments of similar genomic areas from different organisms of the same species [283]. Other motif sequences for possible experiments can be found in the additional materials from here [284, 285]. However, multiple sequence alignments are not always optimal. Some variations of the same DNA motif may be false positives (a result which wrongly indicates that a piece of DNA sequence is a DNA motif), but their overall analysis eliminates these false-positive variations if they are present in a decent proportion.

8.3   Major Functions of DNA Motifs

What is the function of DNA motifs? The general momentum regarding these structures is the association between DNA motifs and DNA-binding proteins. However, DNA motifs can show two major functions. First, a motif can be a binding site for different proteins. Secondly, motifs may be responsible for functional RNA molecules (RNA – ribonucleic acid). Much like proteins, RNA molecules can have different functions too. Their three-dimensional conformation can be dictated by such DNA motifs [286]. The information from various DNA motifs (the order of the nucleotides in the motif) leads to spatial contortions into the RNA transcript that provide specific structures and functions [286, 287]. Another property of a DNA motif is to provide catalytic functions to RNA molecules. For instance, enzyme activity is generally associated with protein structures. Nevertheless, RNA molecules can also exhibit enzyme activity [287, 288]. Catalytic RNAs are involved in RNA processing and protein synthesis. A classic example of a catalytic RNA molecule is the hammerhead ribozymes (self-splicing intron) that catalyze their own excision from mRNA, tRNA, and rRNA precursors.

8.3.1   RNA Splicing and DNA Motifs

In higher eukaryotes, genes are organized into or accompanied by regulatory, coding, and noncoding regions. A regulatory sequence is a DNA area, which can increase or decrease the expression of a gene (low or high synthesis of RNA). An ordinary gene contains two types of interspersed areas. The DNA areas that code for proteins are called exons and the remaining noncoding DNA areas are called introns. To express it, a cell first uses the *Polymerase II* enzyme to copy the gene into a precursor messenger RNA (pre-mRNA). Thus, initially the pre-mRNA contains all the information and has about the same length as the gene (Figure 8.1). Subsequent processing of pre-mRNA involves intron cleavage and the process is called RNA splicing. The pre-mRNA splicing ensures the removal of introns from the nascent premature RNA transcript, which produces the mature form of the protein-coding messenger RNA (mRNA template). The splicing process is made by a spliceosome (a ribonucleoprotein complex – many different proteins and RNAs connected to each other to form a large biochemical machine) that positions the distant ends of the introns into its catalytic center. The large size of the spliceosomes dictates a minimum intron length of 80 nucleotides [289]. As a guide, in eukaryotes, the average exon length is ~200 bp (GC%=50±3), whereas the average intron length is ~2900 bp (GC% = 41 ± 6) [290]. Spliceosomes can catalyze and remove the introns by recognizing RNA motifs on the pre-mRNA transcript. The spliceosome recognition of intron areas is dependent on evolutionary conserved RNA motifs for exon–intron and intron–exon borders (Figure 8.1). Nevertheless, since RNA motifs (5′–3′) are complementary to the template strand (3′–5′) of the gene, they can be studied in their DNA format (5′–3′) for ease. In the DNA version, a few motif variations surrounding these splice sites (borders) are shown in Table 8.2.

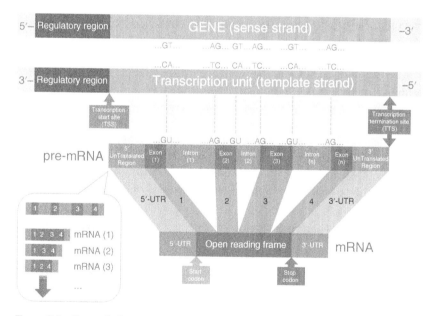

Figure 8.1 Transcription, alternative splicing, and translation. It shows the typical structure of a gene in eukaryotes and reveals the relationship of the two DNA strands with the pre-mRNA sequence. It further shows the relative locations of the splicing sites (i.e. donor "GU" - exon–intron sites and the acceptor "AG" – intron–exon sites) and points out the ability of pre-mRNA to combine different coding regions (exons) into various mRNA isoforms.

Table 8.2 Splicing motifs from human genes.

On DNA – sense strand		On the pre-mRNA	
Exon–intron (donor) sites	Intron–exon (acceptor) sites	Exon–intron (donor) sites	Intron–exon (acceptor) sites
5′-GAA**GT**GAGT-3′	5′-TCTT**AG**GAT-3′	5′-GAA**GU**GAGU-3′	5′-UCUU**AG**GAU-3′
5′-TTC**GT**AAGT-3′	5′-TTTA**AG**CCA-3′	5′-UUC**GU**AAGU-3′	5′-UUUA**AG**CCA-3′
5′-AA**GGT**ACTT-3′	5′-CTGC**AG**CAT-3′	5′-AA**GGU**ACUU-3′	5′-CUGC**AG**CAU-3′
5′-CT**GGT**GAGC-3′	5′-GCAC**AG**GCC-3′	5′-CU**GGU**GAGC-3′	5′-GCAC**AG**GCC-3′
5′-AGA**GT**GAGT-3′	5′-TCTT**AG**GAT-3′	5′-AGA**GU**GAGU-3′	5′-UCUU**AG**GAU-3′
5′-CA**GGT**AGAG-3′	5′-GTTT**AG**CTC-3′	5′-CA**GGU**AGAG-3′	5′-GUUU**AG**CUC-3′
5′-ACT**GT**ACGT-3′	5′-TTCT**AG**GAA-3′	5′-ACU**GU**ACGU-3′	5′-UUCU**AG**GAA-3′
5′-CT**GGT**GAGT-3′	5′-TTGC**AG**TTG-3′	5′-CU**GGU**GAGU-3′	5′-UUGC**AG**UUG-3′
		5′ Splice site	3′ Splice site

The table shows different motif variations known to indicate the exon-intron boundary and the intron–exon boundary in pre-mRNA transcripts of human genes. The motifs are presented in both the DNA format and the RNA format for reference. Note that these motifs are biologically active in the pre-mRNA format.

In the table above, a few motif variations are presented (Table 8.2). These motifs represent real splice site locations determined from different *Homo sapiens* genes (human genes). A simple visual inspection over the list of motifs indicates that GT dinucleotides are dominant (or GU in RNA form) for exon–intron (donor) sites and AG dinucleotides are dominant for intron–exon (acceptor) sites (Table 8.2). Note that each motif variation in the table contains nine positions.

However, such motif variations can be extracted from multiple sequence alignments at arbitrary lengths. Here, nine positions are used for ease. Other types of splice sites in eukaryotes may use different motifs and splice site combinations [291, 292]. However, motifs that use the GT–AG splice site combination are the most frequent (~99%) and well known [284, 291, 292]. The DNA motifs for RNA splice sites are highly conserved across species [292]. This conclusion was based on alignments between sequenced RNAs and the genomes from which they originated [293]. **Intron structure.** On the pre-mRNA molecule, the structure of introns shows three main elements: a GU splice site at one end (5′ splice site), followed by a *branch point,* and a polypyrimidine tract that precedes the AG splice site at the other end (3′ splice site). Branch points are point locations (i.e. an "A" – adenine nucleotide) that actively participate in the intron removal and contain their own set of DNA motifs. The polypyrimidine tracts consist of long areas of U and C nucleotides [285, 294–296]. Genome-wide detection of alternative splicing (AS) in expressed sequences of human genes shows which positions from the DNA motifs are a part of the intron and which are a part of the exon [293, 297–300]. A single pre-mRNA can produce multiple mRNA isoforms by using alternative splice sites (Figure 8.1). AS is a fundamental mechanism of genetic regulation, which modifies the sequence of pre-mRNA transcripts in higher eukaryotes [301–304]. The regulation of AS is made through coupling with transcription [305–307]. This process results in exon shuffling and consequently leads to many types of mRNA molecules, which originate from a single gene. Thus, a gene may encode for several types of proteins. This is the reason why many eukaryotic organisms contain more types of proteins than the observed number of genes. Moreover, many mRNA types originating from noncoding genes have their own functions. Thus, RNA splicing is an essential part of eukaryotic gene expression. The DNA motifs of splice sites are of high importance for understanding the pre-mRNA to mRNA conversion. Consequently, DNA motifs associated with such boundaries will be used as an example in the implementations from the following chapters.

8.4 Conclusions

This chapter described the interplay between DNA-binding proteins and DNA motifs over the course of evolution. Furthermore, the notion of degeneracy was explored in connection with critical and noncritical mutations. To better understand the notion of degeneracy, concrete examples of motifs have been provided from the scientific literature. These examples included known DNA-binding proteins and the motif variations to which they bind (Table 8.1). Once these explanations were given, the major functions of DNA motifs were briefly described. Toward the end, the chapter focused on DNA motifs involved in RNA splicing. A short set of motif variations was given to be further used for illustration and experimentation in the following chapters (Table 8.2). Therefore, this chapter made a short review of these functional DNA units and brought an intuitive view of the biological processes that take place in the cell nucleus.

9

Representation of Motifs (II)

9.1 Introduction

The position-specific scoring matrices (PSSMs) are the most common method for representation of motifs (patterns) in biological sequences [281, 308]. A brief history of the development and application of computer algorithms for the analysis and prediction of DNA-binding sites, can be found here [281]. The approach described in the following implementations revolves around two matrices, namely: matrices s and p (the PSSM). Matrix s will encapsulate the motif variations from the set. Matrix p will contain the analysis made on the data stored in s. Successive steps will transform a matrix p from a *position frequency matrix* (PFM) to a *position probability matrix* (PPM) and then to a *position weight matrix* (PWM). Toward the end of the chapter, a series of additional experiments open new ways to understand the PWM data. The PWM will be later used to scan a long sequence for the presence of elements that resemble those in the initial set. Note that PSSM is a term that includes the PFM, PPM, and the PWM.

9.2 The Training Data

A set of known variations for a specific motif can help with the discovery of other motif variations in new DNA locations. To start this process, a set of known motif sequences (motif variations) is mandatory. Alternatively, any set from Table 8.1 could be used. However, a shorter set is chosen to demonstrate the method (Table 9.1). This set is composed of 10 motif variations and represents the DNA version of the exon–intron boundary observed on pre-mRNA transcripts from different *Homo sapiens* genes.

The motif variations from Table 9.1 are only a small fraction of a larger set not shown here. Two reorganizations are needed to use the above set for an implementation. The motifs from the set are separated by a delimiter and placed

Algorithms in Bioinformatics: Theory and Implementation, First Edition. Paul A. Gagniuc.
© 2021 John Wiley & Sons, Inc. Published 2021 by John Wiley & Sons, Inc.
Companion website: www.wiley.com/go/gagniuc/algorithmsinbioinformatics

Table 9.1 A set of motifs.

Motif set for exon–intron (donor) sites
GAAGTGAGT
TTCGTAAGT
AAGGTACTT
CTGGTGAGC
AGAGTGAGT
CAGGTAGAG
ACTGTACGT
CTGGTGAGT
TATGTAAGT
CGGGTGAGC

The table shows 10 motif variations used for algorithm development and verification. The set shown below is representative for the exon–intron (Donor) splice sites found on the pre-mRNA transcripts in humans. These motifs are generally studied in DNA format for annotation and genome mapping.

in a linear format. For convenience, this delimiter is a comma character. The new string of characters is stored in a variable called c:

c = "GAAGTGAGT,TTCGTAAGT,AAGGTACTT,CTGGTGAGC,AGAGTGAGT, CAGGTAGAG,ACTGTACGT,CTGGTGAGT,TATGTAAGT,CGGGTGAGC"

A PSSM will be obtained based on variable c. After completion, the PSSM will later be used to detect motifs like those from the set. Detection will be performed on a test DNA sequence called z:

z = "AAAAAACAGGTGAGTAAAAAAAA"

The short DNA sequence from z is specially built to confirm the correct operation of the final implementation. Later, a series of experiments will be performed on larger z sequences. It is worth mentioning at this stage that z may contain large pieces from an entire genome. Also, at the end of the chapter, motif sequences shown in Table 8.1 will be processed for experimentation purposes in the same manner as the test set from above (Table 9.1).

9.3 A Visualization Function

First, a special function is needed to observe the results in HTML format. Thus, a general-purpose function is defined for visualization of the contents of any matrix or vector. The function below is called *show matrix content* (SMC) and will be used throughout this chapter to view changes made to these mathematical structures:

Additional algorithm 9.1 Note that the source code is out of context and is intended for explanation of the method.

```
// SHOW MATRIX CONTENT

function SMC(m) {
    var r = "<table border=1>";
    for(var i=0; i<m.length; i++) {
        r += "<tr>";
        for(var j=0; j<m[i].length; j++){
            r += "<td>"+m[i][j]+"</td>";
        }
        r += "</tr>";
    }
    r += "</table>";

    return r;
}
```

The above function builds a table in HTML format in which it associates the cells of a table with the elements of an array. The *SMC* function receives an array *m*, which it traverses element by element. The variable *i* represents each row and the variable *j* represents each column in the matrix *m*. **Note**: by default, matrices and vectors start from index 0. Next, *m.length* represents the total number of rows in the matrix *m* and *m[i].length* represents the total number of columns in the matrix *m* in row *i*. Variable *r* is a string variable. The table is built in variable *r* by using the HTML format for a table structure. In HTML format, a table starts with a "<table>" tag and ends with the "</table>" tag. The "border" property indicates the thickness of the line that separates the cells in the table. The "<tr>" tag means a new row inside the table, while the "<td>" tag represents a new horizontal cell. The *SMC* function returns the content of variable *r*. The content of variable *r* is handed over directly to the browser for interpretation by declaring "*document.write(SMC(m));*". The browser parses the tags built by the *SMC* function and shows them to the user as a simple table. **Note**: all the names of the variables inside the function are independent of the names of the variables outside the function. Thus, variables like *m*, *i*, *j*, *r*, or other variables inside the *SMC* func-

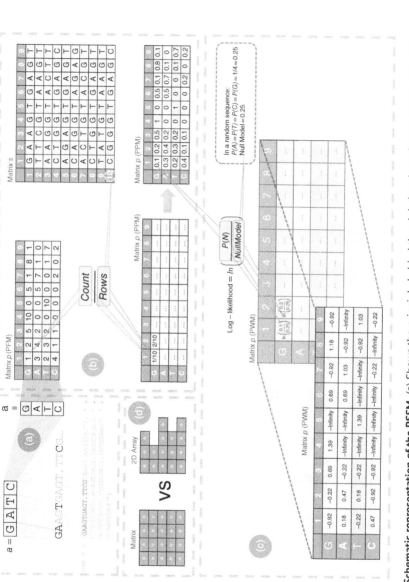

Figure 9.1 **Schematic representation of the PSSM.** (a) Shows the principle behind the alphabet detection module and (b, c) the transformations that matrix p (PSSM) goes through. (d) Shows the difference in terminology between a matrix and a 2D array. (a, b) Shows how the automatic detection sets the order of each symbol from the alphabet on the first column of matrix p.

tion may have different meanings outside the function. The functions described (such as *SMC*) should be seen as separate "boxes" from the main algorithm.

9.4 The Alignment Matrix

At this stage, the input motifs arranged linearly in variable *c* are placed one on top of the other in a matrix *s*. Three main variables are used, namely: *c*, *s*, *m,* and *n*. Variable *c* contains the DNA sequences to be considered. These sequences are placed linearly and are separated by a delimiter (i.e. a comma symbol:','). Variables *s* and *m* are declared initially as one-dimensional arrays ("var s=[]; var m=[];"). Variable *m* is a temporary variable that helps on the construction of matrix *s*. The content of the *c* variable is parsed using the split function (c.split(',');) and the individual sequences are stored in the array variable *m*. The variable *m* holds each individual sequence from *c*. The *n* variable holds an integer signifying the total number of sequences present in the one-dimensional variable *m*. At this stage, the sequences in variable *c* are parsed and placed in a matrix *s* as follows: A traversal is made from the first sequence to the last sequence *n* (i.e. *s*[*first dimension*][*second dimension*][*third dimension*] …). At each step in this traversal, a second dimension is added to the *s* variable. In the second dimension of the *s* variable, the individual characters from sequence *i* are inserted in separate elements (i.e. *s*[0][0] = "G", *s*[0][1] = "A", *s*[0][2] = "A", *s*[0][3] = "G", *s*[0][4] = "T" …). This is done using the split function that helps to associate each element with a letter from sequence *i*. The *s* matrix contains a number of rows equal to the number of sequences in variable *c*, and a number of columns equal to the number of characters in these sequences. The number of characters in any of these sequences is in this case equal. The result of the algorithm can be verified by sending the *s* matrix to the *SMC* function, which in turn will show the structure of the *s* matrix on the main browser. The algorithm below summarizes what is described above:

Additional algorithm 9.2 Note that the source code is out of context and is intended for explanation of the method.

```
//THE ALIGNMENT MATRIX

var c = "GAAGTGAGT,TTCGTAAGT,AAGGTACTT,CTGGTGAGC,AGAGTGAGT," +
        "CAGGTAGAG,ACTGTACGT,CTGGTGAGT,TATGTAAGT,CGGGTGAGC";

var s = [];
var m = [];
```

(Continued)

Additional algorithm 9.2 (Continued)

```
m = c.split(',');
var n = m.length;

for(var i=0; i<n; i++){
    s[i] = [];
    s[i] = m[i].split("");
}

document.write(SMC(s));

Output:
```

G	A	A	G	T	G	A	G	T
T	T	C	G	T	A	A	G	T
A	A	G	G	T	A	C	T	T
C	T	G	G	T	G	A	G	C
A	G	A	G	T	G	A	G	T
C	A	G	G	T	A	G	A	G
A	C	T	G	T	A	C	G	T
C	T	G	G	T	G	A	G	T
T	A	T	G	T	A	A	G	T
C	G	G	G	T	G	A	G	C

Note: any array can be called a matrix, only if each row in the array contains an equal number of columns. A one-dimensional array with a single row and multiple columns can be called a vector (Figure 9.1). Note that the use of variable *m* and *n* can be avoided entirely since the following lines: "`m=c.split(',');` `s[i]=[]; s[i]=m[i].split("");`" can be written in a single line of code as "`s[i] = [] = c.split(',')[i].split("");`". Also, the use of variable *n* can be avoided by replacing the following lines "`var m=[]; m=c.split(',');` `var n=m.length;`" with "`c.split(',').length`". Thus, with a reduction in elegance, the above algorithm can be rewritten as:

Additional algorithm 9.3 Note that the source code is out of context and is intended for explanation of the method.

```
//THE ALIGNMENT MATRIX - Short version

var c = "GAAGTGAGT,TTCGTAAGT,AAGGTACTT,CTGGTGAGC,AGAGTGAGT," +
        "CAGGTAGAG,ACTGTACGT,CTGGTGAGT,TATGTAAGT,CGGGTGAGC";
```

```
var s = [];
for(var i=0; i<c.split(',').length; i++){
    s[i] = [] = c.split(',')[i].split('');
}

document.write(SMC(s));
```

9.5 Alphabet Detection

Detection of the letters that make up an input sequence can be crucial for the adaptability of an algorithm. This allows the possibility to apply the algorithm to any type of sequence, from DNA, RNA to protein sequences or even to normal text for linguistics. For example, in linguistics, this approach can show what is the probability that a misspelled word has a certain meaning. However, we will return to this later in the following chapters. Some examples related to the meaning of the alphabet for different sequences are presented below (Table 9.2).

DNA polymers consist of a total of four types of molecules, ideally represented by four different letters. The alphabet is built of four types of characters, namely: {A, T, C, G}, while in the case of RNA the alphabet contains the {A, U, C, G} characters.

However, uncharacterized nonbiological sequences may have a varying alphabet. In order to build an adaptable algorithm, the alphabet must be automatically detected from the very beginning. To start the detection process, the character sequence stored in variable c is filtered of delimiters ($'$, $'$) and the result is stored in variable t. The filtering process is performed by using the replace function (c.replace(/,/g,"");). The elimination of delimiters is important because the delimiter itself can be detected as a new character type. The total length of the sequence from variable t is represented by the value stored in variable k. Thus, the detection of all unique characters requires a loop over the characters inside the sequence from variable t. This loop is made from the first character in t (index $i = 0$), up to the last character in t, which is represented by the value stored in variable k. (index $i = k$). At each step, a second loop is made over the characters stored in variable a. Initially, variable a is empty; however, the first character in t will be the first character stored in variable a. Before this second loop, a flag variable q is initialized, where its value is set to 1 (Figure 9.1a). During the loop made over variable a, the variable q becomes zero only if the character i in the sequence t is the same as the character j in variable a. A condition is imposed at the end of the loop over variable a. If the flag variable q is zero, then the new unique character is added to variable a. The loop over t moves to the second step $(i+1)$ and the cycle is repeated. At $i = k$, the algorithm ensures the detection of

Table 9.2 Examples of alphabets for different sequences.

Sequence	Sequence length	Sequence alphabet	Characters in alphabet
"AAAAAAAAAAAA"	12	{A}	1
"AABBBABABBAA"	12	{A, B}	2
"GCGGATTCGAAC"	12	{G, C, A, T}	4
"TTAGAAATTTATA"	13	{T, A, G}	3
"ATXXGAAXGTC"	11	{A, T, X, G, C}	5
"GGGGGGGGGTGGGGGGGGG"	18	{G, T}	2
"HSCYUFTAFUGHDHFFG"	17	{H, S, C, Y, U, F, T, A, G, D}	10

The table provides an intuitive view on the notion of alphabet, which will continue to be pivotal in the following steps of the implementation. The detection of the alphabet from the beginning of an analysis allows an algorithm to adapt itself to the input sequence. Thus, any character type can be encountered in the input sequence without additional challenges for the implementation.

all the characters that form the alphabet for the input sequence. Subsequently, variable *a* stores the alphabet of the input sequence and will play a pivotal role in the next stages. The implementation of those discussed above can be made either by using string variables and the substring function or by using array variables and the split function. The first alternative includes string variables and the substring function, as shown below:

Additional algorithm 9.4 Note that the source code is out of context and is intended for explanation of the method.

```
//DETECT ALL SYMBOLS USING THE STRING APPROACH

var c = "GAAGTGAGT,TTCGTAAGT,AAGGTACTT,CTGGTGAGC,AGAGTGAGT," +
        "CAGGTAGAG,ACTGTACGT,CTGGTGAGT,TATGTAAGT,CGGGTGAGC";

var x, y;
var a='';

var t = c.replace(/,/g, '');
var k = t.length;

for(var i=0; i<=k; i++){
    var q = 1;
    for(var j=0; j<=a.length; j++){
        x = t.substr(i, 1);
        y = a.substr(j, 1);
        if (x === y) {q = 0;}
    }
}
```

```
        if (q === 1) {a += x;}
}

document.write(a);

Output:
GATC
```

The result of the above implementation will be stored in variable *a*. Since *a* is a string variable, it can be shown directly in the browser window by using "document.write(a);", without using the *SMC* function. The second approach includes a vector whose components contain unique characters found in the sequence. The second approach based on array variables is presented below:

Additional algorithm 9.5 Note that the source code is out of context and is intended for explanation of the method.

```
//DETECT ALL SYMBOLS USING THE ARRAY APPROACH

var c = "GAAGTGAGT,TTCGTAAGT,AAGGTACTT,CTGGTGAGC,AGAGTGAGT," +
        "CAGGTAGAG,ACTGTACGT,CTGGTGAGT,TATGTAAGT,CGGGTGAGC";

var a = [];

var t = c.replace(/,/g, '').split('');
var k = t.length;

for(var i=0; i<=k; i++){
    var q = 1;
    for(var j=0; j<=a.length; j++){
        if (t[i] === a[j]) {q = 0;}
    }
    if (q === 1) {a.push(t[i]);}
}

document.write(SMC(a));

Output:

G
A
T
C
```

In a complementary way to the first version, the result of the above implementation will be stored in the array variable *a*. Because *a* is an array variable, it cannot be written in a simple way as in the previous implementation. Thus, variable *a* is passed to the *SMC* function, which returns the HTML format of the variable *a* content. The two equivalent approaches are related to the preferred style of implementation rather than a real optimization to increase the execution speed. Nevertheless, the approach that uses arrays will be considered for ease. Thus, up to this point, vector *a* contains the alphabet of the sequence stored in variable *c*. The first letter detected by the above implementations is "G." Thus, the order of the letters in variable *a* is input specific and it is imposed by the order in which the algorithm finds unique characters in the analyzed text. Nevertheless, the order of the characters will not be important in the scheme developed further.

9.6 The Position-Specific Scoring Matrix (PSSM) Initialization

In the next stage, a profile matrix (character counting matrix) can be initialized by using vector *a* (Figure 9.1a,b). A profile matrix contains the number of occurrences for each character from the columns of matrix *s* (Figure 9.1b). However, the profile matrix must be initially constructed. The aim here is the incorporation of letters and their frequencies in the same matrix. Thus, the structure of the matrix should be mixed, namely the matrix will contain elements that store strings or numbers. In the following implementation, this matrix is denoted by p (Additional algorithm 9.6). The below algorithm shows an initialization of matrix *p*, in which the first column contains the unique characters from variable *a* (p [h] [0] =a [h] ;) and the other columns contain zero values (p [h] [i] =0 ;):

Additional algorithm 9.6 Note that the source code is out of context and is intended for explanation of the method.

```
// PROFILE MATRIX INITIALIZATION

var p = [];

for(var h=0; h<a.length; h++){

    p[h] = [];

    for(var i=0; i<=s[0].length; i++) {
        p[h][i] =0;
```

```
        p[h][0]=a[h];
    }
}

document.write(SMC(p));

Output:

G    0    0    0    0    0    0    0    0    0
A    0    0    0    0    0    0    0    0    0
T    0    0    0    0    0    0    0    0    0
C    0    0    0    0    0    0    0    0    0
```

The initialization process starts with the declaration of the first dimension for matrix p (var p=[];). A loop above variable a generates the second dimension and frames the number of rows in the p matrix (p[h]=[];). At the same time, a second loop is added to expand the number of elements on each row. The total number of columns in matrix p is equal to the number of columns found on matrix s plus one column reserved for the unique characters from variable a. Characters from variable a are duplicated on the first column of matrix p, both for optimal visualization and later use. Note that s[0].length assumes that all sequences in the motif set will be the same length as the first.

9.7 The Position Frequency Matrix (PFM)

The PFM shows total counts for each position in the motif set (Figure 9.1b). Counting of individual characters from matrix s is performed post initialization. The implementation below shows how many occurrences of each character from the alphabet are on individual columns of matrix s:

Additional algorithm 9.7 Note that the source code is out of context and is intended for explanation of the method.

```
// THE POSITION FREQUENCY MATRIX

for(var i=0; i<s.length; i++) {
```

(Continued)

Additional algorithm 9.7 (Continued)

```
    for(var j=0; j<s[i].length; j++){

        for(var h=0; h<a.length; h++){

            if (s[i][j] === a[h]) {p[h][j+1]++;}

        }
    }
}

document.write(SMC(p));

Output:

G    1    2    5    10    0    5    1    8    1
A    3    4    2    0     0    5    7    1    0
T    2    3    2    0     10   0    0    1    7
C    4    1    1    0     0    0    2    0    2
```

Thus, the purpose of the above algorithm is to fill the elements of matrix p with integer values by counting the occurrences of each character from the columns of matrix s. The content of each element in matrix s is compared with the content from each component of variable a. The order of the letters is identical in variable a and the first column of matrix p. Such a correspondence allows for a targeted incrementation of the values over matrix p every time there is a match between a character found in matrix s and a character found in variable a. After this operation, matrix p is called a PFM.

9.8 The Position Probability Matrix (PPM)

The same matrix p from above (PFM) can be converted further to a PPM. A PPM shows which character in the alphabet ({A, T, G, C}) has a higher probability of occurrence on a specific position (column). This is, of course, visible in the PFM (Figure 9.1b). However, the values from PFM are transformed into probabilities for a more advanced subsequent evaluation. To make such a transformation to a PPM, the number of occurrences of each character (p[i][j+1]) recorded in matrix p is divided by the number of rows (s.length) found in matrix s (p[i][j+1]/s.length). This division shows the probability of occurrence of each character on a specific column/position:

Additional algorithm 9.8 Note that the source code is out of context and is intended for explanation of the method.

```
// THE POSITION PROBABILITY MATRIX

for(var i=0; i<p.length; i++) {

    for(var j=0; j<p[i].length-1; j++){

        p[i][j+1]=p[i][j+1]/s.length;
    }
}

document.write(SMC(p));

Output:

G    0.1    0.2    0.5    1    0    0.5    0.1    0.8    0.1
A    0.3    0.4    0.2    0    0    0.5    0.7    0.1    0
T    0.2    0.3    0.2    0    1    0      0      0.1    0.7
C    0.4    0.1    0.1    0    0    0      0.2    0      0.2
```

Note that the values from each $j + 1$ column in matrix p add up to 1. Variable j represents the columns and variable i represents the rows of matrix p. The first column ($j = 0$) of matrix p contains the alphabet of the input sequence. To avoid the first column, the loop above matrix p is made from $j + 1$ and ends at the penultimate column (p[i].length-1).

9.8.1 A Kind of PPM Pseudo-Scanner

Notice that at this point the probability of any sequence can be calculated given the PPM. For instance, consider a new sequence equal in length with the number of positions in the PPM: "AAAAAAAAA." The probability of sequence

"AAAAAAAA" given p (the PPM) can be calculated by multiplying together the corresponding (i) probability values from each position ($j + 1$) of the PPM:

$$P(AAAAAAAA \mid p) = 0.3 \times 0.4 \times 0.2 \times 0 \times 0 \times 0.5 \times 0.7 \times 0.1 \times 0 = 0$$

In the same manner, the probability of a sequence "CTAGTAAAT" given p (the PPM) can be calculated as:

$$P(CTAGTAAAT \mid p) = 0.4 \times 0.3 \times 0.2 \times 1 \times 1 \times 0.5 \times 0.7 \times 0.1 \times 0.7$$
$$= 0.000588$$

Even at this early stage, some kind of detector can be built. Please consider a new sequence z = "AACTAGTAAATA." The question that arises is: Which are the areas in z that most closely resemble the motif variations from the set? Such calculations can be made for consecutive segments from z that are equal in length with the motif variations from the set (nine symbols). A probability value will be obtained for each segment:

z = "<u>AACTAGTAA</u>ATA":

$$P(AACTAGTAA \mid p) = 0.3 \times 0.4 \times 0.1 \times 0 \times 0 \times 0.5 \times 0 \times 0.1 \times 0 = 0$$

z = "A<u>ACTAGTAA</u>TA":

$$P(ACTAGTAAA \mid p) = 0.3 \times 0.1 \times 0.2 \times 0 \times 0 \times 0 \times 0.7 \times 0.1 \times 0 = 0$$

z = "AA<u>CTAGTAAAT</u>A":

$$P(CTAGTAAAT \mid p) = 0.4 \times 0.3 \times 0.2 \times 1 \times 1 \times 0.5 \times 0.7 \times 0.1 \times 0.7$$
$$= 0.000588$$

z = "AAC<u>TAGTAAATA</u>":

$$P(TAGTAAATA \mid p) = 0.2 \times 0.4 \times 0.5 \times 0 \times 0 \times 0.5 \times 0.7 \times 0.1 \times 0 = 0$$

A positive value indicates how close the content of a segment is to the motif set. In other words, each probability indicates whether a segment resembles ($P > 0$) or does not resemble ($P = 0$) the variations from the set. Here, the segment "CTAGTAAAT" from z showed the highest value (0.000588). Those described above can be tested by using the implementation from below (Additional algorithm 9.9).

Additional algorithm 9.9 Note that the source code is out of context and is intended for explanation of the method.

```
// POSITION PROBABILITY MATRIX (PPM)

for(var i=0; i<p.length; i++) {
    for(var j=0; j<p[i].length-1; j++){
```

```
            p[i][j+1]=p[i][j+1]/s.length;
    }
}

// A KIND OF PPM PSEUDO SCANNER
var z = "AACTAGTAAATA";

var r = '';
var w = '';
var u = z.length - s.length + 2;

for(var l=0; l<u; l++) {

    w = z.substr(l, s.length-1);
    var d = w.split('');

    var score = 0;
    for(var f=0; f<d.length; f++){
        for(var h=0; h<a.length; h++){
            if(d[f]==p[h][0]) {
                if(f==0) {
                    score = Number(p[h][f+1]);
                } else {
                    score *= Number(p[h][f+1]);
                }
            }
        }
    }
    r += score + ',';
}

document.write(r);

Output:
0,0,0.000588,0
```

Thus, the third segment from *z* contains a sequence close to the motif variations
from the set. However, relevant motif variations present on the segments of z by

chance, may contain a symbol on one of the positions that corresponds to a zero value in the PPM. (e.g. any segment in z that does not contain a "G" on position 4 and a "T" on position 5, and a "G" or an "A" on position 6, and so on). Multiplying any value by zero inactivates the final result (e.g. $0.3 \times 0 \times 0.4 = 0$). Thus, important and valid segments from z can be lost in the detection process. Nevertheless, if matrix p (the PSSM) is initiated by using pseudo-counts instead of zero values, then the impediment related to the multiplication is removed (please see the "*Pseudo-counts and negative infinity*" subchapter). **Important:** Moreover, such a PPM approach does not differentiate between random and nonrandom sequences which are different from the target (the motif set). To make such a differentiation, another important step is needed that opens new possibilities. **Note**: Supplemental information about the design and functionality of the above scanner is described in detail in the implementation of Additional algorithm 10.1.

9.9 The Position Weight Matrix (PWM)

The PWM is a *log-likelihood matrix* (LLM) [308]. The same matrix p from above is further converted to a PWM (Figure 9.1c). A likelihood ratio (LR) is used to compare the fit of two models. The first model is the observation and is represented by the numerical values found on matrix p. The background is the second model. A primitive background model is the *null model,* which assumes that nucleotides succeed each other independently (Bernoulli model). Thus, the second model represents the expectation and considers the probability of a random occurrence for each of the characters in the alphabet:

$$P(G) = P(A) = P(T) = P(C) = \frac{one\ of\ the\ symbols}{symbols\ in\ the\ alphabet} = \frac{1}{4} = 0.25$$

In the context of DNA, these 0.25 values can be obtained by using, for example, a background set and a second PPM. However, to better understand how the value of 0.25 can be reached in the most practical way possible, additional examples are needed (please see the next subchapter – the background model). In the *null model,* the hypothesis is that all characters in the alphabet have the same probability of occurrence on any of the columns of matrix s. Thus, to compare the fit of the two models, the observed values are divided by the expected values:

$$Likelihood = \frac{observed}{expected} = \frac{P_{i,j+1}}{0.25}$$

The natural logarithm is used to calculate the values from the log-likelihoods Matrix. The natural logarithm is the logarithm to the base e (Euler's number), where *e* is an irrational constant approximately equal to 2.718281828. Thus, the log-likelihood ratio (LLR) of observed to expected is:

$$LLR = \ln\left(\frac{observed}{expected}\right) = \ln(observed) - \ln(expected)$$

where in the present example it can be written as:

$$LLR = \ln\left(\frac{observed}{expected}\right) = \ln\left(\frac{p_{i,j+1}}{0.25}\right) = \ln(p_{i,j+1}) - \ln(0.25)$$

Some natural logarithm laws useful in programming:

$$\ln\left(\frac{x}{y}\right) = \ln(x) - \ln(y) \qquad \ln(x \times y) = \ln(x) + \ln(y) \qquad \ln(e) = 1 \qquad \log_n(x) = \frac{\ln(x)}{\ln(n)}$$

The algorithm below divides the values from the *p* matrix to the *null model* (0.25) and it takes the natural logarithm of the result (`Math.log(p[i][j+1]/0.25)`), thus, transforming the *p* matrix into an PWM:

Additional algorithm 9.10 Note that the source code is out of context and is intended for explanation of the method.

```
// THE POSITION WEIGHT MATRIX

for(var i=0; i<p.length; i++) {

    for(var j=0; j<p[i].length-1; j++){

        p[i][j+1]=Math.log(p[i][j+1]/0.25).toFixed(2);
    }
}

document.write(SMC(p));

Output:

G -0.92 -0.22 0.69  1.39     -Infinity 0.69      -0.92     1.16      -0.92
A 0.18   0.47  -0.22 -Infinity -Infinity 0.69      1.03      -0.92     -Infinity
T -0.22 0.18  -0.22 -Infinity 1.39      -Infinity -Infinity -0.92     1.03
C 0.47  -0.92 -0.92 -Infinity -Infinity -Infinity -0.22     -Infinity -0.22
```

In the above case, the natural logarithm was used (Additional algorithm 9.10). However, according to different preferences and needs, a logarithm in any base *n* can be used:

$$\log_n(x) = \frac{\ln(x)}{\ln(n)}$$

The function below (Additional algorithm 9.11) represents an option to calculate a logarithm base *n* by using the above formula:

Additional algorithm 9.11 **Note that the source code is out of context and is intended for explanation of the method.**

```
function Log(n, v) {
   return Math.log(v) / Math.log(n);
}
```

The *Log* function can be used as "Log (base, value)". The example from Additional algorithm 9.10 can be repeated and the PWM can for example be based on a logarithm in base 10 (Log(10,p[i][j+1]/0.25)):

Additional algorithm 9.12 **Note that the source code is out of context and is intended for explanation of the method.**

```
//PROBABILITY MATRIX to LOG 10

for(var i=0; i<p.length; i++) {

    for(var j=0; j<p[i].length-1; j++){

        p[i][j+1]=Log(10,p[i][j+1]/0.25).toFixed(2);
    }
}

function Log(n, v) {
   return Math.log(v) / Math.log(n);
}

document.write(SMC(p));

Output:

G -0.40 -0.10 0.30  0.60      -Infinity 0.30      -0.40    0.51    -0.40
A 0.08  0.20  -0.10 -Infinity -Infinity 0.30      0.45    -0.40    -Infinity
T -0.10 0.08  -0.10 -Infinity 0.60      -Infinity -Infinity -0.40  0.45
C 0.2   -0.40 -0.40 -Infinity -Infinity -Infinity -0.10    -Infinity -0.10
```

A short intuitive conclusion regarding the meaning of the PWM values is in order. The values from the PWM elements are calculated using log likelihoods (Table 9.3). The LR is used to compare the fit of two models. The expected value

Table 9.3 The fate of a PPM value.

PPM value observed (x)	Background model expected (y)	Likelihood ratio (x/y)	PWM value ln(x/y)
0.1	0.25	0.4	−0.91629
0.2	0.25	0.8	−0.22314
0.3	0.25	1.2	0.182322
0.4	0.25	1.6	0.470004
0.5	0.25	2	0.693147
0.6	0.25	2.4	0.875469
0.7	0.25	2.8	1.029619
0.8	0.25	3.2	1.163151
0.9	0.25	3.6	1.280934
1	0.25	4	1.386294

The table shows the logarithmic results given the interplay between different observed values and the expected values. Note that for observed probability values of 1, the maximum value showed by the natural logarithm is 1.38 (i.e. $\ln(1/0.25) = 1.386$), whereas for small probability values (e.g. 1×10^{-28}) the result of the natural logarithm may show extreme negative values (i.e. $\ln(1 \times 10^{-28}/0.25) = -63.08608824$).

represents the prediction (0.25 – the *null model*) and the observed value is represented by the data of the motif set (PPM). In the PWM stage, each observed value is divided by the expected value: LR = (observed/expected). The logarithm of LR is the final result for any of the $j + 1$ elements from the PWM. Some examples showing the interplay between observed and expected values are presented in Table 9.3.

If the observed and expected values are equal, the value of LR will be equal to 1 (i.e. 0.25/0.25 = 1) and the logarithm of 1 will be zero in the PWM (i.e. $\ln(0.25/0.25) = \ln(1) = 0$). However, note that in our PPM the closest results to 0.25 are 0.2 or 0.3 (see the output of Additional algorithm 9.10 and Table 9.3). Therefore, a result closer to zero can be provided only by the elements that contain these two values (i.e. $\ln(0.2/0.25) = \ln(0.8) = -0.22$; $\ln(0.3/0.25) = \ln(1.2) = 0.18$). For more clarity, please see the output of Additional algorithm 9.10.

9.10 The Background Model

The background model (the expectation) is usually anything that is not our target (the observation). However, in the *null model*, the background is uniform and implies an equal frequency between the symbols of the alphabet. To obtain the LRs, all the values from the PPM were divided by 0.25 (please see the previous subchapter). But the consideration was that the values in the PPM (of the motif set) were divided by the corresponding values from a background PPM. However, since such a background PPM can only show 0.25 for each element, we only used the 0.25 value without mentioning a secondary background PPM. In practice, a series of special sequence variations can be used for a secondary background PPM. Such variations should contain as many positions as the motif variations from the motif set (a length match). But what kind of sequence variations could generate a rudimentary *null model* in the form of a PPM? The table below shows four artificial examples of background sets. In each set, there are four sequence variations that ensure an equal frequency for all the symbols of the alphabet (Table 9.4).

Given the equivalence, the same background PPM can be obtained from any of the sets from the table below (Table 9.4). To put things into perspective, the same background PPM will be produced even when the background sets from below are placed one below the other in a continuous bigger background set. Consequently, any of the sets from below (or all of them) can be used for the *null model*. To test this, use the implementation from the "*All in one implementation*" subchapter and change the content of the c variable with any of the sets shown in the below table (Table 9.4). For a brief introduction, Table 9.5 shows how a background PPM looks like for the background set (4) from Table 9.4.

But how many variations can a set contain to always ensure an equal probability of occurrence for each symbol in the alphabet? Here, a rule can easily be observed. Any background set can be built if an equal frequency is allowed on each position. However, such a frequency can obviously be obtained only by using a series of sequence variations that is a multiple of the number of symbols found in the

Table 9.4 Sets of variations for null models.

Background set (1) null model	Background set (2) null model	Background set (3) null model	Background set (4) null model
AAAAAAAAA	ATATATATA	AAAAATTTT	ACAACGTTA
TTTTTTTTT	TATATATAT	TTTTTAAAA	GTTCTTACT
CCCCCCCCC	CGCGCGCGC	CCCCGGGG	CACTACCGC
GGGGGGGGG	GCGCGCGCG	GGGGGCCCC	TGGGGAGAG

The table shows an obvious first set that ensures equal frequencies for each symbol (first column – set 1). Three other less obvious artificial sets are presented in which the symbols are kept with an equal frequency on each position. For an increase in naturalness, sets 2, 3, and 4 show increasing permutations between symbols on each position. The order of these variations on rows is irrelevant. Also, notice that each position contains an "A" character, a "C" character, a "G" character, and a "T" character. Thus, their probability on each column will be $\frac{1}{4}$.

Table 9.5 The null model in PPM format.

Background set (4)	The background position probability matrix (PPM)								
ACAACGTTA	A	0.25	0.25	0.25	0.25	0.25	0.25	0.25	0.25	0.25
GTTCTTACT	C	0.25	0.25	0.25	0.25	0.25	0.25	0.25	0.25	0.25
CACTACCGC	G	0.25	0.25	0.25	0.25	0.25	0.25	0.25	0.25	0.25
TGGGGAGAG	T	0.25	0.25	0.25	0.25	0.25	0.25	0.25	0.25	0.25

```
var c = "ACAACGTTA,GTTCTTACT,CACTACCGC,TGGGGAGAG";
```

The table shows a result based on sequences from the Background set (4) of the previous table. Note that the background sets for the null model generate equal probabilities in all the elements of a PPM and consequently only zero values across the PWM. This table indicates why in the previous subchapter all values in the PPM of the motif set were divided by 0.25. To avoid using a second PPM, the value of 0.25 was used by simple deduction. The last line of the table shows the background set (4) in a linear format for practical usage.

alphabet of the set. For a comparison, the motif set can use an arbitrary number of motif variations; however, the background set for the *null model* can only use a number of sequence variations multiple of n, where n represents the number of symbols in the alphabet of the set:

$$SV = ALF \times n$$

where SV is the number of sequences that make up a background set (the rows for the background set), ALF is the number of symbols in the alphabet, and n is an integer, which represents the number of occurrences for each symbol in the alphabet on a certain position. In other words, on any position, the number of occurrences for any symbol is n. Based on the formula from above, some determinations can be made. For instance, $SV = ALF \times n = 4 \times 1 = 4$, where this result shows that it takes four sequences to position four different symbols on the same column. At $SV = ALF \times n = 4 \times 2 = 8$, the result shows that it takes eight sequences to position four symbols equally on a column (two occurrences for each symbol). We can continue with $SV = ALF \times n = 4 \times 3 = 12$, where the result shows that 12 sequences are needed in a background set to position four symbols equally on a column (three occurrences for each symbol). Of course, this calculation can continue for increasing values of n. Nevertheless, exceeding the minimum number of sequences for the background set is meaningless in practice since the *null model* requires an equal probability of occurrence for each symbol (0.25 for all symbols of DNA). Thus, all discussions from above are for theoretical purposes and point out the origin of "0.25" as an expected value for the LR. Other background models (different from the *null model*) have no restrictions on the number of sequences used for the background set. Note that discussions on alternative background models will be detailed in Chapter 12.

9.11   The Consensus Sequence

A consensus sequence is an ideal motif sequence and the maxima of the motif set. The most frequent characters on each column of matrix s constitute the consensus sequence. Thus, the consensus sequence can be constructed based on any version of the p matrix, either from the PFM, PPM, or with the help of the PWM. The algorithm below makes a maximization between the values present on the columns $j + 1$ of matrix p. The highest value on a column j (p[i][j+1]) dictates the representative character of the column, taken from $j = 0$ (p[i][0]). Variable i represents the rows of the matrix p. The array e contains three indexes, each with a specific role. Index 0 (e[0]) holds the highest value found on a column and is reset when switching to a new column (e[0]=0;). Index 1 (e[1]) holds the representative character for the column and is set each time a higher value is found on the column (e[1]=p[i][0];). Index 2 (e[2]) holds the consensus sequence. The implementation below has the role of extracting the consensus sequence from the p matrix:

Additional algorithm 9.13 Note that the source code is out of context and is intended for explanation of the method.

```
// CONSENSUS SEQUENCE

var e = [];

e[0] = 0;
e[1] = '';
e[2] = '';

for(var j=0; j<p[0].length-1; j++){

    for(var i=0; i<p.length; i++) {

        if(p[i][j+1]>e[0]){
            e[0]=p[i][j+1];
            e[1]=p[i][0];
        }
```

```
    }
    e[0]=0;
    e[2] += e[1];
}

document.write(e[2]);

Output:
CAGGTGAGT
```

Since matrix p contains an equal number of elements on any of the rows, the number of columns can be represented by the number of elements on the first row (p[0].length). To avoid scanning the first column from matrix p, the cycle starts at column $j + 1$ and ends at the last column minus 1 (p[0].length-1). Variable i indicates the rows on column $j + 1$. At each step i, a comparison is made between the value from the matrix element (p[i][j+1]) and the value stored in variable e (e[0]). If the value from the matrix is greater than the value stored in the index 0 of variable e (p[i][j+1]>e[0]), then the index 0 of variable e is updated with the value from the element of the matrix (e[0]=p[i][j+1];). The consensus sequence obtained from a maximization among the values present on the matrix p columns is the most primitive version of motif characterization. For example, two close values that differ insignificantly on one of the columns will generate an uncertainty about the character that should be representative for that position. Thus, two characters would have relatively equal probabilities of being significant in that specific location of the motif sequence. Graphical representations can be used for a more sophisticated evaluation of motif sequences. For instance, matrix p can be used to design motif logos (Sequence logos). In a graphical manner, a sequence logo shows the probability of occurrence for each symbol at specific positions on the motif. Later, in this book (appendices), the implementation of these graphical approaches will be discussed in detail.

9.11.1 The Consensus – Not Necessarily Functional

A consensus sequence is truly an ideal sequence and is not necessarily functional since it is not easy to understand what changes (point mutations) in the DNA motif are accepted in the pre-mRNA transcript by the spliceosome complex [309–311]. In other words, the consensus sequence can only be functional if it is observed in the motif set; otherwise, it may not exist in the biological reality (**Table** 9.6).

Table 9.6 Motif set vs. consensus.

Motif set	Consensus	Distance	Alignment	Mismatches
GAAGTGAGT	CAGGTGAGT	2	5 - [GAAGTGAGT] - 3 5 - [CAGGTGAGT] - 3	2
TTCGTAAGT	CAGGTGAGT	4	5 - [TTCGTAAGT] - 3 5 - [CAGGTGAGT] - 3	4
AAGGTACTT	CAGGTGAGT	4	5 - [AAGGTACTT] - 3 5 - [CAGGTGAGT] - 3	4
CTGGTGAGC	CAGGTGAGT	2	5 - [CTGGTGAGC] - 3 5 - [CAGGTGAGT] - 3	2
AGAGTGAGT	CAGGTGAGT	2	5 - [AGAGTGAGT] - 3 5 - [CAGGTGAGT] - 3	3
CAGGTAGAG	CAGGTGAGT	2	5 - [CAGGTAGAG] - 3 5 - [CAGGTGAGT] - 3	4
ACTGTACGT	CAGGTGAGT	5	5 - [ACTGTACGT] - 3 5 - [CAGGTGAGT] - 3	5
CTGGTGAGT	CAGGTGAGT	1	5 - [CTGGTGAGT] - 3 5 - [CAGGTGAGT] - 3	1
TATGTAAGT	CAGGTGAGT	3	5 - [TATGTAAGT] - 3 5 - [CAGGTGAGT] - 3	3
CGGGTGAGC	CAGGTGAGT	2	5 - [CGGGTGAGC] - 3 5 - [CAGGTGAGT] - 3	2

In order to verify whether the consensus sequence exists, the table shows a comparison between the motif set and the consensus sequence. The motif variations are listed on the first column and the consensus sequence is listed on the second column. The third column shows the Levenshtein distance between sequences found on the first column and the second column. The differences between the two columns are also verified with the help of sequence alignments that objectively show the mismatches between the motif set and the consensus sequence. The number of mismatches is presented on the last column and is visually represented on the first and second columns using underlined font.

On the other hand, a motif set can be subjective if the motif variations originate from different prediction methods and not from experiments, namely sequence alignments made between sequenced mRNA molecules and their genome. Thus, in subjective cases, the presence of the consensus sequence in the motif set propagates a degree of uncertainty over the functionality of the consensus.

The table above shows that the consensus sequence does not exist in the motif set (Table 9.6). The presence of at least one zero value on the last column of the table would have indicated that the consensus sequence existed in the biological sense. Namely, a value of zero would indicate that there is no difference between at least one motif variation from the set and the consensus sequence. A reasonable deduction in this case can be that the consensus sequence is ideal and is not found in real biological sequences. Of course, here the motif set is for exemplification purposes and it has only 10 motif variations. Thus, it may be obvious that for small DNA motifs (i.e. the motif of nine positions used for illustration) the probability of finding a consensus sequence in the motif set is particularly high if the number of motif variations is large. But for longer DNA motifs (i.e. >15 b) the probability of finding the consensus sequence in the motif set is particularly low regardless of the number of motif variations in the set. **Note:** The *Levenshtein* distance between two strings is the number of deletions, insertions, or substitutions required to transform a motif variation into a consensus motif. The rows from Table 9.6 show the *Levenshtein* distance for each motif variation and an alignment between each motif variation from the set and the consensus motif. The computations of the alignments from Table 9.6 were made using the implementations from the sequence alignment chapter. Please see the sequence alignment chapter.

9.12 Mutational Intolerance

A genetic alteration over the DNA area, which corresponds to the exon–intron boundary found in the pre-mRNA transcript, can disrupt RNA splicing. Such disruptions result in the loss of exons or the inclusion of introns in the exon area, which further results in an altered protein-coding sequence [312]. However, in evolution, certain genetic alterations are tolerated in sensitive DNA areas [312]. These mutational tolerances can be seen in any motif variation to a certain degree (Table 9.6). For instance, these can be seen more often in different motifs of DNA-binding sites. Thus, an extremely interesting question would be: Can we detect the motif variations that are not accepted by the DNA-binding domain of a protein? This question deserves an introductory story. In World War II, the number of planes shot down in combat was very high. The returning planes, however, had a series of holes from the cartridges fired by the enemy. Then, the larger military bases decided to reinforce the planes in places where they were riddled with bullets to reduce their chances of being shot down. But the frequency of combat planes shot down by the enemy remained constant. Mathematicians called to inspect aviation wrecks advised the military to make reinforcements in places where there were no bullet holes in the duralumin sheet of the planes. The reasoning was that the combat planes were returning from the fight precisely because those parts of the duralumin sheet remained untouched. The association of motif variations with the returning planes from the above story gives us some important clues. **Important**: For exemplification, consider for a moment that the motif variations used above represent some fictional DNA-binding sites for certain proteins. It is not easy to understand what changes (point mutations) are accepted by a DNA-binding domain of a protein (the 3D structure of the amino acid–binding motif of the protein). A quick look at the DNA sequences from the motif set shows that on columns (positions) 4 and 5 the characters "G" and "T" are constant. In conclusion, a change in the positions 4 and 5 would inactivate the motif, in the sense that the complementarity between the DNA-binding site and the DNA-binding domain of the protein would disappear. But this is an obvious deduction. What is important instead, is what nonessential nucleotide from one position of the motif excludes another nonessential nucleotide from another position to keep the motif functional? In other words, which nucleotides in which positions cannot be present at the same time? This question automatically leads us to the notion of dependency between the motif positions and is discussed extensively in the Markov chains chapter.

9.13 From Motifs to PWMs

A matrix s was constructed using a set of functional DNA sequences. By using the alignment matrix s, a p matrix was initialized. This p matrix was further completed with the number of occurrences of each character on the columns of matrix s. Once completed, the PFM p was converted step by step into a PPM and a log-likelihood PWM. Also, a consensus sequence was determined based on matrix p. The implementation below contains all that has been discussed so far, and can be directly copied and pasted into an HTML file for further experimentation:

Additional algorithm 9.14 Note that the source code is in context and works with copy/paste.

```
<script>

var c = "GAAGTGAGT,TTCGTAAGT,AAGGTACTT,CTGGTGAGC,AGAGTGAGT," +
        "CAGGTAGAG,ACTGTACGT,CTGGTGAGT,TATGTAAGT,CGGGTGAGC";

var s = [];
```

```
var m = [];

m = c.split(',');
var n = m.length;

//THE ALIGNMENT MATRIX
for(var i=0; i<n; i++){
    s[i] = [];
    s[i]=m[i].split('');
}

document.write('<hr>Alignment matrix: ' + SMC(s));

//DETECT ALL LETTERS USING ARRAYS
var a = [];

var t = c.replace(/,/g, '').split('');
var k = t.length;

for(var i=0; i<=k; i++){
    var q = 1;
    for(var j=0; j<=a.length; j++){
        if (t[i] === a[j]) {q = 0;}
    }
    if (q === 1) {a.push(t[i]);}
}

document.write('<hr>Alphabet: ' + SMC(a));

// PROFILE MATRIX INITIALIZATION
var p = [];

for(var h=0; h<a.length; h++){

    p[h]=[];

    for(var i=0; i<=s[0].length; i++) {
        p[h][i]=0;
        p[h][0]=a[h];
    }
}

document.write('<hr>Matrix p initialization: ' + SMC(p));

// THE POSITION FREQUENCY MATRIX
for(var i=0; i<s.length; i++) {

    for(var j=0; j<s[i].length; j++){

        for(var h=0; h<a.length; h++){

            if (s[i][j] === a[h]) {p[h][j+1]++;}
        }
```

(Continued)

Additional algorithm 9.14 (Continued)

```javascript
        }
}

document.write('<hr>Position frequency matrix: ' + SMC(p));

// THE POSITION PROBABILITY MATRIX
for(var i=0; i<p.length; i++) {
    for(var j=0; j<p[i].length-1; j++){

        p[i][j+1]=p[i][j+1]/s.length;

    }
}

document.write('<hr>Position probability matrix: ' + SMC(p));

// CONSENSUS SEQUENCE
var e = [];

e[0] = 0;
e[1] = '';
e[2] = '';

for(var j=0; j<p[0].length-1; j++){

    for(var i=0; i<p.length; i++) {

        if(p[i][j+1]>e[0]){
            e[0]=p[i][j+1];
            e[1]=p[i][0];
        }
    }
    e[0]=0;
    e[2] += e[1];
}

document.write('<hr>Consensus sequence: ' + e[2]);

//THE POSITION WEIGHT MATRIX
for(var i=0; i<p.length; i++) {

    for(var j=0; j<p[i].length-1; j++){

        p[i][j+1]=Math.log(p[i][j+1]/0.25).toFixed(2);
    }
}

document.write('<hr>The Position Weight Matrix: ' + SMC(p));

function Log(x, y) {
  return Math.log(y) / Math.log(x);
}
```

```
// SHOW MATRIX CONTENT
function SMC(m) {
    var r = "<table border=1>";
    for(var i=0; i<m.length; i++) {
        r += "<tr>";
        for(var j=0; j<m[i].length; j++){
            r += "<td>"+m[i][j]+"</td>";
        }
        r += "</tr>";
    }
    r += "</table>";

    return r;
}

</script>
```

Output:

Alignment matrix:

G	A	A	G	T	G	A	G	T
T	T	C	G	T	A	A	G	T
A	A	G	G	T	A	C	T	T
C	T	G	G	T	G	A	G	C
A	G	A	G	T	G	A	G	T
C	A	G	G	T	A	G	A	G
A	C	T	G	T	A	C	G	T
C	T	G	G	T	G	A	G	T
T	A	T	G	T	A	A	G	T
C	G	G	G	T	G	A	G	C

Alphabet:
G
A
T
C

Matrix p initialization:

G	0	0	0	0	0	0	0	0	0
A	0	0	0	0	0	0	0	0	0
T	0	0	0	0	0	0	0	0	0
C	0	0	0	0	0	0	0	0	0

Position frequency matrix:

G	1	2	5	10	0	5	1	8	1
A	3	4	2	0	0	5	7	1	0
T	2	3	2	0	10	0	0	1	7
C	4	1	1	0	0	0	2	0	2

Position probability matrix:

G	0.1	0.2	0.5	1	0	0.5	0.1	0.8	0.1
A	0.3	0.4	0.2	0	0	0.5	0.7	0.1	0
T	0.2	0.3	0.2	0	1	0	0	0.1	0.7
C	0.4	0.1	0.1	0	0	0	0.2	0	0.2

(Continued)

Additional algorithm 9.14 (Continued)

```
Consensus sequence: CAGGTGAGT

Log-likelihood matrix:
G -0.92 -0.22 0.69  1.39     -Infinity 0.69     -0.92    1.16    -0.92
A 0.18  0.47  -0.22 -Infinity -Infinity 0.69     1.03     -0.92   -Infinity
T -0.22 0.18  -0.22 -Infinity 1.39      -Infinity -Infinity -0.92   1.03
C 0.47  -0.92 -0.92 -Infinity -Infinity -Infinity -0.22    -Infinity -0.22
```

9.14 Pseudo-Counts and Negative Infinity

One of the obvious issues when building the PWM is that ln(0) is undefined (i.e. "−Infinity"). This situation is encountered when small datasets are used (only a few motif variations in the set). To avoid such situations, there are several solutions. The first solution is to pass the zero values from the PFM directly to the PWM without any intermediate processing. There is also the unorthodox option to detect the PWM elements containing the string "−Infinity" and replace it with zero. However, such solutions can lead to the detection of false signals when PWM is used. Also, it would be less elegant and would add to the complexity of the implementation. Nevertheless, these approaches are still relevant and useful. The second solution is the use of pseudo-count values. A pseudo-count is a small number (i.e. 0.0000001) that can be added to all positions with zero frequency. The idea behind pseudo-counts is that very small numbers do not visibly affect the final result of an analysis and it can help avoid various calculation errors. The smaller the pseudo-count number, the more negative the logarithm result will be (e.g. ln(0.00001) = −11.51). Some examples showing the relationship between natural logarithms and different pseudo-count values are given in Table 9.7.

Of course, adding pseudo-count values only for positions with zero frequency, it is not a mandatory condition. For instance, consider an arbitrary value such as 0.63. Logarithm of 0.63 will provide a value of −0.46203546. Adding a pseudo-count value to 0.63 would lead to a new value of 0.630000001 (0.63 + 0.000000001 = 0.630000001). Logarithm of 0.630000001 will provide a value of −0.462035458. The difference between "−0.46203546" and "−0.462035458" is small enough to encourage the use of pseudo-count values regardless of frequency, because there are no serious repercussions throughout the analysis. Thus, pseudo-count values can be inserted from the initialization stage of the p matrix. Above, matrix p was initialized by inserting zero values in all the elements (p[h][i]=0;). However, such an initialization can start for instance, with the insertion of values such as "0.00000000000000001" (written as

Table 9.7 Examples of pseudo-counts and their effects on the negative scale.

Pseudo-count value	*ln*(pseudo-count)
0.01	−4.605170186
0.001	−6.907755279
0.0001	−9.210340372
0.00001	−11.51292546
0.000001	−13.81551056
0.0000001	−16.11809565
0.00000001	−18.42068074
0.000000001	−20.72326584

The table shows the relationship between different pseudo-count values and the negative values provided by the natural logarithm. This suggests the possibility of changing the negative scale of the score values in an arbitrary manner.

1×10^{-17}) instead of "0". Inserting pseudo-count values from the beginning does not affect the final result in any way. Untouched pseudo-count values will show extreme negative values in the PWM and the other values will be unchanged to a few decimal places. The implementation below demonstrates the approach using pseudo-count values:

Additional algorithm 9.15 Note that the source code is in context and works with copy/paste.

```
<script>

var c = "GAAGTGAGT,TTCGTAAGT,AAGGTACTT,CTGGTGAGC,AGAGTGAGT," +
        "CAGGTAGAG,ACTGTACGT,CTGGTGAGT,TATGTAAGT,CGGGTGAGC";

var s = [];
var m = [];

m = c.split(',');
var n = m.length;

//THE ALIGNMENT MATRIX
for(var i=0; i<n; i++){
    s[i] = [];
```

(Continued)

Additional algorithm 9.15 (Continued)

```
    s[i]=m[i].split('');
}

// DETECT ALL SYMBOLS USING ARRAYS
var a = [];
var t = c.replace(/,/g, '').split('');
var k = t.length;

for(var i=0; i<=k; i++){
    var q = 1;
    for(var j=0; j<=a.length; j++){
        if (t[i] === a[j]) {q = 0;}
    }
    if (q === 1) {a.push(t[i]);}
}

// PROFILE MATRIX INITIALIZATION
var p = [];

for(var h=0; h<a.length; h++){

    p[h]=[];

    for(var i=0; i<=s[0].length; i++) {
        p[h][i]=0.00000000000000001;
        p[h][0]=a[h];
    }
}

// THE POSITION FREQUENCY MATRIX
for(var i=0; i<s.length; i++) {

    for(var j=0; j<s[i].length; j++){

        for(var h=0; h<a.length; h++){

            if (s[i][j] === a[h]) {p[h][j+1]++;}
        }
    }
}

// THE POSITION PROBABILITY MATRIX
for(var i=0; i<p.length; i++) {
    for(var j=0; j<p[i].length-1; j++){

        p[i][j+1]=p[i][j+1]/s.length;

    }
}

// THE POSITION WEIGHT MATRIX
for(var i=0; i<p.length; i++) {
    for(var j=0; j<p[i].length-1; j++){
```

```
        p[i][j+1]=Math.log(p[i][j+1]/0.25).toFixed(2);
    }
}

document.write(SMC(p));

// SHOW MATRIX CONTENT
function SMC(m) {
    var r = "<table border=1>";
    for(var i=0; i<m.length; i++) {
        r += "<tr>";
        for(var j=0; j<m[i].length; j++){
            r += "<td>"+m[i][j]+"</td>";
        }
        r += "</tr>";
    }
    r += "</table>";

    return r;
}

</script>

Output:

G -0.92   -0.22   0.69    1.39   -40.06   0.69   -0.92   1.16    -0.92
A 0.18    0.47    -0.22  -40.06  -40.06   0.69    1.03   -0.92  -40.06
T -0.22   0.18    -0.22  -40.06   1.39   -40.06  -40.06  -0.92   1.03
C 0.47    -0.92   -0.92  -40.06  -40.06  -40.06  -0.22  -40.06  -0.22
```

It should be noted that all the values in the PWM remained the same as in the previous examples. However, the text "−Infinity" has been replaced with a negative number, namely −40.06 (ln(0.00000000000000001) = −40.06). In the next stages, this implementation will be used for detection of similar motifs in different sequences.

9.15 Conclusions

A mathematical model for representation of motifs was explored in detail. The implementation stage began with a general function for visualization of results, followed by a detection algorithm, which identifies all the unique symbols from an arbitrary sequence. The result of this algorithm was further used to dynamically initiate a PSSM. A series of step by step algorithms described the design of a PSSM. The PSSM steps incorporated the design stages of a PFM, a PPM, and a final PWM. Next, detailed discussions were presented for a better understanding of the notion of background. Based on the PSSM encoding, an algorithm for determination of a consensus motif/pattern was implemented and the biological functionality of the consensus was debated in more detail. These discussions continued with a few extra notions about mutational intolerance explained in connection with the consensus sequence. In a final phase, a recapitulation was made starting from the motif representation up to the PWM phase. This recapitulation led to a full implementation and a discussion related to pseudo-counts and negative infinity. Note that this chapter dealt with the PWM training and from this point on the PWM can be used by a scanner for detections.

10

The Motif Scanner (III)

10.1 Introduction

A motif scanner uses the position weight matrix (PWM) to recognize a DNA motif within a long stretch of DNA. In the following discussions, such a long stretch of DNA will be denoted by z. Up to this point, the PWM has been successfully built and can be used for motif detection in different DNA sequences. In the final stage, consecutive areas from a test sequence (z) are compared to the PWM. These consecutive areas are called *sliding windows* (Figure 10.1). Each comparison leads to the calculation of a score value (a sum of log likelihoods over the PWM). Scores based on *sliding windows* can help with the location of similar motifs/patterns on a given sequence z. A z-sequence can be of arbitrary size, like a DNA sequence of a simple gene or an entire genome file. A *sliding window* represents a virtual portion of constant length above the characters of a z-sequence (Figure 10.1). Such a virtual construct is moved step by step along the sequence (Figure 10.1). A step of 1 means that a window slides with one letter on each step. A step of 2 means that a window moves with two letters on each step and so on. Sliding windows with a step of 1 are regularly used for qualitative analysis. Larger steps jump over n characters and valuable information related to the sequence is lost. Nevertheless, there are special conditions in which larger steps (>1) are used for quantitative analysis. The characters found in the sliding window area at each step represent the contents of a sliding window. For instance, a sliding window with step 1 and a virtual length of four characters above the "ATGATTAT" sequence will show the contents of five sliding windows, namely: "ATGA," "TGAT," "GATT," "ATTA," "TTAT." In our current case, the length of a sliding window is equal to the number of characters of any one sequence from the motif set (nine characters – see the content of variable c from far above).

Algorithms in Bioinformatics: Theory and Implementation, First Edition. Paul A. Gagniuc.
© 2021 John Wiley & Sons, Inc. Published 2021 by John Wiley & Sons, Inc.
Companion website: www.wiley.com/go/gagniuc/algorithmsinbioinformatics

AAAAAACAG|GTGAGTAAAAAAAA

(a)

	A	A	A	A	A	A	C	A	G
G	−0.92	−0.22	0.69	1.39	−40.06	0.69	−0.92	1.16	−0.92
A	0.18	0.47	−0.22	−40.06	−40.06	0.69	1.03	−0.92	−40.06
T	−0.22	0.18	−0.22	−40.06	1.39	−40.06	−40.06	−0.92	1.03
C	0.47	−0.92	−0.92	−40.06	−40.06	−40.06	−0.22	−40.06	−0.22

$= -81.1$

A|AAAAACAGG|TGAGTAAAAAAAA

(b)

	A	A	A	A	A	C	A	G	G
G	−0.92	−0.22	0.69	1.39	−40.06	0.69	−0.92	1.16	−0.92
A	0.18	0.47	−0.22	−40.06	−40.06	0.69	1.03	−0.92	−40.06
T	−0.22	0.18	−0.22	−40.06	1.39	−40.06	−40.06	−0.92	1.03
C	0.47	−0.92	−0.92	−40.06	−40.06	−40.06	−0.22	−40.06	−0.22

$= -118.5$

AA|AAAACAGGT|GAGTAAAAAAAA

(c)

	A	A	A	A	C	A	G	G	T
G	−0.92	−0.22	0.69	1.39	−40.06	0.69	−0.92	1.16	−0.92
A	0.18	0.47	−0.22	−40.06	−40.06	0.69	1.03	−0.92	−40.06
T	−0.22	0.18	−0.22	−40.06	1.39	−40.06	−40.06	−0.92	1.03
C	0.47	−0.92	−0.92	−40.06	−40.06	−40.06	−0.22	−40.06	−0.22

$= -77.7$

Figure 10.1   A diagram of the scanning process. (a) Shows the contents of the first sliding window and the corresponding values from the PWM. (b) It shows the contents of the second sliding window and the correspondence with the PWM values. (c) The same scheme is presented for the third sliding window. The process continues until the last symbol in the z-sequence is incorporated inside a sliding window. For each step, the PWM values corresponding to each symbol from the sliding window are summed and the result is shown next to the PWM. Note that a sliding window of step 1 is used here.

10.2   Looking for Signals

Consecutive score values extracted from sliding windows can determine the formation of a representative signal for sequence z. The score peaks from the signal determine the location of possible motif variations that resemble those in the motif set. For implementation, a special z-sequence will be considered. The sequence contains a series of "A" characters and the consensus sequence ("CAGGTGAGT") inserted in between these characters:

z = "AAAAAACAGGTGAGTAAAAAAAA"

The consensus sequence is known and has already been determined far above. Since the consensus sequence is inserted as a dummy, the analysis of the sequence from variable z should exhibit one score with a maximum positive value above the consensus sequence position.

Three loops are used to detect a motif inside a new sequence stored in variable z. The first loop traverses the content from z by using variable l as an index. The second loop traverses the character positions from the sliding window by using the f variable as an index. The third loop traverses the first column of matrix p by using variable h as an index. The main loop traverses the contents of the z variable from the first character (1=0) to the u character, where u is the length of the sequence in variable z (z.length) minus the length of the sliding window (s.length). Note that the length of the sliding window is represented by the number of columns from matrix s. Initially, sequence z is broken into sliding windows inside the first loop (w=z.substr(1,s.length-1);). The contents of each sliding window are temporarily converted to an array format for a more direct use (d = w.split(");). As a reminder, note that there is a dimensional correspondence between variable a, containing the alphabet, and the first column of matrix p. Also, there is a dimensional correspondence between the f positions from the sliding window and the $f + 1$ columns of matrix p. Inside the main loop, each character from the sliding window (d[f]) is compared with each character from the first column of matrix p (p[h][0]). If there is a match (d[f]==p[h][0]), then variable h will indicate the row on the first column (p[h][0]), corresponding to the character in the sliding window (d[f]). Thus, $f+1$ can be representative of the matrix p column, from where the corresponding value can be extracted (p[h][f+1]). The sum of the log-likelihoods values above matrix p (p[h][f+1]) for a sliding window represents the score. Each sliding window generates a score. The successive score values indicate the candidate positions for the motif in variable z. The implementation below summarizes all that is discussed above (Additional algorithm 10.1):

Additional algorithm 10.1 Note that the source code is out of context and is intended for explanation of the method.

```
// THE MOTIF SCANNER

var z = "AAAAAACAGGTGAGTAAAAAAAA";

document.write('z = "' + z + '"<hr>');

var w = ";
var u = z.length - s.length + 2;
```

(Continued)

Additional algorithm 10.1 (Continued)

```
for(var l=0; l<u; l++) {

    w = z.substr(l, s.length-1);
    var d = w.split(");

    var r = 'score = ';
    var score = 0;
    for(var f=0; f<d.length; f++){

        for(var h=0; h<a.length; h++){

            if(d[f]==p[h][0])
            {
                r += '(' + p[h][f+1] + ')+';
                score += Number(p[h][f+1]);
            }
        }
    }

    score = score.toFixed(1);
    document.write('sw('+(l+1)+') = "'+w+'"<br>'+r+'='+score+'<hr>');
}

Output:

z = "AAAAAACAGGTGAGTAAAAAAAA"

sw(1) = "AAAAAACAG"
score = (0.18)+(0.47)+(-0.22)+(-40.06)+(-40.06)+(0.69)+(-0.22)+(-0.92)
        +(-0.92)+=-81.1

sw(2) = "AAAAACAGG"
score = (0.18)+(0.47)+(-0.22)+(-40.06)+(-40.06)+(-40.06)+(1.03)+(1.16)
        +(-0.92)+=-118.5

sw(3) = "AAAACAGGT"
score = (0.18)+(0.47)+(-0.22)+(-40.06)+(-40.06)+(0.69)+(-0.92)+(1.16)
        +(1.03)+=-77.7

sw(4) = "AAACAGGTG"
score = (0.18)+(0.47)+(-0.22)+(-40.06)+(-40.06)+(0.69)+(-0.92)+(-0.92)
        +(-0.92)+=-81.8

sw(5) = "AACAGGTGA"
score = (0.18)+(0.47)+(-0.92)+(-40.06)+(-40.06)+(0.69)+(-40.06)+(1.16)
        +(-40.06)+=-158.7

sw(6) = "ACAGGTGAG"
score = (0.18)+(-0.92)+(-0.22)+(1.39)+(-40.06)+(-40.06)+(-0.92)+(-0.92)
        +(-0.92)+=-82.5

sw(7) = "CAGGTGAGT"
score = (0.47)+(0.47)+(0.69)+(1.39)+(1.39)+(0.69)+(1.03)+(1.16)+(1.03)
        +=8.3
```

```
sw(8)  = "AGGTGAGTA"
score = (0.18)+(-0.22)+(0.69)+(-40.06)+(-40.06)+(0.69)+(-0.92)+(-0.92)
        +(-40.06)+=-120.7

sw(9)  = "GGTGAGTAA"
score = (-0.92)+(-0.22)+(-0.22)+(1.39)+(-40.06)+(0.69)+(-40.06)+(-0.92)
        +(-40.06)+=-120.4

sw(10) = "GTGAGTAAA"
score = (-0.92)+(0.18)+(0.69)+(-40.06)+(-40.06)+(-40.06)+(1.03)+(-0.92)
        +(-40.06)+=-160.2

sw(11) = "TGAGTAAAA"
score = (-0.22)+(-0.22)+(-0.22)+(1.39)+(1.39)+(0.69)+(1.03)+(-0.92)
        +(-40.06)+=-37.1

sw(12) = "GAGTAAAAA"
score = (-0.92)+(0.47)+(0.69)+(-40.06)+(-40.06)+(0.69)+(1.03)+(-0.92)
        +(-40.06)+=-119.1

sw(13) = "AGTAAAAAA"
score = (0.18)+(-0.22)+(-0.22)+(-40.06)+(-40.06)+(0.69)+(1.03)+(-0.92)
        +(-40.06)+=-119.6

sw(14) = "GTAAAAAAA"
score = (-0.92)+(0.18)+(-0.22)+(-40.06)+(-40.06)+(0.69)+(1.03)+(-0.92)
        +(-40.06)+=-120.3

sw(15) = "TAAAAAAAA"
score = (-0.22)+(0.47)+(-0.22)+(-40.06)+(-40.06)+(0.69)+(1.03)+(-0.92)
        +(-40.06)+=-119.4
```

Thus, a series of values were obtained from each sliding window. The output of the above implementation initially shows the z sequence in clear text, then the data related to each sliding window, such as: (i) the sliding window number and content, (ii) the method of calculating the score, and (iii) the score obtained from the sliding window content. Before describing the significance of the score values, a reconfiguration of the implementation is mandatory because in practice only the series of scores obtained from sliding windows is needed.

10.3 A Functional Scanner

The final reconfiguration includes two main parts. The first part consists of the PWM construction from a total of 10 motifs aligned one below the other. The second part uses the PWM to detect possible new motifs inside an uncharacterized sequence (z). Another short optimization was also added. The PPM and the PWM were merged and a variable g was used to calculate the probabilities and the logarithm in the same nested loop. These simple enhancements can be seen in the implementation below (Additional algorithm 10.2):

Additional algorithm 10.2 Note that the source code is in context and works with copy/paste.

```
<script>

// AN IMPLEMENTATION BASED ON PSEUDO COUNTS

var c = "GAAGTGAGT,TTCGTAAGT,AAGGTACTT,CTGGTGAGC,AGAGTGAGT," +
        "CAGGTAGAG,ACTGTACGT,CTGGTGAGT,TATGTAAGT,CGGGTGAGC";

var s = [];
var m = [];

m = c.split(',');
var n = m.length;

//THE ALIGNMENT MATRIX
for(var i=0; i<n; i++){
    s[i] = [];
    s[i]=m[i].split("");
}

//ALPHABET DETECTION
var a = [];
var t = c.replace(/,/,/g, "").split("");
var k = t.length;

for(var i=0; i<=k; i++){
    var q = 1;
    for(var j=0; j<=a.length; j++){
        if (t[i] === a[j]) {q = 0;}
    }
    if (q === 1) {a.push(t[i]);}
}

// PROFILE MATRIX INITIALIZATION
var p = [];

for(var h=0; h<a.length; h++){
    p[h]=[];
    for(var i=0; i<=s[0].length; i++) {
        p[h][i]=0.00000000000000001;
        p[h][0]=a[h];
```

```
    }
}

// THE POSITION FREQUENCY MATRIX
for(var i=0; i<s.length; i++) {
    for(var j=0; j<s[i].length; j++){
        for(var h=0; h<a.length; h++){
            if (s[i][j] === a[h]) {p[h][j+1]++;}
        }
    }
}

// POSITION PROBABILITY MATRIX & LOG LIKELIHOOD MATRIX
for(var i=0; i<p.length; i++) {
    for(var j=0; j<p[i].length-1; j++){
        var g = p[i][j+1]/s.length;
        p[i][j+1]=Math.log(g/0.25).toFixed(2);
    }
}

// SEARCH MOTIF IN Z
var z = "AAAAAACAGGTGAGTAAAAAAAA";

var r = ";
var w = ";
var u = z.length - s.length + 2;

for(var l=0; l<u; l++) {

    w = z.substr(l, s.length-1);
    var d = w.split(");

    var score = 0;
    for(var f=0; f<d.length; f++){
        for(var h=0; h<a.length; h++){
            if(d[f]==p[h][0]) {score += Number(p[h][f+1]);}
        }
    }
    r += score.toFixed(0) + ',';
}

document.write(r);

</script>

Output:
-81,-118,-78,-82,-159,-82,8,-121,-120,-160,-37,-119,-120,-120,
-119,
```

Note: In the above implementation, the ".toFixed(0)" method rounds all values to integers to make the text and the numbers more intelligible for discussions. In practice, the score values should have at least two decimals (".toFixed(2)"). Additional algorithm 10.2 produces a series of scores based on fifteen sliding window locations. Thus, each number in the output is representative of the contents of a sliding window. In this experiment, the score increases sharply to a positive number on the seventh sliding window, namely to a score value of 8. The seventh sliding window contains the location where the consensus sequence was artificially planted. Thus, the detection of the consensus sequence confirms that the algorithm is running as expected. Some additional experiments based on the above implementation are presented below (Table 10.1):

The table below shows the results obtained from seven specially constructed z-sequences. The first two sequences from the table show a nine-character insert containing the "GT" letters on positions 3 and 4 (a shift to the left – compared to the positions 4 and 5 found in the motif set). The purpose of this pseudo-motif is to observe how the algorithm behaves in such a situation. A detection of the pseudo-motif is made in the third sliding window. Thus, the algorithm considers the letter before the insert because the "GT" characters are found on positions 4 and 5. Nonetheless, the algorithm identifies this pseudo-motif but gives it a relatively low score value (i.e. 3). Sequences three and four are random in nature. For these two random sequences, the signal shows no detection as expected. Sequence

Table 10.1 Additional tests by using pseudo-counts.

No	Sequence *z*	Planted motif	Signal
1	AAA**AAGTAAAGT**AAAAA	AAGTAAAGT	−160,−160,3,−75,−122, −161,−160,−36,−118,
2	AGTC**AAGTAAAGT**AGTCTC	AAGTAAAGT	−122,−162,−161,3,−75,−122, −122,−117,2,−157,−122,
3	ATCAGCGATTACG	None/random	−122,−38,−159,−121,−120,
4	GCGATTACGCATGACGTAA	None/random	−120,−37,−202,−79,−121,−118, −122,−79,−80,−120,−159,
5	CCCCCCCC**CAGGTGAGT**CCC	CAGGTGAGT	−162,−202,−124,−120,−80,−84, −160,−82,8,−81,−120,−161,
6	AAAAAAAGGGGGCCCCCCCCC	None	−78,−80,−80,−80,−38,−78,−78, −120,−120,−161,−163,−163,−162,
7	GTGTGTGTGTGTGTGTGT	None	−123,−35,−123,−35,−123,−35, −123,−35,−123,−35,

Here, different sequence configurations and their signals are presented in detail. The signals corresponding to each z-sequence show a compliant operation for Additional algorithm 10.2.

five is a repetitive sequence with a consensus motif planted in the middle of the sequence. As expected, this sequence shows a strong peak for the ninth sliding window. The last two sequences in the table are meant as a challenge for the algorithm because each z-sequence shows different repetitive patterns. As desired, in these cases, the signals also show no detection. **Note**: for comparative observations, the z-sequences from Table 10.1 will be analyzed repeatedly along several subchapters by using other implementations.

10.4 The Meaning of Scores

A series of scores above the z-sequence can indicate which areas are similar to the motif variations from the set. Each score over z was calculated by adding the relevant log-likelihood values from each position in the PWM. This section describes the meaning of different score values and brings to the fore three ideal sequences: the consensus sequence, the anticonsensus sequence, and a third sequence that mimics the background model.

10.4.1 A Score Value Above Zero

A score value above zero indicates that a sliding window contains a region from the z sequence that is similar to the PWM maxima, namely similar to the motif variations from the set. The logical assumption here is that the PWM represents functional motifs (real). By association, the content of a sliding window that resembles the PWM maxima is therefore also functional. The value of the score indicates how likely is the functionality of the contents from the sliding window (Table 10.2). Note that the PWM maxima are a consensus sequence. Therefore, a maximum

Table 10.2 The meaning of scores.

Score	Indicates	Meaning
Score=0	Random	Equally likely to be functional or random
Score>0	Nonrandom	More likely to be functional
Score<0	Nonrandom	Less likely to be functional

The table describes the significance of the observed values. A score value above zero indicates that the sliding window contains a region from the z sequence that is more likely to be functional. A score value of zero indicates that the sliding window contains a region from the z sequence that is equally likely to be functional or random. A score value below zero indicates that a sliding window contains a region from the z sequence that is less likely to be functional.

Table 10.3 Range of values for scores.

Sequence	Signal	Peak
A maximum score value above zero (score>0)		
z="AAAACAGGTGAGTAAAA"	-78, -82, -159, -82, 8, -121, -120, -160, -37	MAX = 8

G	-0.92	-0.22	0.69	1.39	-40.06	0.69	-0.92	1.16	-0.92
A	0.18	0.47	-0.22	-40.06	-40.06	0.69	1.03	-0.92	-40.06
T	-0.22	0.18	-0.22	-40.06	1.39	-40.06	-40.06	-0.92	1.03
C	0.47	-0.92	-0.92	-40.06	-40.06	-40.06	-0.22	-40.06	-0.22

Sequence	Signal	Peak
A score of zero (score~0)		
z="AAAAAGTTGTAGTCAAAA"	-121,-158,-79,-79,-117,0,-199,-162,-78,-118,-120	Around zero = 0.45~0

G	-0.92	-0.22	0.69	1.39	-40.06	0.69	-0.92	1.16	-0.92
A	0.18	0.47	-0.22	-40.06	-40.06	0.69	-0.92	1.16	-40.06
T	-0.22	0.18	-0.22	-40.06	1.39	-40.06	-40.06	-0.92	1.03
C	0.47	-0.92	-0.92	-40.06	-40.06	-40.06	-0.22	-40.06	-0.22

Sequence	Signal	Peak
A minimum score below zero (score<0)		
z="GTGTGGCCCCTCATGTG"	-120,-120,-119,-159,-124,-243,-122,-81,-119,-124	MIN = -243

G	-0.92	-0.22	0.69	1.39	-40.06	0.69	-0.92	1.16	-0.92
A	0.18	0.47	-0.22	-40.06	-40.06	0.69	1.03	-0.92	-40.06
T	-0.22	0.18	-0.22	-40.06	1.39	-40.06	-40.06	-0.92	1.03
C	0.47	-0.92	0.92	-40.06	-40.06	-40.06	-0.22	-40.06	-0.22

To provide a good understanding of the score values, the table shows three extreme cases. Each case is accompanied by a z sequence, the main signal composed of score values and a PWM that indicates the elements used to calculate the planted sequence. The first case uses a z sequence in which a consensus was planted to obtain a maximum positive score value. The second case uses a z sequence in which a neutral sequence was planted. The neutral sequence guarantees a score of zero inside the signal. The third case uses a z sequence in which an anticonsensus sequence has been introduced to obtain a minimum score value within the signal.

score can be obtained by using the consensus sequence detected by Additional algorithm 9.13. A sequence z that includes a planted consensus sequence (e.g. "CAGGTGAGT") can further point out the significance of positive score values (Table 10.3).

Again, a consensus sequence can be understood as the maxima of the PWM model. The consecutive symbols that show the highest value on each position of the PWM can form a consensus sequence. Any position ($j+1$ column) from the PWM may show more than one maximum value. For instance, the PWM shows 0.69 for "G" and "A" on position 6. This means that characters "G" and "A" have an equal probability of occurrence in this position. Thus, several variations of the consensus sequence can be constructed when there are such cases of equal probability (e.g. for our case both "CAGGT**G**AGT" and "CAGGT**A**AGT" are consensus). However, any variation of the consensus sequence will yield the same maximum score value (i.e. in our case 8.3). To test this, use Additional algorithm 10.2 and change the contents of the z variable alternatively with each variation of the consensus.

10.4.2 A Score Value Below Zero

A score value below zero indicates that a sliding window contains a region from the z sequence that is nonrandom but different from the PWM model. To further understand the significance of such negative score values, one can construct a z sequence that includes an anticonsensus sequence (e.g. "GCCCCCTCA"). By complementarity, an anticonsensus sequence is the minima of the PWM model. The consecutive symbols that show the lowest value on each position of the PWM can form an anticonsensus sequence. In other words, to build an anticonsensus sequence, for each position we choose the symbol that shows the lowest value in the PWM. Also, any position/column from the PWM can show many equal minimum values. Consequently, several variations of the anticonsensus sequence can be constructed. To test the anticonsensus sequence, please replace the z variable content from the Additional algorithm 10.2 with "GTGTGGCCCCCTCATGTGT" (Table 10.3).

The result will show the most negative output (-240) for the sliding window that contains the anticonsensus sequence. An important distinction should be made in the case of negative scores. Score values that are lower than zero are far from both the background model and the target model (motif set). The background model used here is the *null model* and involves randomness. But the background model can also be a nonrandom model. In other words, if the background model would be different from the *null model*, then a score value of zero would indicate no difference between the contents of a sliding window over z and the background set (whatever the background set may represent). This will be exemplified and discussed in detail in the other subchapters.

10.4.3 A Score Value of Zero

A score value of zero indicates that a sliding window contains a region from the z-sequence that is similar to the background model. Zero scores are rare. The reason is that biological sequences are not random in nature. Moreover, different chapters from this book will most likely indicate that true randomness is difficult to obtain on one-dimensional sequences. But what is the origin of a score with a zero value and what is the explanation behind it? Each position in the sliding window has a direct correspondence to the PWM columns. Any symbol from the sliding window has a corresponding row on the PWM. The intersection between a corresponding row and the position of the symbol inside the sliding window indicates the PWM element used for summation. The sum made over these PWM elements leads to a score value. However, the sum over the PWM elements can show a score of zero for two reasons. The first reason would be that the sum is made over the PWM elements that contain only zero values (i.e. observed and expected values are equal for each relevant element from PWM – rare cases). The second reason would be that the sum of positive and negative values from the relevant PWM elements cancel each other out and reach the equilibrium at zero. However, a combination of the two situations is usually true. A sequence that shows a score of zero can be artificially designed (Table 10.3). For instance, such a sequence can be: "GTTGTAGTC." The relevant values from each position of the PWM sum up to:

$$score = (-0.92) + (0.18) + (-0.22) + (1.39) + (1.39) + (0.69)$$
$$+ (-0.92) + (-0.92) + (-0.22) = 0.45$$

Notice that the negative and positive values provided by each position, cancel each other out. Thus, sequence "GTTGTAGTC" shows a score value of ~0 (Table 10.3).

10.5 Conclusions

The operating principle behind a motif scanner was discussed at the beginning of the chapter. A functional motif scanner was implemented, and the results were shown as signals (Additional algorithm 10.2). The meaning of these signals was then explained in the form of successive score values. The motif scanner used a long sequence (e.g. the DNA from a certain genomic area) to formulate a series of scores based on the PWM data. The values of these scores indicated which of the areas of the sequence was similar to the motif encoded in the PWM. In a second phase, the complete meaning of these scores was further explored through experimentation (Table 10.3).

11

Understanding the Parameters (IV)

11.1 Introduction

This chapter makes a parallel analysis between two approaches. Both approaches are variations of the same scanner shown in Additional algorithm 10.2. The implementation steps from Additional algorithm 10.2 are repeated in two alternative versions for experimentation. One experiment explores the importance of Laplace estimators (pseudo-counts). The second experiment explores the results based on the artificial transfer of zero count values between a position frequency matrix (PFM) and a position weight matrix (PWM). Next, the parameters responsible for signal discrimination and signal sensitivity are discussed and modified with the help of a series of experiments. These experiments show the existence of a balance in which the target motifs are optimally detected. The chapter draws a parallel between bioinformatics and computer security to show the applications of different bioinformatics methods in other fields. Later, a scanner that uses previously trained PWMs will be presented in a very short implementation in the final part of the chapter. The idea of such a scanner would be the use of the PWMs that are present in the additional materials of scientific articles. Moreover, a signal-filtering module is added to the same scanner. Based on the filter, score values below a threshold value are eliminated. Toward the end of the chapter, the signal filter is tested using a series of experiments described in detail.

11.2 Experimentation

To make a comparison between signals obtained with the help of pseudo-count values and zero values, two head-to-tail implementations are required. Thus, by using two such implementations, we can highlight the importance of pseudo-counts and further initiate some experiments to understand what a false signal represents. The first implementation is based on pseudo-counts and the

Algorithms in Bioinformatics: Theory and Implementation, First Edition. Paul A. Gagniuc.
© 2021 John Wiley & Sons, Inc. Published 2021 by John Wiley & Sons, Inc.
Companion website: www.wiley.com/go/gagniuc/algorithmsinbioinformatics

second implementation is based on zero values. The two implementations differ only slightly and represent a revision of the methods discussed so far.

11.2.1 A Scanner Implementation Based on Pseudo-Counts

Negative scores provide an intuitive view of the detection process. However, in practice, negative scores can be irrelevant for motif sequence detection. These negative scores can therefore be eliminated with the help of a threshold condition (if(score<0){score=0;}). Such a condition will highlight only the relevant signals for the analysis. Thus, all scores that are below zero are removed from the results and are replaced with scores of zero. The implementation below puts this threshold into practice:

Additional algorithm 11.1 Note that the source code is in context and works with copy/paste.

```
<script>

// PSEUDO COUNTS AND ELIMINATION OF SCORES BELOW ZERO

var c = "GAAGTGAGT,TTCGTAAGT,AAGGTACTT,CTGGTGAGC,AGAGTGAGT," +
        "CAGGTAGAG,ACTGTACGT,CTGGTGAGT,TATGTAAGT,CGGGTGAGC";

var s = [];
var m = [];

m = c.split(',');
var n = m.length;

//THE ALIGNMENT MATRIX
for(var i=0; i<n; i++){
    s[i] = [];
    s[i]=m[i].split("");
}

//ALPHABET DETECTION
var a = [];
var t = c.replace(/,/g, "").split("");
var k = t.length;

for(var i=0; i<=k; i++){
    var q = 1;
    for(var j=0; j<=a.length; j++){
        if (t[i] === a[j]) {q = 0;}
    }
    if (q === 1) {a.push(t[i]);}
}
```

```
// PROFILE MATRIX INITIALIZATION
var p = [];

for(var h=0; h<a.length; h++){
    p[h]=[];
    for(var i=0; i<=s[0].length; i++) {
        p[h][i]=0.00000000000000001;
        p[h][0]=a[h];
    }
}

// THE POSITION FREQUENCY MATRIX
for(var i=0; i<s.length; i++) {
    for(var j=0; j<s[i].length; j++){
        for(var h=0; h<a.length; h++){
            if (s[i][j] === a[h]) {p[h][j+1]++;}
        }
    }
}

// POSITION PROBABILITY MATRIX & POSITION WEIGHT MATRIX
for(var i=0; i<p.length; i++) {
    for(var j=0; j<p[i].length-1; j++){
        var g = p[i][j+1]/s.length;
        p[i][j+1]=Math.log(g/0.25).toFixed(2);
    }
}

// SEARCH MOTIF IN Z
var z = "AAAAAACAGGTGAGTAAAAAAAA";

var r = ";
var w = ";
var u = z.length - s.length + 2;

for(var l=0; l<u; l++) {

    w = z.substr(l, s.length-1);
    var d = w.split(");

    var score = 0;
    for(var f=0; f<d.length; f++){
        for(var h=0; h<a.length; h++){
            if(d[f]==p[h][0]) {score += Number(p[h][f+1]);}
        }
    }
}
```

(Continued)

Additional algorithm 11.1 (Continued)

```
    if(score<0){score=0;}
    r += score.toFixed(0) + ',';
}

document.write(r);

</script>


Output:

0,0,0,0,0,0,8,0,0,0,0,0,0,0,0,0,
```

The above results show that all score values below zero have been removed and the output signal contains a positive score on the expected position, where the test motif was artificially planted.

11.2.2 A Scanner Implementation Based on Propagation of Zero Counts

One question can be: what do the results look like without pseudo-counts? To see this, a few small changes can be made. Two simple changes can be observed in the new implementation from below compared to the implementation based on pseudo-counts (above). The first change is the removal of the pseudo-count value ($p[h][i]= 0.00000000000000001;$) with a replacement of zero ($p[h][i]=0;$). To avoid the logarithm-related issues, a second modification imposes a condition on the construction of the PWM. This condition moves a zero value from the position probability matrix (PPM) to the PWM without any additional calculations. These simple changes can be seen in the implementation below (Additional algorithm 11.2):

Additional algorithm 11.2 Note that the source code is in context and works with copy/paste.

```
<script>

// AN IMPLEMENTATION BASED ON ZERO VALUES

var c = "GAAGTGAGT,TTCGTAAGT,AAGGTACTT,CTGGTGAGC,AGAGTGAGT," +
```

```
              "CAGGTAGAG,ACTGTACGT,CTGGTGAGT,TATGTAAGT,CGGGTGAGC";
var s = [];
var m = [];

m = c.split(',');
var n = m.length;

//THE ALIGNMENT MATRIX
for(var i=0; i<n; i++){
    s[i] = [];
    s[i]=m[i].split("");
}

//ALPHABET DETECTION
var a = [];
var t = c.replace(/,/g, "").split("");
var k = t.length;

for(var i=0; i<=k; i++){
    var q = 1;
    for(var j=0; j<=a.length; j++){
        if (t[i] === a[j]) {q = 0;}
    }
    if (q === 1) {a.push(t[i]);}
}

// PROFILE MATRIX INITIALIZATION
var p = [];

for(var h=0; h<a.length; h++){
    p[h]=[];
    for(var i=0; i<=s[0].length; i++) {
        p[h][i]=0;
        p[h][0]=a[h];
    }
}

// THE POSITION FREQUENCY MATRIX
for(var i=0; i<s.length; i++) {
    for(var j=0; j<s[i].length; j++){
        for(var h=0; h<a.length; h++){
            if (s[i][j] === a[h]) {p[h][j+1]++;}
        }
    }
}

// POSITION PROBABILITY MATRIX & POSITION WEIGHT MATRIX
for(var i=0; i<p.length; i++) {
    for(var j=0; j<p[i].length-1; j++){
```

(Continued)

Additional algorithm 11.2 (Continued)

```
            if(p[i][j+1]==0){
                p[i][j+1]=0;
            } else {
                var g = p[i][j+1]/s.length;
                p[i][j+1]=Math.log(g/0.25).toFixed(2);
            }
        }
}

// SEARCH MOTIF IN Z
var z = "AAAAAACAGGTGAGTAAAAAAAA";

var r = ";
var w = ";
var u = z.length - s.length + 2;

for(var l=0; l<u; l++) {

    w = z.substr(l, s.length-1);
    var d = w.split(");

    var score = 0;
    for(var f=0; f<d.length; f++){
        for(var h=0; h<a.length; h++){
            if(d[f]==p[h][0])
            {
                score += Number(p[h][f+1]);
            }
        }
    }

    if(score<0){score=0;}
    r += score.toFixed(0) + ',';
}

document.write(r);

</script>


Output:

0,2,2,0,2,0,8,0,0,0,3,1,1,0,1,
```

The above results show that all score values below zero have been removed and the output signal contains multiple positive scores. The maximum score value is shown on the expected position, where the test motif was artificially planted. These two tests already point out that the second implementation (Additional algorithm 11.2) contains a series of false-positive score values. To test the level of discrimination of the two scanners, some simple tests are required (see below).

11.3 Signal Discrimination

Perhaps the most important aspect of a detection algorithm is related to the power of discrimination. This mainly refers to the correct detection of a functional/specific target. In other words, how sharp is the algorithm in differentiating between the target and other sequence structures? Some experiments presented below highlight the differences between the implementation that uses pseudo-count values (Additional algorithm 11.1) and the implementation in which the zero counts are propagated to the PWM stage (Additional algorithm 11.2). Seven sequences built specifically for testing are analyzed with both scanners and the results are presented in the table below (Table 11.1).

Table 11.1 Scanner discrimination power.

z-sequence	Planted motif	Signal	Discrimination
AAA**AAGTAAAGT**AAAAA	AAGTAAAGT	PC: 0,0,3,0,0,0,0,0,0,	High
		ZC: 0,1,3,5,0,0,0,4,2,	Low
AGTC**AAGTAAAGT**AGTCTC	AAGTAAAGT	PC: 0,0,0,3,0,0,0,0,2,0,0,	High
		ZC: 0,0,0,3,5,0,0,3,2,3,0,	Low
ATCAGCGATTACG	None/random	PC: 0,0,0,0,0,	High
		ZC: 0,2,1,0,0,	Low
GCGATTACGCATGACGTAA	None/random	PC: 0,0,0,0,0,0,0,0,0,0,0,	High
		ZC: 0,3,0,1,0,3,0,2,1,0,1,	Low
CCCCCCCC**CAGGTGAGT**CCC	CAGGTGAGT	PC: 0,0,0,0,0,0,0,8,0,0,0,	High
		ZC: 0,0,0,0,1,0,0,0,8,0,0,0,	Low
AAAAAAAGGGGGCCCCCCCCC	None	PC: 0,0,0,0,0,0,0,0,0,0,0,0,	High
		ZC: 2,0,0,0,3,2,2,0,0,0,0,0,0,	Low
GTGTGTGTGTGTGTGTGT	None	PC: 0,0,0,0,0,0,0,0,0,0,	High
		ZC: 0,5,0,5,0,5,0,5,0,5,	Low

The table contains four columns. The first column shows the z-sequence taken into consideration for analysis. The second column describes the particularity of each z-sequence. The third column shows the signals provided by both scanners. "PC" is short for pseudo-count and refers to Additional algorithm 11.1, whereas "ZC" is short for zero count and refers to the signal provided by Additional algorithm 11.2. The last column of the table describes the comparison between the two scanners in terms related to discrimination. Note that Additional algorithm 11.2 shows a peak every time a "GT" is encountered on positions 4 and 5 of the sliding window.

The above results are based on a fixed background, namely the *null model*. The table from above indicates that the implementation shown in Additional algorithm 11.1 exhibits a high discrimination power compared to Additional algorithm 11.2, in absolutely all aspects of these experiments. For a more colorful description on discrimination, the two approaches can be associated with the idea of metal detectors. One metal detector (e.g. Additional algorithm 11.2) senses any metal in its range, whereas the other metal detector senses only gold (e.g. Additional algorithm 11.1). To push the associations closer to the information technology area, the scanners described in this subchapter are like the antivirus engines used in computer security. There is a striking similarity between the heuristic parts of an antivirus engine and the methods discussed here. Thus, envisage one antivirus engine (e.g. Additional algorithm 11.2) that detects any malware-like files in its range, generating many false-positive detections in the process, whereas another antivirus engine senses only the true malware files (e.g. Additional algorithm 11.1). Nevertheless, reliable signal/noise discrimination is one of the most important challenges for any detector and needs further discussions.

11.4 False-Positive Results

The possibility of moving the zero values directly from the PFM to the PWM was discussed above as an alternative solution to pseudo-counts. However, such a solution leads to the lowest discrimination power for a scanner. A low discrimination further leads to a correct detection of the target but also to many false-positive results. The importance of pseudo-count values in the PFM and the negative values found in the PWM can be highlighted here. The presence of a pseudo-count in the PFM generates the lowest negative value in the PWM. Such an extreme negative value cannot be canceled out by the sum of the positive values from other positions of the PWM. For example, a value of -40.06 on one of the positions will drag the entire score value below zero. As an example, consider the "AAGAAAAGT" content from a sliding window. The "AAGAAAAGT" content will generate a false-positive score for Additional algorithm 11.2, and it will be correctly discarded by Additional algorithm 11.1. Additional algorithm 11.1 uses pseudo-counts for the PFM. Pseudo-counts lead to negative log likelihood values in the PWM. When the score is calculated, any other character on positions 4 or 5 will carry the lowest negative log-likelihood values. Therefore, if "GT" is missing in these positions, then the negative log-likelihood values will lead to a score value below zero no matter how many positive log-likelihood values will be present on the other positions of the PWM. This simple feature leads to a steep decrease of false positives. Here, the "AAGAAAAGT" content was specially chosen to provide the maximum values from each position of the PWM, except for positions 4 and 5. The content "AAGAAAAGT" shows the "AA" characters on positions 4 and 5 instead of "GT." Thus, Additional algorithm 11.1 will make the following computation based on the PWM values:

G	−0.92	−0.22	0.69	1.39	−40.06	0.69	−0.92	1.16	−0.92
A	0.18	0.47	−0.22	−40.06	−40.06	0.69	1.03	−0.92	−40.06
T	−0.22	0.18	−0.22	−40.06	1.39	−40.06	−40.06	−0.92	1.03
C	0.47	−0.92	−0.92	−40.06	−40.06	−40.06	−0.22	−40.06	−0.22

$$score = (0.18) + (0.47) + (0.69) + (−40.06) + (−40.06) + (0.69) + (1.03)$$
$$+ (1.16) + (1.03) = −74.87$$

A score of −74.87 correctly eliminates the "AAGAAAAGT" sequence from the valid results. On the other hand, Additional algorithm 11.2 will make the same computation on a PWM that contains values of zero instead of "−40.06":

G	−0.92	−0.22	0.69	1.39	0	0.69	−0.92	1.16	−0.92
A	0.18	0.47	−0.22	0	0	0.69	1.03	−0.92	0
T	−0.22	0.18	−0.22	0	1.39	0	0	−0.92	1.03
C	0.47	−0.92	−0.92	0	0	0	−0.22	0	−0.22

$$score = (0.18) + (0.47) + (0.69) + (0) + (0) + (0.69) + (1.03) + (1.16) + (1.03)$$
$$= 5.25$$

It stands to reason that a score with a value of 5.25 for a content such as "AAGAAAAGT" can only be a false-positive result. Note that the "GT" characters are present in the motif set on positions 4 and 5 with a probability of 1 (always present in those positions). Thus, a valid content must contain a "GT" in positions 4 and 5; otherwise, it will generate a false-positive detection. Although the "zero" approach from Additional algorithm 11.2 is not exactly desirable, it can be useful in various experimental conditions.

11.5 Sensitivity Adjustments

But if we were to change the sensitivity of the scanner, how could we do that? The log-likelihood values can be lowered or raised relative to zero by changing the pseudo-count value. For instance, pseudo-count values of 0.00001 or 0.0000001 can have a drastic impact on the negative scale (Table 9.7). A general rule would be that the lower the pseudo-count, the more negative the logarithmic values will be. A high pseudo-count value (e.g. 0.01) leads to a high negative value (e.g. $\ln(0.01) = -4.6$) and vice versa (e.g. 0.000000001, $\ln(0.000000001) = -20.72$). Thus, such negative log-likelihood values have a direct influence over the final detection. This influence is closely related to a balance that can be established in between the false-positive and the false-negative results. The negative values from the PWM generated by pseudo-counts cut off many of the false-positive scores. There is an equilibrium point for a pseudo-count value at which the negativity of the corresponding log likelihood value from the PWM cancels the positive log-likelihood values from the other positions, just in the right amount to avoid the false-positive score values. In this manner, the proper value for the pseudo-count can lead to a near-perfect detection. Consequently, the pseudo-count value can be considered a constant that regulates the sensitivity of the scanner. The more a pseudo-count value tends to zero (0.000 · · · 0001), the stricter the scanner becomes, to the point where only sequences that are almost identical to those from the motif set can generate positive score values. Of course, such a rigidity defeats the purpose of the scanner and is not desired. On the other hand, pseudo-count values that deviate to far from zero on the positive scale (e.g. 0.01), will ensure more permissive results, to the point where nonrelated sequences are detected as possible functional motifs (false positives). This is the type of situation in which it can be said: "*If all you have is a hammer, everything looks like a nail.*" Thus, an appropriate value for the pseudo-count avoids the extremes and leads to a balance in which the scanner is neither too rigid nor too relaxed.

11.6   Beyond Bioinformatics

The theoretical principles and the implementations discussed in this chapter cross the boundaries on applications found in many areas. Probably a much more realistic association could be made between the above scanners and the malware/antivirus scanners. In fact, the principle described here for finding motifs in DNA, RNA, or proteins and the heuristic principles of an antivirus scanner are strikingly similar. Here, our scanners detect small DNA segments. However, in the IT security area, these scanners detect small pieces of files or pieces of information related to the behavior of different applications, such as memory read/write events or events related to the network connections on different ports. Thus, an antivirus/firewall engine may contain a series of heuristic signatures in the form of precalculated PWMs from a collection of malware files or a collection of application behaviors (e.g. log files). For such uses, the sequence variations from the motif set take the shape of string segments specific only to malware files (computer viruses, worms, trojans, backdoors, or metamorphic/polymorphic combinations between these classes of malware). Moreover, the background model for the construction of the PWM would be represented by segments from legitimate files or by pieces of information from the normal behavior of different applications. Of course, in the world of security there are several heuristic methods that can be applied, but the method applied here is a classic and represents a form of machine learning. As a principle of analysis, this method comes long before the modern revolution of computers and the Internet. Thus, the method itself may have been imported in IT security from the nascent field of bioinformatics science in the 1980s and early 1990s.

11.7   A Scanner That Uses a Known PWM

Once a PWM has been calculated, it does not have to be calculated each time the implementation is run. Moreover, many PWMs are already trained and can be found in the additional materials of scientific journals as ready to use. Most of these PWMs are found in Excel format (value + tab + value + tab...). Thus, a simple copy/paste of an Excel table with PWM structure can be directly exploited with the help of an implementation. The idea behind this approach is to load the excel format into a *p* matrix. The principle is presented below (Additional algorithm 11.3):

Additional algorithm 11.3 Note that the source code is out of context and is intended for explanation of the method.

```
// LOADING A KNOWN PWM

c ="|G -0.92   -0.22  0.69   1.39     -40.06   0.69    -0.92  1.16    -0.92" +
  "|A 0.18    0.47   -0.22  -40.06   -40.06   0.69    1.03   -0.92   -40.06" +
  "|T -0.22   0.18   -0.22  -40.06   1.39     -40.06  -40.06 -0.92   1.03" +
  "|C 0.47    -0.92  -0.92  -40.06   -40.06   -40.06  -0.22  -40.06  -0.22"

var n = [];
var m = [];
var p = [];

m = c.split('|');

// LOAD PWM
for(var i=1; i<m.length; i++) {

    p[i-1]=[];
    n = m[i].split('\t');
```

(Continued)

Additional algorithm 11.3 (Continued)

```
    for(var j=0; j<n.length; j++){
        p[i-1][j]=n[j]
    }
}

document.write(SMC(p));

Output:
```

```
G   -0.92   -0.22    0.69    1.39   -40.06   0.69    -0.92    1.16    -0.92
A    0.18    0.47   -0.22  -40.06   -40.06   0.69     1.03   -0.92   -40.06
T   -0.22    0.18   -0.22  -40.06    1.39   -40.06   -40.06  -0.92    1.03
C    0.47   -0.92   -0.92  -40.06   -40.06  -40.06   -0.22   -40.06  -0.22
```

In this version, variable c contains the PWM data, separated by the tab character (in java script the tab character is expressed by "\t"). In Additional algorithm 11.3, the tab character is shown in its native form (i.e. " "). However, optionally, the tab character can be replaced inside the text of the PWM with his JavaScript equivalent (i.e. "\ t"), as shown below (Additional algorithm 11.4):

Additional algorithm 11.4 Note that the source code is out of context and is intended for explanation of the method.

```
// THE SECOND OPTION FOR THE TAB CHARACTER

c = "|G\t-0.92\t-0.22\t0.69\t1.39\t-40.06\t0.69\t-0.92\t1.16\t-0.92" +
    "|A\t0.18\t0.47\t-0.22\t-40.06\t-40.06\t0.69\t1.03\t-0.92\t-40.06" +
    "|T\t-0.22\t0.18\t-0.22\t-40.06\t1.39\t-40.06\t-40.06\t-0.92\t1.03" +
    "|C\t0.47\t-0.92\t-0.92\t-40.06\t-40.06\t-40.06\t-0.22\t-40.06\t-0.22"
```

Moreover, a simple artifice is introduced here to differentiate between the PWM rows. This differentiation is made by using a delimiting character introduced at the beginning of each line in c. The vertical line character was chosen as the delimiter ("|"). Nevertheless, any delimiter can be used as long as it has a low probability of occurring in PWM data and is aesthetically intuitive for the programmer. Compared to the previous implementations, variables c, n, and m take on new roles. In a first phase, the content of the c variable is split, and the resulting segments are stored in the array variable m. Thus, the m variable holds each PWM row as a continuous string. Note: the increment of variable i begins at 1 because the splitting of c into m pieces leaves $m[0]$ empty. Consequently, in the same cycle, the completion of matrix p elements is done from $i - 1$. In the second phase, each row in m is split using the tab character ("\t") and the individual values in the row are kept temporarily in the array variable n. At the same time, the values extracted from variable n are loaded successively on the rows of matrix p. Once loaded, the p matrix is a PWM and can be used directly as such by a scanner. Below is a contextual implementation of a native scanner based on a known PWM (Additional algorithm 11.5):

Additional algorithm 11.5 Note that the source code is in context and works with copy/paste.

```
<script>

// A SCANNER THAT USES A KNOWN PWM

c = "|G -0.92  -0.22   0.69    1.39   -40.06   0.69   -0.92   1.16   -0.92" +
    "|A 0.18    0.47   -0.22  -40.06  -40.06   0.69    1.03  -0.92  -40.06" +
    "|T -0.22   0.18   -0.22  -40.06   1.39   -40.06 -40.06 -0.92   1.03" +
    "|C 0.47   -0.92   -0.92  -40.06  -40.06  -40.06  -0.22  -40.06  -0.22"

var n = [];
var m = [];
var p = [];

m = c.split('|');

// LOAD PWM
for(var i=1; i<m.length; i++) {
    p[i-1]=[];
    n = m[i].split('\t');
    for(var j=0; j<n.length; j++){
        p[i-1][j]=n[j]
    }
}

// SEARCH MOTIF IN Z
var z = "AAAAAACAGGTGAGTAAAAAAAA";

var r = ";
var w = ";
var u = z.length - n.length + 2;

for(var l=0; l<u; l++) {

    w = z.substr(l, n.length-1);
    var d = w.split(");

    var score = 0;
    for(var f=0; f<d.length; f++){
```

(Continued)

Additional algorithm 11.5 (Continued)

```
        for(var h=1; h<m.length; h++){
            if(d[f]==p[h-1][0]) {score += Number(p[h-1][f+1]);}
        }
    }

    if(score<0){score=0;}
    r += score.toFixed(0) + ',';
}

document.write(r);

</script>

Output:
0,0,0,0,0,0,8,0,0,0,0,0,0,0,0,0,
```

The result of the above implementation is the same as the result presented in Additional algorithm 11.1. It may be noted that the above implementation includes two parts. The first part loads the PWM. In the second part, the PWM is used to search for the profile in a sequence z. The first part is already discussed. The second part was also discussed earlier; however, here it undergoes some adaptations. In this case, the s matrix and the a variable do not exist because the p matrix is loaded and not calculated. In previous versions of the scanner, the a variable contained the order of the letters on the first column of matrix p. Here, the p matrix already contains this order by default at loading time. Thus, the length of the content from variable a (a.length) was replaced by the number of rows found on matrix p (m.length). Also, the number representing the columns in matrix s (s.length) has been replaced by the number of columns shown by default in matrix p (n.length). Note that this approach works for all cases where the PWM values are constant. However, PWMs can show dynamic properties which will be discussed in detail in the following subchapters. These dynamic properties imply an adaptation of the PWM values to the z-sequence that is taken into consideration for analysis.

11.8   Signal Thresholds

In research, the score values that indicate a degree of certainty are considered important. The main interest in different experiments is to find real functional targets. Thresholds on signal filters can be imposed for a detection that provides a higher degree of certainty. However, such thresholds are subjective and are imposed by the researcher based on his personal confidence in the PWM data. The maximum score value inside a signal may vary from case to case. Thus, threshold values are set in percentages. The significance of a score shown by the contents of a sliding window was discussed above. Therefore, scores below zero indicate the presence of a nonrandom segment that is different from the target model (the motif set). On the other hand, the maximum score has shown a value of 8 for a perfect match. The distance between a score with a value of zero and a score with a maximum value may represent a range between 0% and 100%. Thus, 0% would represent a score with a value of zero and 100% would represent a score with a maximum value (Additional algorithms 11.6 and 11.7). A complete implementation for imposing thresholds for a motif scanner is shown in Additional algorithm 11.8. Initially, the main signal (r) is split into independent score values, which are then stored in variable v (Additional algorithm 11.6). Accordingly, variable v is an array that holds all the scores from the signal (r). In order to detect the maximum score value in the signal, the values from variable v are traversed only once. Next, the maximum score value is stored in the *max* variable. In a second cycle, each value in v is converted to percentages using the formula shown below (`pro = (100/max)*v[i];`):

$$pro = \frac{100}{max\,(v)} \times v_i$$

The percentage (*pro*) shown by each score value is compared to the threshold variable *th*. The percentage values from variable pro that are higher than the value from the threshold variable (*th*), are retained, while values that are less than *th* are replaced in this case by a line character "-". Variable *pro* is a temporary variable that keeps the percentage associated with the score value only until the comparison with the threshold variable (*th*) is made.

Additional algorithm 11.6 Note that the source code is out of context and is intended for explanation of the method.

```
var v = r.split(',');
var rez = ";
var th = 40;
var max=0;

// FIND MAX
for(var i=0; i<v.length; i++) {
    if(v[i]>max){max=v[i];}
}
```

(Continued)

Additional algorithm 11.6 (Continued)

```
for(var i=0; i<v.length; i++) {
    var pro = (100 / max) * v[i];
    document.write(pro + ',');
    if(pro>th){rez += v[i];} else {rez += '-';}
}

document.write(rez);
```

The percentages that exceed the value from the threshold variable and the line characters, are placed in variable rez, in the same order in which they were encountered. The content of variable *rez* represents the result of the implementation and contains the filtered signal. Moreover, the above implementation can be greatly shortened by using the JavaScript math method, as follows:

Additional algorithm 11.7 Note that the source code is out of context and is intended for explanation of the method.

```
var v = r.split(',');
var rez = '';
var th = 40;

for(var i=0; i<v.length; i++) {
    var pro = (100 / Math.max.apply(Math, v)) * v[i];
    if(pro>th){rez += v[i];} else {rez += '-';}
}
```

The above algorithm contains a threshold set to 40% and the input is taken from variable *r*, which contains the final signal as it did in all the previous implementations. Additional algorithm 11.6 concludes the theory behind the filter algorithm and it can be inserted in context.

11.8.1   Implementation and Filter Testing

To test different thresholds on the current filter, a special z-sequence is constructed to allow a smooth verification of the results. To observe several score values in the same signal, three motif sequences are planted in the z-sequence as follows:

z = "AAAATCAGTGAATAAAAACAGGTGAGTAAAAATAAGTGAATAAAA"

Each of the three motifs planted inside *z* provides a different score value in the following order: 3, 8, 5. The goal here is to eliminate certain scores based on a threshold value. Additional algorithm 11.5 is a scanner with the shortest source code and it was previously discussed and tested. Here, it will be used to test the new implementation for filtering score values below an arbitrary threshold. In short, the new implementation from Additional algorithm 11.6 will be attached at the end of Additional algorithm 11.5, as follows (Additional algorithm 11.8):

Additional algorithm 11.8 Note that the source code is in context and works with copy/paste.

```
<script>

// A SCANNER THAT USES THRESHOLDS AND A KNOWN PWM

c = "|G\t-0.92\t-0.22\t0.69\t1.39\t-40.06\t0.69\t-0.92\t1.16\t-0.92" +
    "|A\t0.18\t0.47\t-0.22\t-40.06\t-40.06\t0.69\t1.03\t-0.92\t-40.06" +
    "|T\t-0.22\t0.18\t-0.22\t-40.06\t1.39\t-40.06\t-40.06\t-0.92\t1.03" +
    "|C\t0.47\t-0.92\t-0.92\t-40.06\t-40.06\t-40.06\t-0.22\t-40.06\t-0.22"

var n = [];
var m = [];
var p = [];

m = c.split('|');

// LOAD PWM
for(var i=1; i<m.length; i++) {
    p[i-1]=[];
    n = m[i].split('\t');
    for(var j=0; j<n.length; j++){
        p[i-1][j]=n[j]
    }
}

// SEARCH MOTIF IN Z
var z = "AAAATCAGTGAATAAAAACAGGTGAGTAAAAATAAGTGAATAAAA";

var r = ";
var w = ";
var u = z.length - n.length + 2;

for(var l=0; l<u; l++) {

    w = z.substr(l, n.length-1);
    var d = w.split(");

    var score = 0;
    for(var f=0; f<d.length; f++){
```

(Continued)

Additional algorithm 11.8 (Continued)

```
        for(var h=1; h<m.length; h++){
            if(d[f]==p[h-1][0]) {score += Number(p[h-1][f+1]);}
        }
    }

    if(score<0){score=0;}
    r += score.toFixed(0) + ',';
}

document.write('Signal by score values:<br>' + r + '<hr>');
document.write('Signal by percent:<br>');

var v = r.split(',');
var rez = ";
var th = 40;
var max=0;

// FIND MAX
for(var i=0; i<v.length; i++) {
    if(v[i]>max){max=v[i];}
}
//max = Math.max(v);

for(var i=0; i<v.length; i++) {
    var pro = (100/max)*v[i];
    document.write(pro + ',');
    if(pro>th){rez += v[i];} else {rez += '-';}
}

document.write('<hr>Signal by threshold [' + th + '%]<br>' + rez);

</script>

Output:
Signal by score values:
0,0,0,0,3,0,0,0,0,0,0,0,0,0,0,0,0,0,8,0,0,0,0,0,0,0,0,0,0,0,0,0,5,0,0,0,0,

Signal by percent:
0,0,0,0,37.5,0,0,0,0,0,0,0,0,0,0,0,0,0,100,0,0,0,0,0,0,0,0,0,0,0,0,0,62.5,
0,0,0,0,0,

Signal by threshold [40%]
---------------8----------5-----
```

The results of the above implementation show a maximum score value of 8. There-fore, the value 8 will represent 100%. By using the above formula, one can calculate the position of each score in the range 0–100%:

$$pro = \frac{100}{max\,(v)} \times v_i = \frac{100}{8} \times 3 = 37.5\%$$

Table 11.2 Signal filters and thresholds.

Test	Result	Threshold (%)
1	----3------------8------------5-----	10
2	----3------------8------------5-----	20
3	---3------------8-----------5-----	30
4	------------------8-----------5-----	40
5	------------------8-----------5-----	50
6	------------------8----------5-----	60
7	------------------8-----------------	70
8	------------------8-----------------	80
9	------------------8-----------------	90

The table shows nine tests with different threshold values, performed on a z-sequence containing three artificially planted motifs. The first motif is designed to show a score of 3. The second motif is designed to show a maximum score of 8 and the third motif is designed to show an intermediate score of 5.

$$pro = \frac{100}{max\,(v)} \times v_i = \frac{100}{8} \times 5 = 62.5\%$$

$$pro = \frac{100}{max\,(v)} \times v_i = \frac{100}{8} \times 8 = 100\%$$

Thus, we notice that a threshold higher than 37.5% eliminates the first score value from the signal and a threshold above 62.5% eliminates the last score value (Table 11.2). Tests conducted by using the above implementation were performed by changing the threshold value (th) of the filter. In Table 11.2, each character (line or digit) represents a sliding window above the z-sequence.

The data show that the filter algorithm behaves as expected and it can be used in its current form. In practice, the filter should not be crucial for conducting an experiment on short z-sequences. The true value of a filter of this kind can be observed when a detection is conducted above an entire genome. For large-scale experiments, the filter can be set to a minimum threshold value (e.g. 10–20%) to avoid the elimination of scores that may reveal important functional areas. The filter algorithm should not be understood as a discrimination method. The detection algorithm itself must provide a high discrimination power. Thus, a low discrimination power of the scanner algorithm (e.g. a poor training for the PWM) should not be supplemented by such a filter. For instance, the scanner from Additional algorithm 11.2 shows a rather low discrimination power and a filter would eliminate small values but would not improve the process itself.

11.9 Conclusions

In the first instance, this chapter followed a series of experiments based on two implementations that used the *null model* as the background. These two designs represented different versions of the same scanner. Each version used a different approach for dealing with logarithm of zero (i.e. undefined; "-Infinity"). The first scanner was based on the use of pseudo-count values inserted at the PSSM initialization stage. The second scanner was based on the propagation of zero count values from the PFM, straight to the PWM. The importance of the pseudo-count values was clearly revealed by comparing the results of the two scanners in a series of experiments. In the second part of the chapter, different types of adjustments have been discussed. Adjustments like signal discrimination and sensitivity were determined by experimentation. Toward the end of the chapter, a scanner version that used a precalculated PWM was implemented and fully tested. Moreover, a signal filter was further developed and added to this scanner to set a threshold for score values.

12

Dynamic Backgrounds (V)

12.1 Introduction

The environment of the algorithms described in these chapters is represented by a z-sequence. Such a z-sequence is a simple text, which may represent either a DNA/RNA molecule or a protein, or, any other type of nonbiological information. Thus, the implementations developed here can therefore be used outside the field of bioinformatics. However, the idea of a background model, which is different from the *null model*, is explored in detail. Moreover, a dynamic background is discussed, implemented, and later tested by experimentation. A dynamic background continuously adapts the algorithm to the environment, giving it a much higher detection specificity. The implementations presented here bring more complexity to this method but also a stronger discrimination between the background and the target motif. Many of those discussed in previous chapters remain valid in this chapter as well. The difference is the approach regarding the background position frequency matrix (PFM - matrix b). The experiments described in this chapter were previously used in the same format and will show a set of sequential detections built specifically for operational checks. The new results are further compared with the experiments from the previous chapters to demonstrate the usefulness of the new approach.

12.2 Toward a Scanner with Two PFMs

In the chapters above, a reference was made to the possibility of using a background that is different from the *null model*. In the null model, a second matrix was considered, whose elements contained a single value, namely 0.25. Provided with this particularity, all values in the p matrix elements were simply divided by 0.25

Algorithms in Bioinformatics: Theory and Implementation, First Edition. Paul A. Gagniuc.
© 2021 John Wiley & Sons, Inc. Published 2021 by John Wiley & Sons, Inc.
Companion website: www.wiley.com/go/gagniuc/algorithmsinbioinformatics

without a special need to use a second matrix in the implementation strategy. Nevertheless, different values for the background require the introduction of a second matrix in the implementation (Figure 12.1). One of the ways in which the presence of a functional motif on a certain sequence z can be appreciated more accurately, is represented by the adaptation of the algorithm to the local environment. The local environment can be represented either by the whole sequence z or by smaller parts from sequence z (buffers/segments in the case of large genomes). Thus, the local environment is the frequency of each symbol on a certain area over sequence z. To extract these frequencies in the format of a PFM (similar to the PFM based on the motif set), the same sliding window strategy is adopted.

The difference in this case is that the content of each sliding window will be a sequence variation for the background set (t). In other words, the background set (t) contains consecutive segments from the z sequence before the z sequence itself is analyzed (Figure 12.1). Thus, the motif set (c) and the background set (t) lead to the calculation of two independent PFMs (Figure 12.1a,b). Both PFMs are then directly used to calculate the PWM, which in turn is used by the scanner to find motifs in sequence z (Figure 12.1c).

12.2.1 The Implementation of Dynamic PWMs

In the new algorithm, the z sequence is declared at the beginning of the implementation (Additional algorithm 12.1). The background set is collected from the z sequence, and then it is stored in the t variable in the same format as the motif set (i.e. variable c). Both sets are each loaded into a matrix. The motif set (c) is loaded in the sp matrix and the background set (t) is loaded in the sb matrix (Figure 12.1a,b). For a position correspondence between the sp matrix and the sb matrix, each sequence in the background set is equal in length to the motif variations from the motif set (`mp[0].length`). Each background sequence is extracted from z, starting from i up to i plus the length observed in the first motif sequence from variable c (`z.substr(i, mp[0].length)`). The total number of sequences in the background set is variable and is represented by the number of symbols from z minus the number of symbols in the first motif sequence found in variable c (`z.length - mp[0].length`). In previous implementations, the p matrix was used constantly, and it was successively transformed from a PFM to a PPM and then to a PWM. The same role is maintained in this case as well. The same as before, the values from the p matrix elements are computed based on the counts found on the columns of the s matrix, renamed here as the sp matrix. However, a second PFM is added, namely the b matrix. The b matrix is the background PFM. Previously, a background PFM was not used because

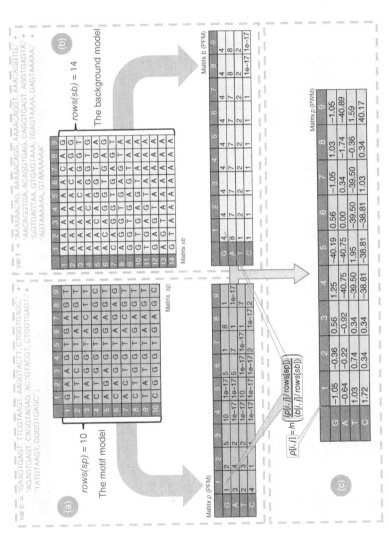

Figure 12.1 Dynamic PWMs. (a) Shows the motif set and the transformation of this set into a PFM stored in matrix p, (b) shows the background set and the transformation of this set into a PFM stored in matrix b. (c) It shows the construction of the PWM with the help of the two PFMs. Note that in the lower right corner of the PWM there is a value of +40.17. This value occurs due to an insufficient variation in the background set. In other words, the z sequence in our example is either too short or poor in information content.

all elements would have contained a single value dictated by the *null model*, namely 0.25. Here, a background PFM is constructed from sequences extracted consecutively from *z*. Thus, the values from the *b* matrix elements are computed based on the counts found on the columns of the *sb* matrix. In an interesting manner, the *z* sequence becomes particularly important for its own analysis. Note that the PWM is calculated by using the two PFMs without an intermediary step for the calculation of PPMs (Figure 12.1). To calculate the PWM, a series of steps are taken in a single cycle. For the elements of each PFM, the frequency values are divided by the number of sequences in the corresponding set. Namely, the values from the *p* matrix are divided by the number of sequences found in the motif set (sp.length) and the values in the *b* matrix are divided by the total number of sliding windows (sb.length – the total number of sequences in the background set):

$$gp = \frac{p_{i,j}}{rows(sp)}$$

$$gb = \frac{b_{i,j}}{rows(sb)}$$

where *gp* represents the observed values and *gb* represents the expected values. Then the values resulting from the homologous elements of the two PFMs divide each other and the logarithm of the result provides the final PWM value:

$$PWM[i,j] = ln\left(\frac{gp}{gb}\right)$$

In this manner, the relationship can be written as:

$$PWM[i,j] = ln\left(\frac{p_{i,j}/rows(sp)}{b_{i,j}/rows(sb)}\right)$$

where $p_{i,j}$ is the PFM of the motif set and $b_{i,j}$ is the PFM of the background set. The variable *rows(sp)* represents the number of motif variations in the motif set (sp.length) and *rows(sb)* represents the number of sequences that make up the background set (sb.length). It is worth mentioning again that the *p* matrix is finally transformed into a PWM. Thus, the relationship can be written more directly as (Figure 12.1c):

$$p_{i,j} = ln\left(\frac{p_{i,j}/rows(sp)}{b_{i,j}/rows(sb)}\right)$$

The above discussion leads the stage for an implementation strategy. The idea behind the implementation is a concomitant processing of two matrices (*p* and *b*)

to the point where the PWM is calculated. The implementation below indicates all
that has been discussed so far:

**Additional algorithm 12.1 Note that the source code is in context and works
with copy/paste.**

```
<script>

// A SCANNER WITH TWO PFMs

var c = "GAAGTGAGT,TTCGTAAGT,AAGGTACTT,CTGGTGAGC,AGAGTGAGT," +
        "CAGGTAGAG,ACTGTACGT,CTGGTGAGT,TATGTAAGT,CGGGTGAGC";

var z = "AAAAAACAGGTGAGTAAAAAAAA";

var sp = [];
var mp = [];
var sb = [];
var mb = [];

mp = c.split(',');
var np = mp.length;

var t = ";
var o = z.length-mp[0].length;

// MAKE BACKGROUND SET t
for(var i=0; i<o; i++) {
    t += z.substr(i, mp[0].length);
    if(i<o-1){t += ','}
}

document.write('t="' + t + '"');

mb = t.split(',');
var nb = mb.length;

// THE ALIGNMENT MATRIX FOR p
for(var i=0; i<np; i++){
    sp[i] = [];
    sp[i] = mp[i].split(");
}

document.write('<hr>' + SMC(sp));
```

(Continued)

Additional algorithm 12.1 (Continued)

```
// THE ALIGNMENT MATRIX FOR b
for(var i=0; i<nb; i++){
    sb[i] = [];
    sb[i] = mb[i].split("");
}

document.write('<hr>' + SMC(sb));

// ALPHABET DETECTION
var a = [];
var t = c.replace(/,/g, "").split("");
var k = t.length;

for(var i=0; i<=k; i++){
    var q = 1;
    for(var j=0; j<=a.length; j++){
        if (t[i] === a[j]) {q = 0;}
    }
    if (q === 1) {a.push(t[i]);}
}

// SIMULTANEOUS MATRIX INITIALIZATION
var p = [];
var b = [];

for(var h=0; h<a.length; h++){
    p[h]=[];
    b[h]=[];
    for(var i=0; i<=sp[0].length; i++) {
        p[h][i]=0.00000000000000001;
        b[h][i]=0.00000000000000001;
        p[h][0]=a[h];
        b[h][0]=a[h];
    }
}

// THE POSITION FREQUENCY MATRIX p
for(var i=0; i<sp.length; i++) {
    for(var j=0; j<sp[i].length; j++){
        for(var h=0; h<a.length; h++){
            if (sp[i][j] === a[h]) {p[h][j+1]++;}
        }
    }
}

document.write('<hr>' + SMC(p));
```

```
// THE POSITION FREQUENCY MATRIX b
for(var i=0; i<sb.length; i++) {
    for(var j=0; j<sb[i].length; j++){
        for(var h=0; h<a.length; h++){
            if (sb[i][j] === a[h]) {b[h][j+1]++;}
        }
    }
}

document.write('<hr>' + SMC(b));

// POSITION PROBABILITY MATRIX & POSITION WEIGHT MATRIX
for(var i=0; i<p.length; i++) {
    for(var j=0; j<p[i].length-1; j++){
        var gp = p[i][j+1]/sp.length;
        var gb = b[i][j+1]/sb.length;
        p[i][j+1]=Math.log(gp/gb).toFixed(2);
    }
}

document.write('<hr>' + SMC(p));

// SCANNER - SEARCH MOTIF IN Z
var r = ";
var w = ";
var u = z.length - sp.length + 2;

for(var l=0; l<u; l++) {

    w = z.substr(l, sp.length-1);
    var d = w.split(");

    var score = 0;
    for(var f=0; f<d.length; f++){
        for(var h=0; h<a.length; h++){
            if(d[f]==p[h][0]) {score += Number(p[h][f+1]);}
        }
    }

    if(score<0){score=0;}
    r += score.toFixed(0) + ',';
}

document.write('<hr>' + r);

function SMC(m) {
    var r = "<table border=1>";
```

(Continued)

Additional algorithm 12.1 (Continued)

```
    for(var i=0; i<m.length; i++) {
        r += "<tr>";
        for(var j=0; j<m[i].length; j++){
            r += "<td>"+m[i][j]+"</td>";
        }
        r += "</tr>";
    }
    r += "</table>";

    return r;
}

</script>

Output:
t="AAAAAACAG,AAAAACAGG,AAAACAGGT,AAACAGGTG,AACAGGTGA,ACAGGTGAG,
CAGGTGAGT,AGGTGAGTA,GGTGAGTAA,GTGAGTAAA,TGAGTAAAA,GAGTAAAAA,
AGTAAAAAA,GTAAAAAAA"
```

```
 G A A G T G A G T
 T T C G T A A G T
 A A G G T A C T T
 C T G G T G A G C
 A G A G T G A G T
 C A G G T A G A G
 A C T G T A C G T
 C T G G T G A G T
 T A T G T A A G T
 C G G G T G A G C

 A A A A A A C A G
 A A A A A C A G G
 A A A A C A G G T
 A A A C A G G T G
 A A C A G G T G A
 A C A G G T G A G
 C A G G T G A G T
 A G G T G A G T A
 G G T G A G T A A
 G T G A G T A A A
 T G A G T A A A A
 G A G T A A A A A
 A G T A A A A A A
 G T A A A A A A A
```

```
 G 1 2 5 10    1e-17 5     1     8     1
 A 3 4 2 1e-17 1e-17 5     7     1     1e-17
 T 2 3 2 1e-17 10    1e-17 1e-17 1     7
 C 4 1 1 1e-17 1e-17 1e-17 2     1e-17 2
```

```
G   4    4    4    4    4    4    4    4        4
A   8    7    7    7    7    7    7    8        8
T   1    2    2    2    2    2    2    2        2
C   1    1    1    1    1    1    1    1e-17    1e-17


G -1.05  -0.36  0.56   1.25   -40.19  0.56    -1.05   1.03   -1.05
A -0.64  -0.22 -0.92  -40.75  -40.75  0.00     0.34  -1.74  -40.89
T  1.03   0.74  0.34  -39.50   1.95  -39.50  -39.50  -0.36   1.59
C  1.72   0.34  0.34  -38.81  -38.81 -38.81   1.03   0.34   40.17


0,0,0,0,0,0,9,0,0,0,0,0,0,0,0,
```

Note that the above implementation can be much shortened; however, to be intelligible for interpretation, it is less optimized in terms of code compression. Additional algorithm 12.1 presents a whole set of results: (1) the background variations extracted from sequence z, (2) the sp matrix containing the motif set, (3) the sb matrix containing the sequence variations (t) for the background set, (4) the PFM based on the motif set, (5) the PFM based on the background set, (6) the PWM calculated using the two PFMs and (7) the signal obtained following the use of the new PWM.

12.2.2 Issues and Corrections for Dynamic PWMs

But is there any signal improvement? We can see that the score for the planted motif has increased from a value of 8 to a value of 9 (Additional algorithm 12.2). This should not raise suspicions given that the background is different from the *null model*. However, tests made on a few special z sequences are indicative of the issues that the implementation may encounter. The table below shows that one of the sequences exhibits a disproportionate score (i.e. 37), well above the maximum value of 8 detected in previous experiments for a genuine motif (Table 12.1). This score is the result of a background set that lacks enough sequence variations.

The size of the background set is directly proportional to the length of the z-sequence. A short z-sequence leads to disproportionate results. This is evident at a simple glance above the PWM where some of the *Log Likelihood* values are quite different from the values previously obtained (Figure 12.1c and Additional algorithm 12.2). For instance, a value of +40.17 can be observed in the lower right corner of the PWM (Figure 12.1c). Note that such score values (i.e. 37) exist due to deficiencies present in the reference system (the background model). To

Table 12.1 Tests performed with pseudo-counts and two PFMs (real background).

Sequence z	Planted motif	Signal	Result
AAA<u>AAGTAAAGT</u>AAAAA	AAGTAAAGT	0,0,0,0,0,0,0,0,0,	Ok
AGTC<u>AAGTAAAGT</u>AGTCTC	AAGTAAAGT	0,0,0,2,0,0,0,0,0,0,0,	Ok
ATCAGCGATTACG	None/random	0,0,0,0,37,	False positive
GCGATTACGCATGACGTAA	None/random	0,0,0,0,0,0,0,0,0,0,	Ok
CCCCCCCC<u>CAGGTGAGT</u>CCC	CAGGTGAGT	0,0,0,0,0,0,0,9,0,0,0,	Ok
AAAAAAAGGGGGCCCCCCCCC	None	0,0,0,0,0,0,0,0,0,0,0,0,	Ok
GTGTGTGTGTGTGTGTGT	None	0,0,0,0,0,0,0,0,0,0,	Ok

The table shows the implementation behavior for seven special z sequences. Note: based on the new background, the algorithm shows that there is no significant score for the first sequence from the table.

avoid disproportionate scores on the detection side, the z sequence should have a minimum size that allows enough variations for the background set. To test this assumption, a longer z sequence can be considered. Such a test sequence can be constructed from a random sequence and a planted motif (i.e. "CAGGTGAGT"), as shown below:

z = "ATCAGCGATTATCTGCTACTGC<u>CAGGTGAGT</u>TATGCGTGTCATTATTTATTAGCGC"

The above sequence is inserted into the z variable from Additional algorithm 12.2 and the implementation is run. The signal returned by the implementation shows a detection of the motif in the expected location:

0,0,0,0,0,0,0,0,0,0,0,0,0,0,0,0,0,0,0,0,0,0,9,0,0,0,0,0,0,0,0,0,0,0,0,0,0,0,0,0,0,0, 0,0,0,0,0,0,

The longer the z-sequence, the higher the number of background variations. This simple approach should avoid frequencies with zero counts for the background PFM (the b matrix). But will it be so? Consider the following z-sequence:

AAAAAAAAAAAAAAAACAGGTGAGT

The signal returned by the implementation shows a detection of the motif in the expected location:

0,0,0,0,0,0,0,0,0,0,0,0,0,0,0,131,

Additional algorithm 12.1 returns an unusually high score value (i.e. 131). Based on this observation we can also conclude that: the lower the information content (high redundancies/repetitions) the more pseudo-counts start to appear in the elements of the background PFM. Thus, we observe that the appearance of frequencies with zero counts is dictated by the content of the z variable. The PFMs and the PWM of this experiment are shown below:

```
Output:
Motif set - matrix p (motif PFM):
G   1    2     5     10      1e-17    5      1       8       1
A   3    4     2     1e-17   1e-17    5      7       1       1e-17
T   2    3     2     1e-17   10       1e-17  1e-17   1       7
C   4    1     1     1e-17   1e-17    1e-17  2       1e-17   2
```

```
Background set - matrix b (background PFM):
G   1e-17   1e-17   1e-17   1       2       2    3    3    4
A   15      14      14      13      12      11   10   10   9
T   1e-17   1e-17   1e-17   1e-17   1e-17   1    1    1    1
C   1e-17   1       1       1       1       1    1    1    1
```

```
The PWM (matrix p):
G   39.55   40.24   41.16   2.71     -39.43   1.32     -0.69    1.39     -0.98
A   -1.20   -0.85   -1.54   -41.30   -41.22   -0.38    0.05     -1.90    -40.94
T   40.24   40.65   40.24   0.41     41.85    -38.74   -38.74   0.41     2.35
C   40.94   0.41    0.41    -38.74   -38.74   -38.74   1.10     -38.74   1.10
```

```
Signal:
0,0,0,0,0,0,0,0,0,0,0,0,0,0,0,131,
```

In the above results, a series of exaggerated log likelihood values in the PWM can be observed (i.e. 40.24). A natural question is: Why do aberrant log-likelihood positive values occur? Suppose the existence of two values, each from a different PPM, which are responsible for the occurrence of a new value in one of the PWM elements. A 0.25 value for the motif model and the pseudo-count 0.00000000000000001 for the background model. Consider the equation below for one of the elements of a PWM:

$$p_{i,j} = ln\left(\frac{p_{i,j}/rows(sp)}{b_{i,j}/rows(sb)}\right)$$

$$p_{i,j} = ln\left(\frac{0.25}{0.00000000000000001}\right)$$

$$p_{i,j} = ln(25000000000000000)$$

$$p_{i,j} = 37.75$$

Thus, zero counts in the background PFM lead to the propagation of the pseudo-count to the stage where the PWM is computed. When the score is calculated, such high log-likelihood values cannot be counteracted by log-likelihood negative values found on the other positions of the PWM. Therefore, exaggerated log-likelihood values in the PWM can lead to false-positive scores. For a direct reference to the scanner, the explanations can be put into perspective as follows: A short z-sequence leads to a poor background set, which is unable to provide at least one count for each element in the background PFM. Therefore, a short z-sequence increases the number of pseudo-count values over the elements of the background PFM (matrix b). The pseudo-count values cause an issue when present in the background PFM. In the background PPM, the pseudo-count from the background PFM is divided by the number of sequences found in the background set (matrix sb). This division results in a value even closer to zero than the initial value of the pseudo-count. Then, the result further divides a corresponding value from the PPM of the motif set (matrix p). The logarithm of this new value may show an exaggerated *log-likelihood* value in the PWM, as seen in the results of Additional algorithm 12.1. Thus, the solution would be the initialization of the background PFM elements with an integer like 1 instead of zero or the pseudo-count value.

12.2.3 Solutions for Aberrant Positive Likelihood Values

To avoid these situations, overly positive log-likelihood values from the PWM can be prevented either at the initialization of the PFM or directly by artificial additions to the background set. It is worth mentioning again that these exaggerated positive values are not necessarily wrong; however, they can lead to false-positive score values (Table 12.1).

Virtual Additions to the Background Set

It takes a minimum of four sequences to ensure one occurrence for each symbol on each position from the background PFM (Table 9.4). Consequently, the first approach assumes the addition of four sequences to the background set (Figure 12.2a). Here, each of the elements from the background PFM (matrix b) has one count by default from the initialization stage (b[h][i]=1). In other words, the method provides a uniform distribution over the background PFM in an artificial manner by initializing the background PFM with values of 1 (Figure 12.2b). This leads to a division of the frequencies in the background PFM to the number of rows in the sb matrix, plus the four sequences considered (sb.length+4). Note: these four sequences are not present in the background set. However, the background set is treated as if these variations exist. The difference between Additional algorithm 12.1 and a new implementation would consist in the following changes:

(a) In the *simultaneous initiation* of the two PFMs, the line:

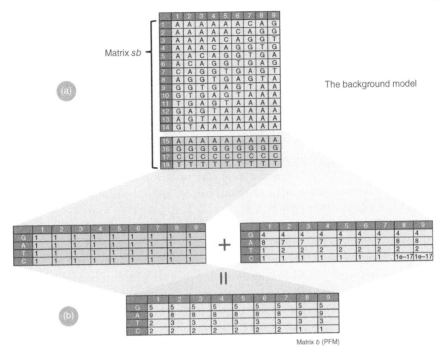

Figure 12.2 Merging a null model into the existing background PFM. (a) Shows the *sb* matrix composed of sequences taken from *z* and the sequences that generate the *null model*. (b) The counts generated by the four sequences of the *null model* are added to the existing background PFM. The addition of the *null model* increases the counts evenly in the new PFM. Notice that this action preserves the proportionality between the original frequencies.

```
"b[h] [i]=0.0000000000000001"
```

is changed to:

```
"b[h] [i]=1".
```

(b) In the *position probability matrix & position weight matrix*, the line:

```
"var gb = b[i] [j+1]/sb.length;"
```

is changed to:

```
"var gb = b[i] [j+1]/(sb.length + 4);".
```

These simple changes allow the application to avoid zero counts in the elements of the background PFM (matrix *b*). In this manner, a new implementation can avoid the aberrant log-likelihood positive values from the PWM, and consequently the false-positive scores. Interestingly, just as the *null model* is added to the background PFM, so can any other model be added in the same manner. Thus, it is possible to make combinations between several existing models according to the same principle described above (Figure 12.2).

Real Additions to the Background Set

There is a second possibility whose approach is much more natural, and which will be used in the next implementation. The sequences for the background set are extracted from z by using the sliding window method. Each sequence variation from z is stored in variable t, according to the same format observed in variable c. For our purposes, subsequent additions can be made to variable t. Four additional sequence variations can be added to the background set to provide at least one count for each of the elements from the background PFM (Figure 12.3). Thus, the method provides a uniform distribution over the background PFM in a natural manner by counting the actual symbols present in the background set. These additional variations can take many forms and they are widely presented in Table 9.4. The simplest version of these four sequences is presented below (Table 12.2).

The table below shows a set of four sequences that ensure equal frequencies for each symbol. Notice that each position contains an "A" character, a "C" character, a "G" character, and a "T" character. Thus, their probability of occurrence on each column will be $1/4$. These four sequences are added to the background set, namely at the end of the t variable, as follows:

```
"t += ',AAAAAAAAA,TTTTTTTTT,GGGGGGGGG,CCCCCCCCC';
```

In this manner, the four sequences are added each time the implementation runs (Figure 12.3b). The introduction of the four sequences in the background set will allow the background PFM to avoid the pseudo-count values, and last but not least to avoid the aberrant log likelihood positive values from the PWM, responsible for signals with false-positive scores (Figure 12.3c).

Figure 12.3 points out the additions brought to this approach. The only change takes place in the background set where the t variable is supplemented with four new sequences designed to avoid the aberrant log-likelihood positive values from the PWM. The number of sequences in the new configuration of the background set (sb.length) will always be equal to the number of symbols in z (z.length) minus the number of positions in the sliding window (sb[0].length), plus four (the four sequences from the t variable). The addition of the four sequences to the background set obviously changed the old values in the PWM just slightly (Figures 12.1c and 12.3c). However, radical changes were made only to the aberrant values of the PWM. This last action clearly improved the impact on the final detection (Additional algorithm 12.3 and Table 12.3).

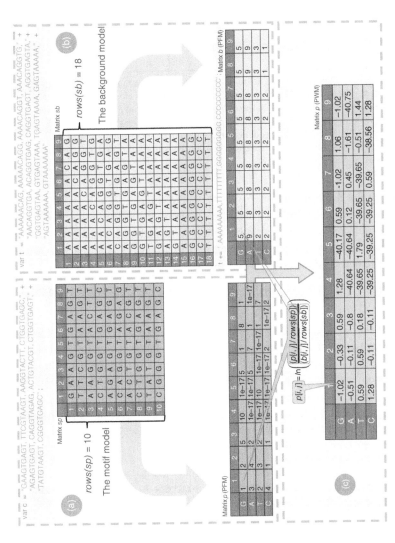

Figure 12.3 Dynamic PWMs with real additions to the background set. The example here refers to a z-sequence presented earlier, namely "AAAAACAGGTGAGTAAAAAAA." (a) Shows the motif set and the transformation of this set into a PFM stored in matrix p, (b) shows the background set and the transformation of this set into a PFM stored in matrix b. Notice the constant additions of the four sequences (t) to the background set and the transformation of this set into a PFM stored in matrix b. (c) It shows the construction of the PWM with the help of the two PFMs.

Table 12.2 Additions to the background set.

Neutral additions to a dynamic background set
AAAAAAAAA
TTTTTTTTT
CCCCCCCCC
GGGGGGGGG

The table shows four sequences that have the role of providing a count for each element in the background PFM. The set of sequences presented in the table was previously discussed in the explanations related to the *null model*. In the *null model*, such a set had the role of ensuring an equal probability of occurrence of all the symbols in the alphabet of the sequence, on each position of a PPM. Note that the four sequences allow for a uniform distribution over the background PFM. Thus, adding the four sequences to the background set does not change the overall properties of the background set.

Verification of the Two Methods

The approaches discussed in the *Virtual additions to the background set* and the *Real additions to the background set* subchapters are tested here. These approaches represent variations for the same solution and their results are identical. Both methods were tested on the *z*-sequence from below:

z="ATCAGCGATTATCTGCTACTGC<u>CAGGTGAGT</u>TATGCGTGTCATTATT TATTAGCGC"

Note that the above *z*-sequence has been used before and is constructed from a random sequence and a motif sequence planted in the middle. To avoid a redundant action, the new implementation will not be shown here. Repeat the experiment by using the same implementation shown in Additional algorithm 12.1. Insert the contents of the above *z*-sequence into the *z* variable from Additional algorithm 12.1. In the "*make background set t*" section of Additional algorithm 12.1, insert the code line from the second method (*real additions to the background set*) as shown below:

Additional algorithm 12.2 Note that the source code is out of context and is intended for explanation of the method.

```
// MAKE BACKGROUND SET t

for(var i=0; i<o; i++) {
    t += z.substr(i, mp[0].length);
```

```
    if(i<o-1){t += ','}
}

t += ',AAAAAAAAA,TTTTTTTTT,GGGGGGGGG,CCCCCCCCC';

document.write('t="' + t + '"');
```

Next, run the algorithm. The z-sequence from above and the new change made to Additional algorithm 12.1 should yield the following results:

```
Output:
Motif set - matrix p (motif PFM):
  G   1    2    5    10       1e-17    5        1        8        1
  A   3    4    2    1e-17    1e-17    5        7        1        1e-17
  T   2    3    2    1e-17    10       1e-17    1e-17    1        7
  C   4    1    1    1e-17    1e-17    1e-17    2        1e-17    2

Background set - matrix b (background PFM):
  G   12   12   12   12   12   11   12   11   12
  A   11   10   11   11   10   11   11   11   10
  T   18   19   18   19   20   20   20   20   20
  C   10   10   10   9    9    9    8    9    9

The PWM (matrix p):
  G   -0.86    -0.16    0.75     1.45     -40.00    0.84     -0.86     1.31      -0.86
  A   0.33     0.71     -0.08    -39.91   -39.82    0.84     1.18      -0.77     -39.82
  T   -0.57    -0.22    -0.57    -40.46   0.94      -40.51   -40.51    -1.37     0.58
  C   0.71     -0.67    -0.67    -39.71   -39.71    -39.71   0.24      -39.71    0.13

Signal:
0,0,0,0,0,0,0,0,0,0,0,0,0,0,0,0,0,0,0,0,0,8,0,0,0,0,0,0,0,0,0,0,0,0,0,0,0,
0,0,0,0,0,0,0,0,0,0,
```

Note that in the output from above the elements in the background PFM (matrix b) contain at least one count and the aberrant positive values in the PWM have been avoided. In conclusion, the lack of counts in any of the elements from the background PFM leads to exaggerated log-likelihood positive values in the PWM. This impediment was solved by uniform additions in the background set that allowed the avoidance of zero counts in the background PFM.

12.3   A Scanner with Two PFMs

A short implementation that includes all that has been discussed so far is shown below (Additional algorithm 12.3). Unlike the previous implementation,

Additional algorithm 12.3 is slightly more optimal and contains the second solution (additions to the *t* variable) for eliminating aberrant log-likelihood positive values from the PWM. Note that the signal extracted from the *z*-sequence is the only output of this implementation, as the content of the matrices is no longer important at this stage.

Additional algorithm 12.3 Note that the source code is in context and works with copy/paste.

```
<script>

// A SCANNER WITH TWO PFMs - SHORT VERSION

var c = "GAAGTGAGT,TTCGTAAGT,AAGGTACTT,CTGGTGAGC,AGAGTGAGT," +
        "CAGGTAGAG,ACTGTACGT,CTGGTGAGT,TATGTAAGT,CGGGTGAGC";

var z = "ATCAGCGATTATCTGCTACTGCCAGGTGAGTTATGCGTGTCATTATT-
TATTAGCGC";

var sp=[], mp=[];
var sb=[], mb=[];

mp = c.split(',');
var np = mp.length;

var t = ";
var o = z.length-mp[0].length;

// MAKE BACKGROUND SET t
for(var i=0; i<o; i++) {
    t += z.substr(i, mp[0].length);
    if(i<o-1){t += ','}
}

t += ',AAAAAAAAA,TTTTTTTTT,GGGGGGGGG,CCCCCCCCC';

mb = t.split(',');
var nb = mb.length;

// THE ALIGNMENT MATRIX FOR p
for(var i=0; i<np; i++){
    sp[i] = [];
    sp[i] = mp[i].split(");
}

// THE ALIGNMENT MATRIX FOR b
for(var i=0; i<nb; i++){
```

```
    sb[i] = [];
    sb[i] = mb[i].split("");
}

// ALPHABET DETECTION
var a = [];
var t = c.replace(/,/g, "").split("");
var k = t.length;

for(var i=0; i<=k; i++){
    var q = 1;
    for(var j=0; j<=a.length; j++){
        if (t[i] === a[j]) {q = 0;}
    }
    if (q === 1) {a.push(t[i]);}
}

// INITIALIZATION
var p = [], b = [];
for(var h=0; h<a.length; h++){
    p[h]=[];
    b[h]=[];
    for(var i=0; i<=sp[0].length; i++) {
        p[h][i]=b[h][i]=0.00000000000000001;
        p[h][0]=b[h][0]=a[h];
    }
}

// THE MOTIF PFM p
for(var i=0; i<sp.length; i++) {
    for(var j=0; j<sp[i].length; j++){
        for(var h=0; h<a.length; h++){
            if (sp[i][j] === a[h]) {p[h][j+1]++;}
        }
    }
}

// THE BACKGROUND PFM b
for(var i=0; i<sb.length; i++) {
    for(var j=0; j<sb[i].length; j++){
        for(var h=0; h<a.length; h++){
            if (sb[i][j] === a[h]) {b[h][j+1]++;}
        }
    }
}

// PPM & PWM
for(var i=0; i<p.length; i++) {
    for(var j=0; j<p[i].length-1; j++){
```

(Continued)

Additional algorithm 12.3 (Continued)

```
            var gp = p[i][j+1]/sp.length;
            var gb = b[i][j+1]/sb.length;
            p[i][j+1]=Math.log(gp/gb).toFixed(2);
        }
}

// THE SCANNER
var r = ", w = ";
var u = z.length - sp.length + 2;

for(var l=0; l<u; l++) {

    w = z.substr(l, sp.length-1);
    var d = w.split(");
    var score = 0;

    for(var f=0; f<d.length; f++){
        for(var h=0; h<a.length; h++){
            if(d[f]==p[h][0]) {
            score += Number(p[h][f+1]);
            }
        }
    }

    if(score<0){score=0;}
    r += score.toFixed(0) + ',';
}

document.write(r);

</script>

Output:
0,0,0,0,0,0,0,0,0,0,0,0,0,0,0,0,0,0,0,0,0,0,0,8,0,0,0,0,0,0,0,0,
0,0,0,0,0,0,0,0,0,0,0,0,0,0,0,0,0,
```

Additional algorithm 12.3 represents the final version of a scanner with two PFMs. Tests performed in Table 12.1 are repeated here to verify the optimal operation of the algorithm. The table below shows perfect detection rates for each of the seven z-sequences taken into consideration (Table 12.3).

The behavior of Additional algorithm 12.3 can be further tested on arbitrary or specially designed z-sequences. A scanner with two PFMs can be slightly confusing, but the clarity appears only through additional experimentation. In addition to the algorithms discussed here, other more advanced implementations and a series

Table 12.3 Tests performed on a scanner with two PFMs.

Sequence z	Planted motif	Signal	Result
AAAA<u>AAGTAAAGT</u>AAAAA	AAGTAAAGT	0,0,0,0,0,0,0,0,0,	Ok
AGTC<u>AAGTAAAGT</u>AGTCTC	AAGTAAAGT	0,0,0,2,0,0,0,0,0,0,	Ok
ATCAGCGATTACG	None/random	0,0,0,0,0,	Ok
GCGATTACGCATGACGTAA	None/random	0,0,0,0,0,0,0,0,0,0,0,	Ok
CCCCCCCC<u>CAGGTGAGT</u>CCC	CAGGTGAGT	0,0,0,0,0,0,0,0,9,0,0,0,	Ok
AAAAAAAGGGGGCCCCCCCCC	None	0,0,0,0,0,0,0,0,0,0,0,0,0,	Ok
GTGTGTGTGTGTGTGTGT	None	0,0,0,0,0,0,0,0,0,0,	Ok

The table shows the behavior of Additional algorithm 12.3 for seven special z-sequences. The results show a perfect discrimination in which only the targets of interest are identified, namely the functional motifs.

of additional experiments can be found online. The online additional experiments are designed to clarify the theoretical principles behind each implementation.

12.4 Information and Background Frequencies on Score Values

The frequency of each symbol in z plays an important role on detection. An increase in the frequency of some symbols over the others is somewhat inversely proportional to the information content of z (Table 12.4 and 12.5). An increase in frequency for a certain symbol can only be achieved by decreasing the entropy inside the z-sequence. To avoid confusion, note that low entropy means low information content and high entropy means high information content. For an intuitive description on the meaning of "information content," consider the following: Information decreases when repetitions (redundancy) inside a sequence increase. Lack of redundancy leads to an increase of the information content (please read the chapter self-sequence alignment). As the information content increases, the frequencies of the symbols become more and more uniform and tend toward equality (Table 12.5). In other words, as the information content of z increases, the background model based on z begins to look more and more like the *null model*. In contrast, as the information content decreases, the symbol frequencies begin to deviate from the uniform distribution of the *null model* (Table 12.4). The background effects on detection are shown more clearly in Table 12.4. For instance, the score value for the same planted motif increases if the frequency of a symbol is gradually strengthened in the background set

Table 12.4 Detection and contrast.

z-Sequence	Signal
AAAAAAAAAAAAAAAACAGGTGAGT	0,0,0,0,0,0,0,0,0,0,0,0,0,0,13,
AAAAAAAAAAAAAAAAACAGGTGAGT	0,0,0,0,0,0,0,0,0,0,0,0,0,0,0,14,
AAAAAAAAAAAAAAAAAACAGGTGAGT	0,0,0,0,0,0,0,0,0,0,0,0,0,0,0,0,14,
AAAAAAAAAAAAAAAAAAACAGGTGAGT	0,0,0,0,0,0,0,0,0,0,0,0,0,0,0,0,0,14,
AAAAAAAAAAAAAAAAAAAACAGGTGAGT	0,0,0,0,0,0,0,0,0,0,0,0,0,0,0,0,0,0,15,
AAAAAAAAAAAAAAAAAAAAACAGGTGAGT	0,0,0,0,0,0,0,0,0,0,0,0,0,0,0,0,0,0,0,15,

Six z-sequences are shown, each with an extra "A" character to increase its frequency in the background PFM. The increase in the frequency of the "A" character leads to an increase in the contrast between the motif set and the background set. Thus, each increase in the frequency of the "A" character leads to an increase of the score value for the planted motif.

(Table 12.4). On the other hand, z-sequences with high information content (few repetitions) tend to equalize the frequencies in the background model (Table 12.5). Thus, the score value for a planted motif tends to decrease because the contrast between the motif set and the background set also decreases. In other words, the information content of the motif begins to be closer and closer in value to the information content shown by other areas from the z-sequence. Note the difference between the z-sequences on the first and last rows of Table 12.5. On the first row, the planted motif is easily distinguishable, while on the last row, it is already difficult to visually identify the same planted motif.

Interestingly, if a z-sequence contains a high information content the frequencies of the symbols can approach a probability of occurrence of 0.25 on any of the positions of the background PPM by pure chance. Thus, a high information content in z generates a background PPM similar to the PPM of the *null model* (Table 9.5). To test the results discussed above, please use Additional algorithm 12.3.

12.5 Dynamic Background vs. Null Model

A dynamic background enables a higher sensitivity for a complete range of possible z-sequences. Although particularly useful, the *null model* is preferred for both simplicity and universality; however, it is not necessarily the ideal answer for complex and highly sensitive analyzes. In previous cases, the *null model* was considered as a fixed point of reference. Prior to the analysis, the contrast was adjusted only by increasing or decreasing the counts of different symbols on each

Table 12.5 Information content and score values.

z-Sequence	A	T	C	G	Signal
AAAAAAAAAAAAACAGGTGAGT	14	2	1	4	0,0,0,0,0,0,0,0,0,0,0,0,12,
ATATATATATATCAGGTGAGT	8	8	1	4	0,0,0,0,0,0,0,0,0,0,0,0,11,
ATCATCATCATCCAGGTGAGT	6	6	5	4	0,0,0,0,0,0,0,0,0,0,0,0,11,
ATCGATCGATCGCAGGTGAGT	5	5	4	7	0,0,0,0,0,0,0,0,0,0,0,0,9,

The table shows four z-sequences whose information content is artificially increased to gain an intuition about the values shown by different score values. The z-sequence with the lowest information content is shown on the first line of the table. The examples continue progressively until the sequence on the last line shows the highest information content. Note that a decrease of the information content in z leads to an increase in the contrast between the motif set and the background set. A higher contrast further leads to higher score values for the same target motif. In parallel, note the relationship between the frequencies of the symbols for each z-sequence and the score values.

position from the motif set (matrix s). Here, the contrast is self-adjusted by frequencies found in the z-sequence itself and the motif set remains in this case a fixed point of reference. Considering the above reasoning, we can naturally ask ourselves: which is the real background in this case? Is it the motif set or the background set? The background model based on sliding windows over the z-sequence provides important information related to local frequencies. On the other hand, the motif variations from the motif set were taken from multiple alignments made between different human genes. However, the background frequencies of each gene may be particularly different from case to case. Each cell type contains a specific chromatin configuration. The predisposition for point mutations follows a distribution molded by euchromatin and heterochromatin areas in the cell nucleus [31]. During evolution, point mutations that occur inside a motif also occur in the proximity of a motif [31]. Proximity is a subjective appreciation that can refer to areas immediately adjacent to the motif or to areas that incorporate the entire gene or even much larger DNA areas [31]. The idea here is that the motif variations from the set come from gene regions with different background frequencies. The local background frequency above a region of a gene can also be reflected on the motifs found there. Nevertheless, these types of considerations can bear important nuances but are often disregarded to avoid a level of complexity that goes beyond desirable. Moreover, too much complexity can lead to an unexpected psychological outcome, namely the researcher becomes doubtful of the end result. For this reason, in bioinformatics the methods are often refined, and experimentation is the decisive factor in adopting different methods for specific purposes.

12.6 Conclusions

A motif is the interface between DNA or RNA and the biochemical processes of the cell. The detection of these interaction points is of crucial importance for understanding the fundamental processes that underlie life. Here, a series of detection approaches were presented in detail through theory, implementation, and experimentation. The background model (the expectation) is the reference system according to which the motif detection is made. The importance of a dynamic background that leads to the adaptation of the algorithm to the environment was discussed in detail. The principle was applied in a new format, namely a scanner with two PFMs. A specific issue related to background PFMs and aberrant positive log-likelihood values was met with two solutions. Both solutions were based on melting a *null model* inside a dynamic background (the background PFM). Experiments based on the new method presented a series of sharp results compared to previous implementations that used the *null model* as a background. Toward the end of the chapter, the differences between a dynamic background and a *null model* were discussed, as well as other considerations with less impact on the final results. One of these considerations was the connection between information and background frequencies, which showed an impact on the score values.

13

Markov Chains: The Machine (I)

13.1 Introduction

Up to this point, the detection of patterns was performed on the basis of a set of motif sequences of a relatively short and constant length. Each position in the pattern was considered independent, i.e. a probability of a letter appearing on a certain position was not dependent on the adjacent positions. Thus, PSSMs can be used successfully for relatively short sequences. But for variable-length sequences, a new method can be applied that is based on the transition probabilities between the letters of an alphabet. Transition probabilities manage to capture the essence of a sequence of symbols. The "essence" is represented by the behavior of the system that produced the sequence of symbols/observations. The meaning of the term "system" is also explained in detail. The first part of the chapter presents a series of discussions about the methodology involved in the computation of transition probabilities and the transition matrices by which these numerical values are represented. Next, a series of implementations describe an algorithm called the *discrete probability detector* (DPD). At this stage, the chapter describes the main steps by which the DPD algorithm is able to transform any sequence of symbols (letters representative of DNA in this case) into a first-order transition matrix. Toward the end of the chapter, the transition matrices produced by the DPD algorithm are used by a Markov chains generator (MCG) for predictions.

13.2 Transition Matrices

Without a doubt, Markov chain is the most important mathematical approach ever developed and the implications of the method extend in the context of all fields, especially in informatics, physics, and philosophy. It was first developed and used by Andrei Markov to debate the lack of free will [249]. Markov chain in

Algorithms in Bioinformatics: Theory and Implementation, First Edition. Paul A. Gagniuc.
© 2021 John Wiley & Sons, Inc. Published 2021 by John Wiley & Sons, Inc.
Companion website: www.wiley.com/go/gagniuc/algorithmsinbioinformatics

itself can be considered a "kind of" primitive yet very powerful neural network. A stochastic matrix (also called a transition matrix) is used to describe the transitions of a Markov chain. That is, a *transition matrix* is par excellence *a memory unit*. With transition matrices, predictions can be made in a Markov chain over several steps, or, these memory units can be compared by a ratio in order to differentiate between two models. Note that transition matrices are also known under a number of alternative names, such as stochastic matrices, Markov matrices, or probability matrices. Transition matrices are square matrices that can take three different structures (Figure 13.1a). The first structure and the one used throughout this chapter is known as the right transition matrix. The sum of the values on the rows of this type of matrix adds up to 1. A second type of matrix is the left transition matrix in which the sum of the values on the columns adds up to 1. A third type of matrix is the double transition matrix in which the values on the columns add up to 1 and the values on the rows also add up to 1. The last type of transition matrix is used in special conditions not described here. Nevertheless, the right stochastic

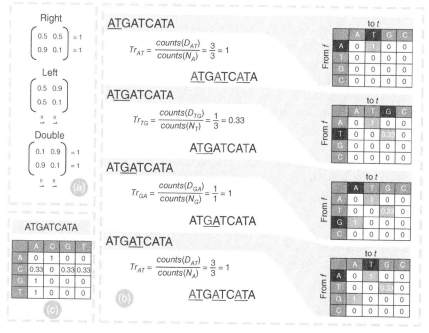

Figure 13.1 Computation of transition matrices. (a) Shows the types of transition matrices, namely the right stochastic matrix, the left stochastic matrix, and the double stochastic matrix. Note that a transition matrix and a stochastic matrix are two names that represent the same thing. (c) Shows just the first four steps in computing a right transition matrix from a sequence "ATGATCATA.". (b) Shows the final transition matrix for sequence "ATGATCATA."

matrices are used in the present examples (the values on the rows add up to 1). The Markov chains method uses a transition matrix to predict the transition from a current state to a future state. However, before this method can be used, a transition matrix must be constructed. The construction of the transition matrix is a training process, which uses a sequence of past observations to capture the behavior of the system that generated the respective sequence. A sequence of past observations can be represented by a succession of different symbols in a string/text, each symbol with a chosen significance. For instance, a sentence written in English can be considered such a sequence, where each type of letter represents a distinct state and every two adjacent letters represent a transition. The system or machine that produced the sentence is the writer (the human brain). To extend this intuition, if the sequence of past observations was produced by a weather-measuring device, the machine is the planetary atmosphere. In the case of DNA, RNA, or protein sequences, the machine is represented by the evolutionary process. Back to the main point, this training process consists of three steps. In the first step, the transition frequencies (counts) are extracted from a sequence of observations (i.e. DNA sequences). In the second step, the frequencies of the states (i.e. the four letters of the DNA alphabet) are determined from the same sequence. In a third step, the transition frequencies (counts) are divided by the frequencies of the states (also counts). The result of this division indicates the value of the transition probability.

$$Tr_{ft} = \frac{counts(D_{ft})}{counts(N_f)}$$

where Tr_{ft} represents the transition probability from one state f to another state t. The meaning of the subscript notation is: f, which means "from" and t, which means "to" (Figure 13.1b). In a sequence of states, D_{ft} represents the frequency of transitions between states f and t, and N_f represents the frequency of state f in the same sequence. Notations "Tr," "D," and "N" also have their meanings, where "Tr" refers to the word transition, "D" indicates the frequency of dinucleotides or two-letter combinations, and "N" refers to nucleotides or the frequency of independent letters. However, a simple example is probably the most eloquent in this case (Figure 13.1b,c). Consider the following sequence:

"ATGATCATA"

A transition probability is calculated by dividing the frequency of transitions by the number of occurrences of the letter that represents the current state. In the above sequence, the first transition is made from state "A" (f) to state "T" (t). Therefore, the transition probability from the current state "A" to the next state "T" in this sequence must be calculated ("ATGATCATA"). In the above sequence, there are three "AT" transitions ("**AT**GA**TC**A**T**A"; $D_{ft} = 3$) and two "A" states ($N_f = 3$). Notice that the last state in the sequence is not counted and the transition *ft* is underlined and the *f* state counts are thickened.

$$Tr_{AT} = \frac{counts(D_{AT})}{counts(N_A)} = \frac{3}{3} = 1$$

Therefore, the probability of a transition from state "A" to state "T" shows a value of 1, which indicates that once the current state is "A," the next state will certainly be "T". Next, a transition between "T" (*f*) and "G" (*t*) is observed (i.e. "A**T**GATCATA"). The transition probability from the current state "T" to the next state "G" must be calculated. In the above sequence, there is only one "TG" pair ("A**TG**ATCATA"; $D_{ft}=1$) and three "T" states ($N_f = 3$). Thus, the substitutions in the equation indicate:

$$Tr_{TG} = \frac{counts(D_{TG})}{counts(N_T)} = \frac{1}{3} = 0.33$$

Therefore, the probability of a transition from state "T" to state "G" shows a value of 0.33, which indicates that once the current state is "T," there is a 0.33 probability that the next state is "G." Next, a transition between the current state "G" (*f*) and the following state "A" (*t*) is observed (i.e. "ATGATCATA"). In the above sequence, there is only one "GA" transition ("AT**GA**TCATA"; $D_{ft} = 1$) and one "G" state over the entire sequence ($N_f = 1$). Thus, the substitutions in the equation indicate:

$$Tr_{GA} = \frac{counts(D_{GA})}{counts(N_G)} = \frac{1}{1} = 1$$

Therefore, the probability of a transition from state "G" to state "A" shows a value of 1, which indicates that once the current state is "G," the next state will certainly be "A." Next, a transition between state "A" (*f*) and state "T" (*t*) is observed again (i.e. "ATG**AT**CATA"). For exemplification, this will be recalculated:

$$Tr_{AT} = \frac{counts(D_{AT})}{counts(N_A)} = \frac{3}{3} = 1$$

Of course, these calculations must be conducted consecutively for all the transitions in the sequence. Next, a transition between state "T" (*f*) and state "C" (*t*) is observed (i.e. "ATGA**TC**ATA"):

$$Tr_{TC} = \frac{counts(D_{TC})}{counts(N_T)} = \frac{1}{3} = 0.33$$

Next, a transition between "C" (*f*) and "A" (*t*) is observed (i.e. "ATGAT**CA**TA"):

$$Tr_{CA} = \frac{counts(D_{CA})}{counts(N_C)} = \frac{1}{1} = 1$$

Next, a transition between "A" (*f*) and "T" (*t*) is observed again (i.e. "ATGATC**AT**A"):

$$Tr_{AT} = \frac{counts(D_{AT})}{counts(N_A)} = \frac{3}{3} = 1$$

The final transition is made between state "T" and state "A" (i.e. "ATGATCA TA"):

$$Tr_{TA} = \frac{counts(D_{TA})}{counts(N_T)} = \frac{1}{3} = 0.33$$

But how are these transition probabilities organized on a transition matrix? Again, f signifies the row of the matrix, which means "from" and t is the column of the matrix which means "to." Thus, each transition probability value is positioned at the intersection of row f with column t (Figure 13.1b). In other words, in a right stochastic matrix, the probability of transition from one state to another state is represented from rows to columns (Figure 13.1b). The rows of a transition matrix represent the states from which the transition is made (f) and the columns of a transition matrix represent the states to which the transition is made (t). The value from the element at the intersection of a row with a column represents the transition probability from the state that represents the row to the state that represents the column. To formulate a transition matrix, the alphabet of the sequence must be determined in advance. In the case of DNA or RNA sequences, the total number of transitions for a four-symbol alphabet ($n = 4$) is 16 ($4 \times 4 = 16$). For proteins, the alphabet can be made from a total of 20 letters ($n = 20$) and a first-order transition matrix may have a maximum of 400 transitions ($n \times n = 20 \times 20 = 400$) arranged in 20 rows and 20 columns. Of course, the composition of amino acids is different from protein to protein and 20 letters in the alphabet represent an upper-bound limit. For other types of sequences, the alphabet may be unknown and must be detected prior to the computation stage. Thus, a transition matrix contains a number of columns equal to the number of rows ($n \times n$; where n is the number of symbols in the alphabet of the sequence). However, even in the case of DNA/RNA, a short s-sequence may not necessarily contain all 16 types of transitions (as can be seen above). For instance, in the example from above, there are only six types of transitions from a total of 16 possible, namely: "AT," "TG," "GA," "TC," "CA," "TA." Also, notice that the "AT" transition repeats and the insertion of the same transition probability value for this transition into the elements of the matrix is no longer necessary (Figure 13.1b).

Of course, the transitions between two letters/states represent the first-order Markov chains. Second-order Markov chains or higher-order Markov chains can be used to more accurately capture the behavior of a system. These higher-order approaches have the ability to capture the relationships between the succession of several states; however, the level of complexity of the transition matrix increases.

13.3 Discrete Probability Detector

To calculate the transition probabilities for any sequence of symbols, a dynamic method must be built; otherwise, a rigid algorithm may be difficult to manage for other types of sequences besides strings that represent DNA molecules. Such an algorithm exists and it is called the DPD. The DPD algorithm has been described in detail in the additional materials of the book *Markov Chains: From Theory to Implementation and Experimentation*, and is especially valuable when the alphabet of the sequence is unknown [249]. The DPD algorithm uses an *s*-sequence to construct a transition matrix. The *s*-sequence can be a variable-length sequence representative for a known genomic region. The *s*-sequence may represent a promoter region, a gene region, a CpG region, and so on. Here, a short *s*-sequence is considered for illustration, namely:

s = "ATCGATTCGATATCATACACGTAT"

The original DPD algorithm is organized in four phases: (i) alphabet detection, (ii) matrix initialization, (iii) frequency detection, and (iv) calculation of transition probabilities. The final result of this algorithm is represented by a transition matrix, which is built in the matrix initialization phase and then fully completed in the last phase, namely in the calculation of transition probabilities phase.

13.3.1 Alphabet Detection

In a first phase, the alphabet of the sequence is detected and the order of the letters in the transition matrix is also established (Additional algorithm 13.1). The alphabet of the sequence sets the number of states in a Markov chain. Each new letter found in *s* is appended to the string forming in variable *a*. Thus, variable *a* gradually increases its content until all types of letters from *s* are identified. The implementation from Additional algorithm 13.1 includes the alphabet detection phase:

Additional algorithm 13.1 Note that the source code is out of context and is intended for explanation of the method.

```
var s = 'ATCGATTCGATATCATACACGTAT';
var a = [];
var k = s.length;

for(var i=0; i<=k; i++){
    var q = 1;
    for(var j=0; j<=a.length; j++){
        if (s[i] === a[j]) {q = 0;}
    }
    if (q === 1) {a.push(s[i]);}
}
```

```
document.write(SMC(a) + '<br>');

Output:

A
T
C
G
```

Above, the *s* variable contains the string that is being analyzed and *k* represents the length of the *s*-sequence. Variable *q* is a flag variable with an initial value of 1. The value of *q* becomes zero only if a letter *i* in the *s*-sequence corresponds with a letter *j* in the *a* variable. Variable *a* holds the letters representing the states. The content of the variable gradually increases in length as the s-sequence is traversed by the main loop. At each step in the loop, a new letter is added to variable *a*, only if the value of *q* becomes zero. Thus, once the *s*-sequence is traversed by the main loop, the number of letters in the *a* variable is equal to the total number of states.

13.3.2 Matrix Initialization

In the second phase, a number of new variables are declared, which are used in the current stage and throughout the following stages (Additional algorithm 13.2). Variable *d* indicates the total number of states and it represents the length of variable *a*. Variable *m* is the main probability matrix, which the function produces. The elements of a matrix *m* are filled with zero values for later use.

Additional algorithm 13.2 Note that the source code is out of context and is intended for explanation of the method.

```
var d = a.length-1;
var m = [];
var e = [];
var l = [];

for(var i=0; i<=d; i++){
    m[i]=[];
```

(Continued)

Additional algorithm 13.2 (Continued)

```
      e[i]=[];
      for(var j=0; j<=d; j++){
          m[i][j]=0;

          if (j === 0) {
              e[i][0]=a[i];
              e[i][1]=0;
          }
      }
}

document.write(SMC(e) + '<br>');
document.write(SMC(m) + '<br>');

Output:

    A     0
    T     0
    C     0
    G     0

    0     0     0     0
    0     0     0     0
    0     0     0     0
    0     0     0     0
```

Moreover, e is a matrix with two columns, namely column 0 and 1. Column 0 stores all the letters found in a. Column 1 stores the number of appearances for each type of letter in s. The first column of matrix e is filled with letters found in variable a, and the second column of matrix e is filled with zero values for later use. Variable l is a vector with two components. Vector l contains the i and $i+1$ letters from s, and it is used in the next stage.

13.3.3 Frequency Detection

In the third phase, the transitions between the letters of *s* are counted and stored in matrix *m* (Additional algorithm 13.3). The strategy in this particular case is to fill matrix *m* with transition counts before the last letter in *s* is reached. In this case, the first column of matrix *e* already contains the letters from variable *a*. The two components of vector *l* contain the *i* and *i* + 1 letters from *s*. The count of individual transitions between letters is made by a comparison between vector *l* and the elements from the first column of matrix *e*. The number of rows in matrix *m* and matrix *e* is the same, namely *d*. Therefore, an extra loop can be avoided by mapping matrix *m* through a coordinate system.

Additional algorithm 13.3 Note that the source code is out of context and is intended for explanation of the method.

```
l[0]="";
l[1]="";

for(var i=0; i<s.length-1; i++){

    l[0] = s.substr(i, 1);
    l[1] = s.substr(i + 1, 1);

    for(var j=0; j<=d; j++){
        if (l[0] === e[j][0]) {
            e[j][1] = e[j][1] + 1;
            r = j;
        }
        if (l[1] === e[j][0]) {c=j;}
    }
    m[r][c] = m[r][c] + 1;
}

document.write(SMC(e) + '<br>');
document.write(SMC(m) + '<br>');
```

(Continued)

Additional algorithm 13.3 (Continued)			
Output:			

A	8		
T	7		
C	5		
G	3		

0	6	2	0
3	1	3	0
2	0	0	3
2	1	0	0

For instance, if the letter from position i in s stored in $l[0]$ and the letter from j row in matrix e (e[j][0]) are the same, then variable r is equal to j. Likewise, if the letter $i+1$ stored in $l[1]$ and the letter from $e[j][0]$ are the same, then variable c is equal to j. Variable r represents the rows of matrix m, whereas variable c represents the columns of matrix m (m[r][c]). Thus, at each step through s, an element of matrix m is always incremented according to the coordinates received from r and c. This "coordinate" approach greatly increases the processing speed of the algorithm. The number of loops is:

$$o = (k - 1) \times d$$

where o represents the number of loops/steps, d represents the number of states (or letter types), and k is the number of letters in s. When the letter stored in $l[0]$ and the letter from the j row in matrix e are the same, the second column of matrix e is also incremented. The second column of matrix e stores the number of appearances for each type of letter in s. Moreover, to make a connection between the formula presented above in subchapter 13.2 and the approach used here, one can make the correspondence between the implementation variables and the transition probability equation:

$$Tr_{ft} = \frac{counts(D_{ft})}{counts(N_f)}$$

In the present case, $l[0]$ represents the letter in f found in s and $l[1]$ represents the letter in t found in s. In other words, f is represented by letter i found in s and t is represented by letter $i+1$ found in s. Thus, D_{ft} represents the counts of the letter combination made by $l[0]$ and $l[1]$, and N_f represents the counts of the letter found in $l[0]$. This concludes the association between the implementation and the main equation.

13.3.4 Calculation of Transition Probabilities

In the fourth phase, the transition probabilities are finally computed. The counts from matrix m elements are divided by the counts from the second column of matrix e (Additional algorithm 13.4):

$$m_{ij} = \frac{m_{ij}}{e_{i1}}$$

which is equivalent to the standard equation previously used in subchapter 13.2:

$$Tr_{ft} = \frac{counts(D_{ft})}{counts(N_f)}$$

where at this stage m_{ij} represents the D_{ft} counts and e_{i1} represents the N_f counts. The results of these divisions are stored in the same positions in matrix m and represent transition probability values. The implementation below makes these computations by using a nested loop, which transforms matrix m from a count matrix into a probability matrix, as follows:

Additional algorithm 13.4 Note that the source code is out of context and is intended for explanation of the method.

```
for (var i=0; i<=d; i++) {
    for (var j=0; j<=d; j++) {
        if (e[i][1] > 0) {
            m[i][j] = (m[i][j]/e[i][1]).toFixed(2);
        }
    }
}

Output:

    0.00        0.75        0.25        0.00
    0.43        0.14        0.43        0.00
    0.40        0.00        0.00        0.60
    0.67        0.33        0.00        0.00
```

Notice, of course, that the resulting matrix *m* is a right stochastic matrix in which the sum of the transition probabilities on any of the rows is 1. Up to this point, a transition matrix *m* has been obtained from a sequence *s*. However, this matrix is not perfectly independent of the *a* variable. Note that the rows and columns of matrix *m* are not associated with any letters. To locate the value of the transition probability from one letter to another, both matrix *m* and the content of variable *a* are needed. To avoid this inconvenience, a transition matrix can be represented by a more complete version, which contains both the transition probability values and the letters associated with the rows and columns of the matrix (Additional algorithm 13.5).

Additional algorithm 13.5 Note that the source code is out of context and is intended for explanation of the method.

```
var t = [];

for(var i=0; i<=d+1; i++){

    t[i]=[];

    for(var j=0; j<=d+1; j++){

            if (j==0 && i==0) {t[i][j] = '(p)';}
            if (i>0 && j>0) {t[i][j] = m[i-1][j-1];}
            if (i==0 && j>0 && j<=d+1) {t[i][j]=a[j-1];}
            if (j==0 && i>0 && i<=d+1) {t[i][j]=a[i-1];}
    }
}
```

Output:

(p)	A	T	C	G
A	0.00	0.75	0.25	0.00
T	0.43	0.14	0.43	0.00
C	0.40	0.00	0.00	0.60
G	0.67	0.33	0.00	0.00

In this way, the result of the algorithm can be a matrix that contains all the data necessary for a correct use. Matrix *m* contains the transition probabilities and the rows and columns represent the letters found in the variable *a*. To obtain the most useful visual representation, matrix *m* can be used to construct a matrix *t* containing both the numerical values and the representative letters for these transitions (Additional algorithm 13.5). Above, a matrix *t* is constructed with an extra column and row to accommodate the letters indicating the transitions from variable *a*. The transition probability values from matrix *m* are rewritten in matrix *t* as they are, starting from the second column and the second row. As a final conclusion, the implementation below links all the above steps discussed so far:

Additional algorithm 13.6 Note that the source code is in context and works with copy/paste.

```
<script>

// DISCRETE PROBABILITY DETECTOR

var s = 'ATCGATTCGATATCATACACGTAT';

//--------------[ Phase one ]--------------/
var a = [];
var k = s.length;

for(var i=0; i<=k; i++){
    var q = 1;
    for(var j=0; j<=a.length; j++){
        if (s[i] === a[j]) {q = 0;}
    }
    if (q === 1) {a.push(s[i]);}
}

document.write('a(i):<br>' + SMC(a) + '<hr>');

//--------------[ Phase two ]--------------/
var d = a.length-1;
var m = [];
var e = [];
var l = [];

for(var i=0; i<=d; i++){
```

(Continued)

Additional algorithm 13.6 (Continued)

```
    m[i]=[];
    e[i]=[];
    for(var j=0; j<=d; j++){

        m[i][j]=0;

        if (j === 0) {
            e[i][0]=a[i];
            e[i][1]=0;
        }
    }
}

document.write('e(i,j):<br>' + SMC(e) + '<hr>');

//-------------[ Phase three ]-------------/
l[0]="";
l[1]="";

for(var i=0; i<s.length-1; i++){

    l[0] = s.substr(i, 1);
    l[1] = s.substr(i + 1, 1);

    for(var j=0; j<=d; j++){
        if (l[0] === e[j][0]) {
            e[j][1] = e[j][1] + 1;
            r = j;
        }
        if (l[1] === e[j][0]) {c=j;}
    }
    m[r][c] = m[r][c] + 1;
}

document.write('e(i,j):<br>' + SMC(e) + '<hr>');
document.write('m(i,j):<br>' + SMC(m) + '<hr>');

//-------------[ Phase four ]-------------/
for(var i=0; i<=d; i++){
```

```
    for(var j=0; j<=d; j++){
        if (e[i][1] > 0) {
            m[i][j]=(m[i][j]/e[i][1]).toFixed(2);
        }
    }
}

document.write('m(i,j):<br>' + SMC(m) + '<hr>');

//----------------[ END ]------------------/
var t = [];

for(var i=0; i<=d+1; i++){

    t[i]=[];

    for(var j=0; j<=d+1; j++){

        if (j == 0 && i == 0) {t[i][j] = '(p)';}
        if (i>0 && j>0) {t[i][j] = m[i-1][j-1];}
        if (i == 0 && j>0 && j<=d+1) {t[i][j]=a[j-1];}
        if (j == 0 && i>0 && i<=d+1) {t[i][j]=a[i-1];}
    }
}

document.write('t(i,j):<br>' + SMC(t));

// SHOW MATRIX CONTENT
function SMC(m) {
    var r = "<table border=1>";
    for(var i=0; i<m.length; i++) {
        r += "<tr>";
        for(var j=0; j<m[i].length; j++){
            r += "<td>"+m[i][j]+"</td>";
        }
        r += "</tr>";
    }
    r += "</table>";
```

(Continued)

Additional algorithm 13.6 (Continued)

```
      return r;
}

</script>

Output:

a(i):
A
T
C
G

e(i,j):
A      0
T      0
C      0
G      0

e(i,j):
A      8
T      7
C      5
G      3

m(i,j):
0      6         2         0
3      1         3         0
2      0         0         3
2      1         0         0

m(i,j):
0.00      0.75      0.25      0.00
0.43      0.14      0.43      0.00
0.40      0.00      0.00      0.60
0.67      0.33      0.00      0.00
```

```
t(i,j):
(p)        A        T        C        G
A        0.00     0.75     0.25     0.00
T        0.43     0.14     0.43     0.00
C        0.40     0.00     0.00     0.60
G        0.67     0.33     0.00     0.00
```

The output of Additional algorithm 13.6 shows the content of matrix *m* and *e* at each phase. At the end, the output shows a special version of the transition matrix (matrix *t*). Note that the *SMC* function was presented in the previous chapters and only serves to show the content of any matrix. To be called repeatedly for multiple matrices, the DPD algorithm is inserted into a function (Additional algorithm 13.7). This function is simply called "DPD" for convenience.

Additional algorithm 13.7 Note that the source code is in context and works with copy/paste.

```
<script>

// DISCRETE PROBABILITY DETECTOR

var s = 'ATCGATTCGATATCATACACGTAT';

document.write(SMC(DPD(s, '(+)')));

function DPD(s, n){

    //-------------[ Phase one ]-------------/
    var a = [];
    var k = s.length;

    for(var i=0; i<=k; i++){
        var q = 1;
        for(var j=0; j<=a.length; j++){
            if (s[i] === a[j]) {q = 0;}
        }
        if (q === 1) {a.push(s[i]);}
    }
```

(Continued)

Additional algorithm 13.7 (Continued)

```
//--------------[ Phase two ]--------------/
var d = a.length-1;
var m = [];
var e = [];
var l = [];

for(var i=0; i<=d; i++){
    m[i]=[];
    e[i]=[];
    for(var j=0; j<=d; j++){

        m[i][j]=0;

        if (j === 0) {
            e[i][0]=a[i];
            e[i][1]=0;
        }
    }
}

//-------------[ Phase three ]--------------/
l[0]="";
l[1]="";

for(var i=0; i<s.length-1; i++){

    l[0] = s.substr(i, 1);
    l[1] = s.substr(i + 1, 1);

    for(var j=0; j<=d; j++){
        if (l[0] === e[j][0]) {
            e[j][1] = e[j][1] + 1;
            r = j;
        }
        if (l[1] === e[j][0]) {c=j;}
    }
    m[r][c] = m[r][c] + 1;
}

//-------------[ Phase four ]--------------/
for(var i=0; i<=d; i++){
    for(var j=0; j<=d; j++){
```

```
                  if (e[i][1] > 0) {
                        m[i][j]=(m[i][j]/e[i][1]).toFixed(2);
                  }
          }
    }

    //----------------[ END ]------------------/
    var t = [];

    for(var i=0; i<=d+1; i++){

        t[i]=[];

        for(var j=0; j<=d+1; j++){

            if (j == 0 && i == 0) {t[i][j] = n;}
            if (i>0 && j>0) {t[i][j] = m[i-1][j-1];}
            if (i == 0 && j>0 && j<=d+1) {t[i][j]=a[j-1];}
            if (j == 0 && i>0 && i<=d+1) {t[i][j]=a[i-1];}
        }
    }

return t;

}

// SHOW MATRIX CONTENT
function SMC(m) {
    var r = "<table border=1>";
    for(var i=0; i<m.length; i++) {
        r += "<tr>";
        for(var j=0; j<m[i].length; j++){
            r += "<td>"+m[i][j]+"</td>";
        }
        r += "</tr>";
    }
    r += "</table>";

    return r;
}

</script>
```

(Continued)

Additional algorithm 13.7 (Continued)

Output:

```
(+)     A       T       C       G
 A     0.00    0.75    0.25    0.00
 T     0.43    0.14    0.43    0.00
 C     0.40    0.00    0.00    0.60
 G     0.67    0.33    0.00    0.00
```

Also, to differentiate between the returned transition matrices, the *DPD* function allows each matrix *t* to be labeled in the upper left corner. Thus, the *DPD* function receives two inputs, namely the content of the *s* variable and the identifying label (' (+) ') for the matrix. In turn, the function returns the transition matrix (*t*) as can be seen in the output of Additional algorithm 13.7. This concludes the DPD algorithm.

13.3.5 Particularities in Calculating the Transition Probabilities

A DNA or RNA sequence will always show a total of four possible states (four types of letters in the alphabet) and a protein sequence will contain a maximum of 20 states (the 20 letters representative for amino acids). However, note that at this point the sequence may contain any type of symbol, not only the letters representative of DNA. Specially constructed sequences may show various interesting features that may appear in a transition matrix. For instance, consider the following sequence:

$$s = \text{``HAHAAAHQ''}$$

where "H", "A," and "Q" are three states. Note that Additional algorithm 13.7 can be used by simply replacing the sequence in *s* with the new sequences described herein. If a new state at the end of *s* does not occur in the rest of *s*, then matrix *m* will contain a row with all elements on zero (Table 13.1). Since it is at the end of *s*, the new letter does not make a transition to anything. In contrast, consider a second sequence:

$$s = \text{``QHAHAAAH''}$$

where "H", "A," and "Q" are again the same three states. If a state from the beginning of *s* does not occur in the rest of *s*, then matrix *m* will contain a column

Table 13.1 Transition matrices and specially constructed sequences.

s = "HAHAAAHQ"				s = "QHAHAAAH"			
(+)	H	A	Q	(+)	Q	H	A
H	0.00	0.67	0.33	Q	0.00	1.00	0.00
A	0.50	0.50	0.00	H	0.00	0.00	1.00
Q	0.00	0.00	0.00	A	0.00	0.50	0.50

Shows the transition matrices for two specially constructed sequences, namely the sequence "HAHAAAHQ" on the left of the table that induces a line with zero values and the sequence "QHAHAAAH" on the right, which induces a column with elements with zero values. The table also shows the behavior of the alphabet detector in setting the order of the states in the matrix.

with all elements on zero (Table 13.1). Since the first letter it is only seen at the beginning of s, no other letter makes a transition to it.

Moreover, Table 13.1 shows the behavior of the alphabet detector method. In the first sequence (i.e. "HAHAAAHQ"), the "Q" state appears at the end of the sequence and it is positioned on the last row and last column in the transition matrix. On the other hand, state "Q" appears first in the second sequence (i.e. "QHAHAAAH"), and the alphabet detector sets this state on the first column and the first row of the matrix.

13.4 Markov Chains Generators

But how well a transition matrix captures the essence of a sequence ? Moreover, what can be done with a transition matrix once it is built? To answer these questions, a MCG is introduced (Additional algorithm 13.8). A transition matrix is calculated based on a training sequence (Additional algorithm 13.7). A MCG is a prediction machine that uses a transition matrix to generate sequences that are similar to the training sequence. Thus, the output of a MCG mimics the training sequence and the process itself represents a prediction. The MCG can also be used to verify the correct operation of the DPD algorithm. Once the DPD algorithm produces a transition matrix (called here the "original" transition matrix) using a training sequence, that transition matrix can be used by the MCG to predict a similar sequence. In turn, the sequence produced by the MCG can be used by the DPD algorithm to produce a new transition matrix. If the original transition matrix and the transition matrix of the predicted sequence contain close transition probability values, then the DPD algorithm and the MCG machine work as expected.

13.4.1   The Experiment

How can the probabilities in the transition matrix be formulated? Namely, how can a machine provide one letter/symbol after another according to a certain probability value? These are the main questions for implementing an MCG. To answer these questions, an experiment may be imagined (Figure 13.2). Consider four jars labeled the same as the letters of the DNA alphabet, namely jar "A," jar "T," "C," and "G" (Figure 13.2b). Each of these jars should contain a number of types of balls equal to the number of types of jars, namely balls of type "A," "T," "C," and "G." The proportion of these balls in a jar must reflect the proportion

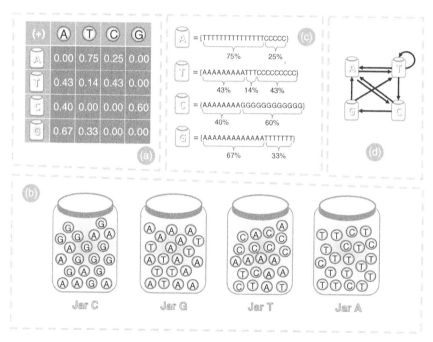

Figure 13.2   A Markov Chain experiment. (a) The probability values present inside a 4×4 transition matrix (P) are directly used for an automatic generation of (b) the letter composition that make up the representation of four jars. Each row in the transition matrix is associated with a state (a jar/state "A," "T," "C," or "G"). (c) The values on each row of a transition matrix are mirrored in a string. The four letter sequences have a calculated proportion of "A," "T," "C," and "G" letters according to the values on the rows of the matrix. Thus, the composition of a string reflects the probability values on a row. The chance of a letter selected at random from one of the four sequences is dictated by the proportions of "A," "T," "C," and "G" letters. (d) A set of rules is imposed on this system. Initially, the random selection of a letter can be done arbitrarily from one of the four strings. The randomly selected letter is noted and added to the output sequence and it also indicates the next string from which the random selection will be made.

indicated on the row of a transition matrix, where the row in the transition matrix signifies the jar and a ball chosen at random from that jar represents the jar from which the next extraction is made (next state). These probabilities can be transformed into proportions. For example, any of the jars can be filled with a constant number of balls, say, 20 balls. But the proportion of ball types out of a total of 20 can be calculated from the values on the row of the matrix. Note that the sum of the probabilities on a row in the transition matrix is 1. Thus, 1 (unity) can represent the total number of balls in a jar, namely 20 balls in the current example. Let us take a concrete example from the transition matrix present in the Additional algorithm 13.7 output. The second row in the matrix represents the "from" state "A." Therefore, this row represents the "A" jar. But what would be the proportion of "A," "T," "C," and "G" balls in this "A" jar? A simple look at the second row of the transition matrix indicates that from state "A" there is a probability of transition of 0.75 to state "T," a probability of 0.25 for a transition to state "C" and a probability of zero for the transition to state "A" or "G." Returning to the "A" jar, it can be intuitively deduced that in this jar there are only two types of balls out of the four possible types, namely "T" type balls and "C" type balls. But what should be the proportion for the two types of balls? The unit (number 1) is represented by the 20 balls and the value of the probabilities can be transformed into a percentage by multiplying the value of the transition probability to 100. It is now clear that of the 20 balls, 75% must be of the "T" type and 25% must be of the "C" type. The problem becomes one related to proportions, namely how many "T" balls represent 75% of a total of 20 balls? Well, the 20 balls represent 100% and 75% is:

$$\text{"T" } balls \, in \, jar \, A = \frac{20 \, balls}{100\%} \times 75\% = 15 \, balls$$

Of course, this calculation is also formulated for the other types of balls:

$$\text{"C" } balls \, in \, jar \, A = \frac{20 \, balls}{100\%} \times 25\% = 5 \, balls$$

$$\text{"A" } balls \, in \, jar \, A = \frac{20 \, balls}{100\%} \times 0\% = 0 \, balls$$

$$\text{"G" } balls \, in \, jar \, A = \frac{20 \, balls}{100\%} \times 0\% = 0 \, balls$$

Note that each result must be rounded because in this experiment the balls are represented by integers. For a more direct calculation of the ball proportions in each jar, a transition probability value can be multiplied by the total number of balls and the same result is obtained:

$$\text{"T" } balls \, in \, jar \, A = 0.75 \times 20 \, balls = 15 \, balls$$

$$\text{"C" } balls \, in \, jar \, A = 0.25 \times 20 \, balls = 5 \, balls$$

"A" *balls in jar* A $= 0 \times 20$ *balls* $= 0$ *balls*

"G" *balls in jar* A $= 0 \times 20$ *balls* $= 0$ *balls*

Thus, the number of balls can be calculated in all four types of jars by using the above method (Figure 13.2a–c). Note that the order in which the calculations are made is not important. Again, the proportion of balls in the "A" jar:

"T" *balls in jar* A $= 0.75 \times 20$ *balls* $= 15$ *balls*

"C" *balls in jar* A $= 0.25 \times 20$ *balls* $= 5$ *balls*

"A" *balls in jar* A $= 0 \times 20$ *balls* $= 0$ *balls*

"G" *balls in jar* A $= 0 \times 20$ *balls* $= 0$ *balls*

The proportion of balls in thé "T" jar:

"T" *balls in jar* T $= 0.14 \times 20$ *balls* $= 2.8 = 3$ *balls*

"C" *balls in jar* T $= 0.43 \times 20$ *balls* $= 8.6 = 9$ *balls*

"A" *balls in jar* T $= 0.43 \times 20$ *balls* $= 8.6 = 9$ *balls*

"G" *balls in jar* T $= 0 \times 20$ *balls* $= 0$ *balls*

The proportion of balls in the "C" jar:

"T" *balls in jar* C $= 0 \times 20$ *balls* $= 0$ *balls*

"C" *balls in jar* C $= 0 \times 20$ *balls* $= 0$ *balls*

"A" *balls in jar* C $= 0.4 \times 20$ *balls* $= 8$ *balls*

"G" *balls in jar* C $= 0.6 \times 20$ *balls* $= 12$ *balls*

The proportion of balls in the "G" jar:

"T" *balls in jar* G $= 0.33 \times 20$ *balls* $= 6.6 = 7$ *balls*

"C" *balls in jar* G $= 0 \times 20$ *balls* $= 0$ *balls*

"A" *balls in jar* G $= 0.67 \times 20$ *balls* $= 13.4 = 13$ *balls*

"G" *balls in jar* G $= 0 \times 20$ *balls* $= 0$ *balls*

It can be concluded that the "A" jar contains 15 balls "T" and 5 balls "C"; the "T" jar contains 3 balls type "T," 9 balls type "C," and 9 balls type "A"; the "C" jar contains 8 balls type "A" and 12 balls type "G"; and the "G" jar contains 7 balls type "T" and 13 balls type "A" (Figure 13.2b). Up to this point, the jars are filled with the required ball types according to the values indicated by the transition matrix (Figure 13.2a–c). But how would such a Markov machine work? Ball draws with replacement are made according to predetermined rules (Figure 13.2d). The types of balls ("A," "C," "T," or "G") selected at random are noted successively as a sequence (Figure 13.3). This sequence represents the output of the machine. Commonly, the first extraction can be made from a random jar. The type of ball is then noted and it represents the first letter in the sequence produced by this machine (Figure 13.3). If the ball type coincides with the current jar type (e.g. an "A" ball is extracted from the "A" jar), then the next extraction is made from the same jar. If the ball type indicates another type of jar, the next draw is made from that jar and the letter indicating the type of ball is added to the output sequence.

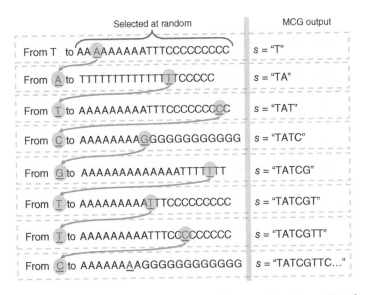

Figure 13.3 The Markov chains generator. The principle of operation of a Markov chains generator (MCG) is shown for the first eight steps. Starting at the top of the figure, the *s*-sequence from the MCG output begins randomly with one of the four letters. In this case the first letter was "T." The next random selection is made from the string associated with state "T" (i.e. Jar[2]) and the randomly selected letter in this case was "A." Thus, the next random selection is made from the string corresponding to state "A" (i.e. Jar[1]) and the randomly selected letter in this case was "T." Thus, the next random selection is made from the string corresponding to state "T" (i.e. Jar[2]). These steps continue until the machine is stopped.

These draws continue and the representative letters for the balls are added to the output at each step. However, whether it is a random extraction of balls from a jar or a random extraction of balls from a box in which the balls are placed linearly, the process is the same. Such a linear arrangement can even be represented by a sequence of letters from which a random selection of a letter can be made (Figure 13.2a–c). Thus, a random selection in a string is equivalent to a random selection of a ball in a jar (Figure 13.2b,c). The above thought experiment can be implemented quite directly with the help of an algorithm that will be discussed further. This concludes the principles behind a MCG and opens up the possibility for practical applications in bioinformatics and in other fields of research.

13.4.2 The Implementation

However, first the MCG implementation is shown and discussed below (Additional algorithm 13.8). **In a first stage** of the implementation, the transition matrix is set and the transition probabilities are declared exactly as these values were calculated by the DPD algorithm in the output of Additional algorithm 13.7. As a general rule, a right transition matrix is composed of horizontal probability vectors arranged one on top of the other. In a probability vector, the values of the components add up to 1. Therefore, the transition matrix from above (Figure 13.2a) is declared in the form of vectors, each represented by a group of statements. Each group represents a state from which the system starts and the elements in the group represent the probabilities of the states to which the transition can be made.

Additional algorithm 13.8 Note that the source code is in context and works with copy/paste.

```
<script>

// MARKOV CHAINS GENERATOR
var P = [];
var Jar = [];

for(var i=0; i<=4; i++){P[i]=[];}
for(var i=1; i<=4; i++){Jar[i]=[];}

P[0][0] = "A";
P[0][1] = "T";
P[0][2] = "C";
P[0][3] = "G";
```

```
// From A
P[1][0] = 0;     // to A
P[1][1] = 0.75;  // to T
P[1][2] = 0.25;  // to C
P[1][3] = 0;     // to G

//From T
P[2][0] = 0.43;  // to A
P[2][1] = 0.14;  // to T
P[2][2] = 0.43;  // to C
P[2][3] = 0;     // to G

//From C
P[3][0] = 0.40;  // to A
P[3][1] = 0;     // to T
P[3][2] = 0;     // to C
P[3][3] = 0.60;  // to G

//From G
P[4][0] = 0.67;  // to A
P[4][1] = 0.33;  // to T
P[4][2] = 0;     // to C
P[4][3] = 0;     // to G

for(var j=1; j<=4; j++){
    Jar[j] = Fill_Jar(j);

    document.write('From '+P[0][j-1] + ' to ' + Jar[j] + "</br>");
}

var draws = 40;
var q = "";
var z = "";
var a;

a = Draw(1);

for (var i=1; i<=draws; i++){
    for (var j=0; j<=3; j++){
        if (a === P[0][j]){
            a = Draw(j + 1);
            q = q + P[0][j];
            j=3;
        }
```

(Continued)

Additional algorithm 13.8 (Continued)

```
        }
    }

document.write("s = " + q + "</br>");

function Draw(S) {
    var rc = Math.floor((Math.random() * Jar[S].length));
    var b = Jar[S].substr(rc, 1);
    return b;
}

function Fill_Jar(S){
var Ltot = 20;
var a = 1;
var b = "";
    for (var i=0; i<=3; i++){
        a = Math.round(Ltot * P[S][i]);
        for (var j=1; j<=a; j++){
            b = b + P[0][i];
        }
    }
return b;
}

</script>

Output:

From A to TTTTTTTTTTTTTTTTCCCCC
From T to AAAAAAAAATTTCCCCCCCCC
From C to AAAAAAAAGGGGGGGGGGGG
From G to AAAAAAAAAAAAATTTTTTT
s = TATCGTTCATATATCACGACATACATATACACGATTACAT
```

Notice that the sum of the probabilities in each group is 1 (Additional algorithm 13.8). Also note that the order and letters from the original transition matrix are stored in the first group and are especially important for the second phase. **In the second phase** of the implementation, four strings are generated to simulate the jars from the above thought experiment (Figure 13.2c). Each string is representative of one state (one jar):

Jar "A" is *Jar*[1] = "TTTTTTTTTTTTTTTTCCCCC"

Jar "T" is *Jar*[2] = "AAAAAAAAATTTCCCCCCCCC"

Jar "C" is *Jar*[3] = "AAAAAAAAGGGGGGGGGGGG"

Jar "G" is *Jar*[4] = "AAAAAAAAAAAAAAATTTTTTT"

The composition of the string is determined by the *Fill_Jar* function using the transition probabilities found in the corresponding row of the transition matrix (Figure 13.2a,c). Therefore, the *Fill_jar* function is called repeatedly for each row *j* of the matrix by using a loop and the results of the function are stored in a vector (Jar[j]) whose components (j) represent the jars discussed in the example from above (Figure 13.2c).

13.4.3 Simulation of Transition Probabilities

When called, the *Fill_jar* function reads only the transition probability values from a specific row (S) of the transition matrix (P). Inside the function, the length of a string that is representative for a jar is set by default to a constant size using the *Ltot* variable. In the example from above, the chosen length was 20 letters (Ltot = 20). Next, function *Fill_jar* initiates a loop that traverses each position *i* in the row (S). On each step of the loop, the proportion for a specific type of letter (a) is determined by multiplying the transition probability value (P[S][i]) to the total number of letters (Ltot). Thus, the result of the multiplication (Ltot * P[S][i]) provides the proportion (a) for a specific type of letter (P[0][i]). Next, variable *a* sets the upper bound for a loop in which the letter (P[0][i]) is added repeatedly one by one to a string *b*. Note that the character of the letter is taken repeatedly from the same *i* position of row zero (P[0][i]) of the matrix. Once the corresponding proportions of the four letters have been added to the *b* string, the *Fill_jar* function returns the content of the *b* string to the caller (Jar[j] = Fill_Jar(j)).

13.4.4 The Markov machine

Note that the above output is different at each run (Additional algorithm 13.8). But how does the above algorithm work? Up to this point, strings have been constructed for each component of vector *Jar* and the core of the MCG can be further developed. **In the third phase,** the core of the MCG is built. The number of steps (the total number of letters in the MCG output) is set by using the *draws* variable (draws = 40). Next, a variable *a* is used to temporarily store each current state (always the last letter in the output). This variable is first initialized. Its content is set by calling the *Draw* function only once for a random selection from the first component of the Jar vector (representing jar "A"; a = Draw(1)). A random selection over a linear string will always result in a specific probability for any type of letter due to the difference in proportions/composition (Figure 13.2c).

Then, a loop sets the number of random selections from the components of vector *Jar*. At each step inside this cycle, a second loop checks if the content of variable a is the same as one of the letters in column j of line zero in the matrix (a === P[0] [j]). If this condition is true, then the new letter in a is set by the *Draw* function using a random selection from the string stored in the component $j + 1$ from vector *Jar* (a = Draw(j + 1)). In other words, the *Draw* function makes a random selection of a letter from the string in the $j + 1$ component of the *Jar* vector and returns this letter to the caller. Note that in Additional algorithm 13.8 the columns of the transition matrix start from index 0 and the components of the *Jar* vector start from index 1. Thus, in the second loop, the increment variable j starts from 0 and the *Draw* function is called using the $j + 1$ value as an input parameter. The randomly selected letter from the $j + 1$ component of vector *Jar* is then added to the variable containing the MCG output (q = q + P[0] [j]). The last letter from variable a becomes the current state. Again, the algorithm goes to the next step in the loop in which it makes the same check between the new content of variable a and the letters in column j of line zero in the matrix. As before, if the condition is true then the new value in a is set by the *Drow* function using a random selection from the string stored in the component $j + 1$ from vector *Jar*. The randomly selected letter from the $j + 1$ component of the *Jar* vector is then again added to the variable containing the MCG output (q = q + P[0] [j]). This cycle continues until the upper bound is reached (i<=draws). Once the upper bound is reached (draws), the complete sequence (q) is obtained in the output of the MCG (Figure 13.3). Notice that each string from the components of the *Jar* vector is intentionally organized into letter domains/clusters for exemplification ("A" repetitions, "T" repetitions, and so on). But these domains are not important, namely as long as the proportion of the letters is consistent, the order in the string does not matter. For instance, consider the content of Jar "A" (i.e. Jar[1]):

Jar "A" = "TTTTTTTTTTTTTTTCCCCC"

The above content can also be written as:

Jar "A" = "TCTTCTTCTTTTTTTCTTCT"

Notice that the proportion between letters is the same in both versions. A random selection of a letter from either *Jar* "A" version from above will always be made with a probability of 0.25 for letter "C" and a 0.75 probability for letter "A." Thus, the order of the letters does not matter, but the proportions are important. To conclude, each state has an associated row in the transition matrix. Each row in the transition matrix can be considered a jar filled with balls of different types. Such a jar can be represented by a string, and the letters from the string can represent balls of different types (Figure 13.2a–d).

13.4.5   Result Verification

Note that an MCG is already a prediction machine. Once an observation sequence has been molded into a transition matrix, this transition matrix can be further used to generate a sequence of observations similar to the original sequence. Thus, an MCG has the role of mimicking the output of the original machine. Up to this point, MCG has produced a series of characters, presumably similar to the original sequence of observations. After all, it is just a sequence that at first glance may or may not look like the original. But how can this be tested? Well, first a training sequence was used to calculate a transition matrix by using the DPD algorithm:

s = "ATCGATTCGATATCATACACGTAT"

(+)	A	T	C	G
A	0.00	0.75	0.25	0.00
T	0.43	0.14	0.43	0.00
C	0.40	0.00	0.00	0.60
G	0.67	0.33	0.00	0.00

Next, the above transition matrix was used by an MCG to mimic the properties of the training sequence by generating a new sequence in the output, namely:

s = "TATCGTTCATATATCACGACATACATATACACGATTACAT"

Notice that the output, of course, is not the same as the initial training sequence. However, the MCG (Additional algorithm 13.8) should be able to copy the behavior shown by the training sequence. Thus, when the above MCG output sequence is processed by the *DPD* function, a new transition matrix can be generated. This new transition matrix should contain close transition probability values to the original transition matrix. The close values of the transition probabilities between the two transition matrices can be a validation method for both the operation of the MCG and the implementation of the DPD algorithm. Thus, the s sequence generated by the MCG can in turn generate a transition matrix, as can be seen below:

(+)	T	A	C	G
T	0.17	0.58	0.25	0.00
A	0.60	0.00	0.40	0.00
C	0.00	0.67	0.00	0.33
G	0.33	0.67	0.00	0.00

Of course, due to the alphabet detection module in the DPD function, the order of the states on the new transition matrix differs. But, in principle, the original transition matrix and the transition matrix of the MCG sequence are the same. Moreover, the longer the MCG sequence, the more the transition probabilities will converge to the transition probabilities of the original transition matrix. But how long these MCG sequences can be? Of course, sequences of "unlimited" length can be produced by an MCG (namely the *draws* variable can be set to any integer). However, any region from the MCG sequence will generate a transition matrix similar to the original transition matrix. This statement can be verified by experiment using Additional algorithms 13.7 and 13.8. For instance, Additional algorithm 13.8 can be used to generate a sequence of 1000 letters. Repeated random samples of 100 letters (sampling length that can also vary) above the 1000 letters can be made to rationalize this experiment. The introduction of each sample in the DPD algorithm (Additional algorithm 13.7) will generate a transition matrix with transition probability values similar to those of the original transition matrix. Thus, it can be understood that the distribution of the transition probabilities between the letters produced by the MCG is relatively uniform because it is modeled by the probabilities from the original transition matrix. Here, a first-order Markov chain was described. Nevertheless, regardless of the first-, second-, third-, or other higher-order Markov chains, an MCG will have an answer to the following questions: If a sequence of observations exists, can a similar sequence of observations be created? How close can one get to the style and meaning of the initial sequence?

13.5 Conclusions

In this chapter, an introduction was made to Markov chains. In the first part of the chapter, the basic notions related to transition probabilities and transition matrices were described. The methods by which the transition matrices can be obtained from a sequence of observations have been described using the DPD algorithm. The DPD algorithm has been described in detail and compartmentalized into four main phases, namely: (i) alphabet detection, (ii) matrix initialization, (iii) frequency detection, and (iv) calculation of transition probabilities. Next, the implementation of the DPD algorithm was transformed into a function. This approach allows the function to be called repeatedly by other implementations that will be discussed further. Toward the end of the chapter, a MCG was described, implemented, and tested for predictions. The MCG used a transition matrix produced by the DPD algorithm. Thus, the MCG represented a machine mimicking/predicting the output of the system that generated the sequence of observations initially used in the construction of the transition matrix.

14

Markov Chains: Log Likelihood (II)

14.1 Introduction

Transition matrices are memory units that store the meaning and structure of a sequence of observations. Often, it is not a classical prediction that is needed for a meaningful detection of biological functions, but only a comparison between two models, each represented by a transition matrix. A comparison between two models is always very useful for signal detection in regions with unknown meaning along a variable length sequence (i.e. a genome). These models can be largely represented by a DNA/RNA or protein sequence with known biological function/meaning (model "+") and a background model (model "−"). Two types of background models are explored here. The first background model ("−") uses the *null model* and the second background model (also "−") is represented by a sequence different from that of the "+" model. Here, the discrete probability detector (DPD) algorithm is implemented as a function, which is called repeatedly to generate two transition matrices, namely a transition matrix that represents a model of interest ("+"; observation) and a transition matrix that represents the background model ("−"; expectation). Then, the two matrices are combined to generate a single log-likelihood matrix (LLM). Toward the end of the chapter, the LLM is used to build a scanner that detects regions within a z-sequence that resemble the model of interest ("+").

14.2 The Log-Likelihood Matrix

Up to this point, the DPD function allows the generation of a transition matrix from any sequence s (see the previous chapter). The transition matrix returned by the DPD function can be used to calculate a LLM. The LLM requires two matrices, a representative matrix for observations (labeled here as ' (+) ') and a

Algorithms in Bioinformatics: Theory and Implementation, First Edition. Paul A. Gagniuc.
© 2021 John Wiley & Sons, Inc. Published 2021 by John Wiley & Sons, Inc.
Companion website: www.wiley.com/go/gagniuc/algorithmsinbioinformatics

representative matrix for expectations (labeled here as ' (-) '). For the observation model (+), the *DPD* function returns a matrix based on sequence *s*, as can be seen in the output of Additional algorithm 13.7. However, for the expectation model (−), there are two possibilities. The second model (−) can either be calculated by using the *null model* or by using another model, which is different from the *null model*.

14.2.1 A Log-Likelihood Matrix Based on the Null Model

The LLM is a model that shows how different a target sequence can be from a random sequence. The *null model* dictates an equal probability of transition between the symbols of the alphabet of a sequence. In the case of DNA and RNA, this expected probability is 0.25 (four letters in the alphabet; 1/4 = 0.25).

$$P(G) = P(A) = P(T) = P(C) = \frac{unity}{symbols\ in\ the\ alphabet} = \frac{1}{4} = 0.25$$

Note that the methodology for calculating the LLM is similar to that described in the previous chapters. In the previous chapters, the LLM was constructed based on two PPMs (position probability matrices). However, here stochastic matrices are used instead. The implementation below uses sequence *s* and the *null model* to generate a LLM (Additional algorithm 14.1).

Additional algorithm 14.1 Note that the source code is out of context and is intended for explanation of the method.

```
var s = 'ATCGATTCGATATCATACACGTAT';

var q
var p = [] = DPD(s, '(+)');

p[0][0] = '(L)';
for(var i=0; i<p.length; i++){
    for(var j=0; j<p[i].length; j++){
        if (i>0 && j>0) {
            if(p[i][j]==0){p[i][j]=0.000000000000000001;}
            q = p[i][j]/0.25;
            p[i][j] = Log(2,q).toFixed(2);
        }
    }
}

document.write(SMC(p));
```

```
Output:

log likelihood:
(L)      A       T       C       G
A     -57.79   1.58    0.00   -57.79
T      0.78   -0.84    0.78   -57.79
C      0.68  -57.79  -57.79    1.26
G      1.42    0.40  -57.79  -57.79
```

The above implementation divides each numeric value from matrix p by 0.25. To avoid division by zero, the above module checks the value of each element in matrix p. If this value is zero, then it is replaced with a pseudo-count value prior to the division. Of course, the replacement of zero values can be done in the DPD function. However, if this replacement of zero values with pseudo-count values is done in the transition matrix, then the probability values can no longer be rounded to two decimal places, and the transition matrices would be devoid of elegance. Therefore, the replacement is made in the implementation from above (Additional algorithm 14.1). Next, the logarithm (base two) of this result is stored in the same element. Thus, matrix p is transformed into a LLM. To compute a logarithm in base n, the following equation is used:

$$\log_n(x) = \frac{\ln(x)}{\ln(n)}$$

The particularities and the experiments related to logarithms have been described in detail in the previous chapters (please also see Chapter 9). Note: In order to observe the behavior of the pseudo-count value in the context of an LLM, please also see Table 9.7 from Chapter 9. The function below (Additional algorithm 14.2) represents the implementation for the computation of a logarithm base n by using the above formula:

Additional algorithm 14.2 Note that the source code is out of context and is intended for explanation of the method.

```
function Log(n, v) {
  return Math.log(v) / Math.log(n);
}
```

Note that the *log* method in Javascript represents the natural logarithm (*ln*). This concludes the methodology related to the computation of logarithms. For the background model, this approach considers the use of a second matrix whose elements contain only values of 0.25. However, instead of using a second matrix, the division is done directly by 0.25 because the value is constant.

$$LLM = \log_2 \left(\frac{observed}{expected} \right) = \log_2 \left(\frac{p_{i+1,j+1}}{0.25} \right)$$

where the LLM is the log likelihood matrix. Notice that the calculation starts at $i+1$ and $j+1$ because the letters signifying the transitions are present on the first column and the first row. The order of the letters on the columns and rows of matrix p is not important for the moment because all the values from matrix p are divided by a constant value, namely 0.25. Therefore, the order provided by the alphabet detection algorithm is also valid for the LLM. Also, the base of the logarithm is not important in the grand scheme of the method, as can be seen in the following subchapters. Here, a logarithm in base two was used for a change of scenery. Moreover, the natural logarithm as described in the previous chapters can be used just as well (please see the explanations and the experiments from the previous chapters).

14.2.2 A Log-Likelihood Matrix Based on Two Models

A background other than the *null model* involves a second *s*-sequence for a background matrix (−). Thus, the DPD function can be called twice, both for the "+" and "−" transition matrices (Additional algorithm 14.3). To distinguish between the sequence of the "+" model and the sequence of the "−" model, a new notation can be used. Namely, a variable *s0* can store the sequence for the "+" model and a variable *s1* may contain the sequence for the background model (−):

s0 = "ATCGATTCGATATCATACACGTAT"
s1 = "AATCCTATCTTACTATTCTACTCAGTCCC"

The two sequences from above are just two unrelated sequences. The overall idea here is to capture the difference between the two sequences (*s0* and *s1*) by calculating a LLM (Figure 14.1a–c). The calculation of the LLM involves a division of the observed values from the *p* matrix (model "+") by the corresponding expected values ("−" model) from the *b* matrix (the background matrix), as follows (Figure 14.1c):

$$LLR = \log_2 \left(\frac{observed}{expected} \right) = \log_2 \left(\frac{p_{i+1,j+1}}{b_{i+1,j+1}} \right)$$

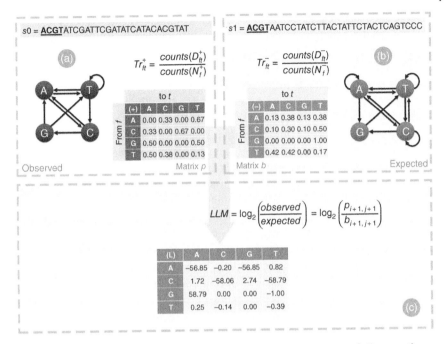

Figure 14.1 The log-likelihood matrix. (a) The panel shows sequence *s0*, the equation by which the transition probabilities are computed, and also it shows the organization of these values in the transition matrix of the "+" model. (b) In contrast, panel b shows sequence *s1*, the equation by which the transition probabilities are computed, and the organization of these values in the transition matrix of the "−" model. The "+" and "−" signs in the superscript of the equations indicate the model to which the equation belongs. Although it is not relevant in this case, every transition matrix contains an associated Markov diagram for illustration. The arrows that are shown in each Markov diagram represent a transition probability value greater than zero. Note the presence of the "ACGT" sequence at the beginning of both *s0* and *s1* (bolded and underlined), which establishes the same order of states on the rows and columns of both transition matrices. (c) It shows the equation that combines the transition matrices from the "+" model and the "−" model, and the final result represented by the log-likelihood matrix (LLM).

The *LLM* is the log-likelihood matrix and can be represented by the same matrix *p*. Thus, matrix *p* is transformed further into a LLM. Consequently, a more informative equation would be:

$$p_{i+1,j+1} = \log_2\left(\frac{p_{i+1,j+1}}{b_{i+1,j+1}}\right)$$

Again, the format of the transition matrices returned by the *DPD* function imposes a calculation that starts from the element $i+1$ and $j+1$. The first column and the first row in matrix *t* returned by the function are occupied by the associated letters/states. Of course, to avoid confusion on the implementation side, note that a *t* matrix returned by the *DPD* function can become either a *p* matrix or a *b* matrix. Moreover, to divide the observed values (i.e. from the "+" model; matrix *p*) by the expected values (i.e. from the "−" model; matrix *b*) on the implementation side, both transition matrices must contain the same order of the letters on columns and rows. Thus, this is where the order of the states on the transition matrix matters. Notice that the order of the letters on columns and rows of matrix *p* or *b* is determined by their respective *s*-sequences (i.e. *s0* and/or *s1*). Since *s0* and *s1* contain different sequences, there is a good chance that the letters of the alphabet may be detected in a different order from case to case. If the order differs between the two transition probability matrices ("+" and "−"), then the LLM cannot be calculated directly. Such a difference between the transition matrices could greatly complicate the calculation of the LLM. For instance, if the observed values ("+") are divided by the expected values ("−") in a tête-à-tête maner, the approach requires a correlation/mapping between the positions of the letters in the two matrices ("+" and "−"). This mandatory correlation would be undesirable, as it increases the level of complexity for the implementation.

Nevertheless, what if we wish a predefined order of the states on the columns and rows of the two transition matrices? To maintain an order of the letters/states in all transition matrices returned by the *DPD* function, a preferred order (*h*) of the letters can be added at the beginning of *s* (e.g. *h* = "ATCG"). In this manner, the alphabet detection method identifies the order inserted at the beginning of the *s*-sequence (i.e. *h* + *s0* and *h* + *s1*). Thus, variable *h* sets the order of the states on the transition matrices returned by the *DPD* function. For very short sequences, the introduction of supplemental symbols (i.e. "ATCG") in *s0* and *s1* can lead to additions in the transition counts for "AT," "TC," and "CG," and an increase in the values of their transition probabilities. However, these added values can be insignificant for longer *s*-sequences. Moreover, since this order (*h*) is added for both models, the end result does not change.

Additional algorithm 14.3 Note that the source code is out of context and is intended for explanation of the method.

```
var h = 'ACGT';

var s0 = h+'ATCGATTCGATATCATACACGTAT';
var s1 = h+'AATCCTATCTTACTATTCTACTCAGTCCC';


var q
var p = [] = DPD(s0, '(+)');
```

```
var b = [] = DPD(s1, '(-)');

document.write('Model (+):<br>'+SMC(p)+'<hr>');
document.write('Model (-):<br>'+SMC(b)+'<hr>');

p[0][0] = '(L)';
for(var i=0; i<p.length; i++){
    for(var j=0; j<p[i].length; j++){
        if (i>0 && j>0) {

            if(p[i][j]==0){p[i][j]=0.000000000000000001;}
            if(b[i][j]==0){b[i][j]=0.000000000000000001;}

            q = p[i][j]/b[i][j];
            p[i][j] = Log(2,q).toFixed(2);
        }
    }
}

document.write('log likelihood:<br>'+SMC(p));

Output:

Model (+):
(+)     A         C         G         T
 A      0.00      0.33      0.00      0.67
 C      0.33      0.00      0.67      0.00
 G      0.50      0.00      0.00      0.50
 T      0.50      0.38      0.00      0.13

Model (-):
(-)     A         C         G         T
 A      0.13      0.38      0.13      0.38
 C      0.10      0.30      0.10      0.50
 G      0.00      0.00      0.00      1.00
 T      0.42      0.42      0.00      0.17

log likelihood:
(L)     A         C         G         T
 A     -56.85    -0.20     -56.85    0.82
 C      1.72     -58.06    2.74      -58.79
 G      58.79     0.00     0.00      -1.00
 T      0.25     -0.14     0.00      -0.39
```

But what is the end result? The immediate result is represented by the LLM; however, the end result is represented by the use of the LLM (please see the next subchapter). Note that the two sequences used in the training of matrix p and b can show variable lengths, while in the PSSM method of the previous chapters, all motif sequences had a constant length.

14.3 Interpretation and Use of the Log-Likelihood Matrix

A LLM has been calculated either by using the *null model* or by using another type of model for the background. Of course, the use of two different models can be particularly useful for different experiments. Thus, the two-sequence training (*s0* and *s1*) approach is used further. But what can be done with the LLM? Once built, the LLM can be used as in the case of the PSSM method described in the previous chapters. Namely, in order to calculate a significance score, a z-sequence may be analyzed by consulting the LLM at each step. However, here each transition in z is checked against the LLM and the sum of the values picked up from the LLM represents the significance score. Consider the *s0* and *s1* sequences from the previous examples and the following z-sequence:

z = "ATCGATTCGATATCAT"

In the first instance, the implementation below calls the *DPD* function to generate the transition matrices for the two models (Additional algorithm 14.4). Then, the two transition matrices (p and b) are used for the LLM computation. Once calculated, the LLM is kept in the memory of the application during the execution session.

Additional algorithm 14.4 Note that the source code is out of context and is intended for explanation of the method.

```
// MAIN SEQUENCE
var z = 'ATCGATTCGATATCAT';

// ORDER
var h = 'ACGT';

// TRAINING MODELS
var s0 = h+'ATCGATTCGATATCATACACGTAT';
var s1 = h+'AATCCTATCTTACTATTCTACTCAGTCCC';
```

```
var q;

// get  the two transition matrices
var p = [] = DPD(s0, '(+)');
var b = [] = DPD(s1, '(-)');

for(var i=0; i<p.length; i++){
    for(var j=0; j<p[i].length; j++){
        if (i>0 && j>0) {

            if(p[i][j]==0){p[i][j]=0.000000000000000001;}
            if(b[i][j]==0){b[i][j]=0.000000000000000001;}

            q = p[i][j]/b[i][j];
            p[i][j] = Log(2,q).toFixed(2);
        }
    }
}

document.write('log likelihood:<br>'+SMC(p)+'<hr>');

var l = [];
var f='';
var t=0;

l[0]="";
l[1]="";

for(var u=0; u<z.length; u++){

    l[0] = z.substr(u, 1);
    l[1] = z.substr(u + 1, 1);

    for(var i=0; i<p.length; i++){
        for(var j=0; j<p[i].length; j++){
            if(p[i][0]==l[0] && p[0][j]==l[1]){
                f+='(' + Number(p[i][j]) + ')+';
                t+=Number(p[i][j]);
            }
        }
    }
}
document.write(z+'<br>');
document.write(f+' = '+t);
```

(Continued)

Additional algorithm 14.4 (Continued)

```
Output:

log likelihood:
(+)     A         C         G         T
A     -56.85    -0.20    -56.85     0.82
C      1.72    -58.06     2.74    -58.79
G     58.79     0.00      0.00     -1.00
T      0.25    -0.14      0.00     -0.39

ATCGATTCGATATCAT
(0.82)+(-0.14)+(2.74)+(58.79)+(0.82)+(-0.39)+(-0.14)
+(2.74)+(58.79)+(0.82)+(0.25)+(0.82)+(-0.14)+(1.72)
+(0.82)  = 128.32
```

The question now is whether the z-sequence is similar to the "+" model or the "−" model. To find the answer to this question, each transition in z is checked by the algorithm against the LLM. For example, the first transition between "A" and "T" shows a value of 0.82 in the LLM. The second transition is made from "T" to "C" and shows a value of −0.14 in the LLM. The next transition is from "C" to "G" and shows a value of 2.74. These verifications continue until the last transition in z is reached, namely the transition between "A" and "T," which shows a value of 0.82. All these values are summed and the result represents the significance score for the z-sequence (Additional algorithm 14.4):

$$Score = (0.82) + (-0.14) + (2.74) + (58.79) + (0.82) + (-0.39) + (-0.14)$$
$$+ (2.74) + (58.79) + (0.82) + (0.25) + (0.82) + (-0.14) + (1.72)$$
$$+ (0.82) = 128.32$$

The output from Additional algorithm 14.4 shows the values from the LLM for each transition found in z. The sum of the values in the LLM leads to a final score for the z-sequence (i.e. 128.32). Up to this point, a global score has been calculated for the entire content found in z.

14.4 Construction of a Markov Scanner

A global score has been obtained for a z-sequence. But for longer z-sequences, the global score loses its value for determining similarity (please see the additional explanations in the chapters related to the PSSM method). The longer the z-sequence, the lower the possibility of variation for the score value, up to the point where it can reach a narrow plateau (for very long z-sequences). On the other

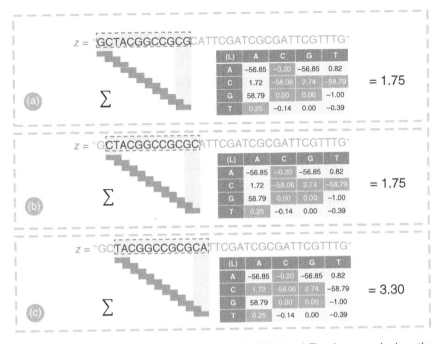

Figure 14.2 **Local score computation by using the LLM.** (a–c) The three panels show the first steps in the scanning process. A window slides above the z-sequence with a step of 1. In any of the three panels (a–c), the ladders indicate the successive transitions over the sliding window. The rows and columns in the LLM have the same meaning as described for transition matrices, namely the rows indicate the current state (the first letter in the combination) and the columns indicate the state to which the transition is made (the second letter in the combination). The value in the element at the intersection of the two positions is the log-likelihood ratio value added to the summation. The sum sign indicates that the log-likelihood ratio values corresponding to each transition are summed together to compute the local score of the sliding window content.

hand, local scores can be much more useful because the variation of the score values for shorter z-sequences can be abrupt and noticeable when plotted on a chart (Figure 14.2a–c). For a change in scenery, please consider the new $s0$ sequence for the "+" model and the $s1$ sequence for the "−" model:

$s0$ = "ATGCGCGCGGCATTCGTACGATGCGTACGGTCTA"

$s1$ = "ATTACTATATTATTTCGCGTAATACTGGTCTATATCGGCATT"

Please also consider the following z-sequence:

z = "GCTACGGCCGCGCATTCGATCGCGATTCGTTTG"

The implementation from Additional algorithm 14.5 computes the local scores for the above z-sequence and includes the new $s0$ and $s1$ sequences for the two models. The Additional algorithm 14.5 contains two functions, the LLM calculation module and a new scan module. The *DPD* function generates the two transition matrices (*p* and *b*) as discussed in the previous examples. The two transition matrices are then used by the LLM calculation module. Once the LLM was obtained, the scan module uses a 12-position sliding window and calls an *SW* function to process its contents (w).

Note that the length of the sliding window can be arbitrary and 12 positions were chosen only to illustrate the method. This processing stage consists of checking the letter transitions in the content of the sliding window against the LLM positions. The *SW* function makes a summation of the appropriate LLM values to compute the significance score (f). Thus, the *SW* function receives the sliding window content (w) and returns the significance score (f).

Additional algorithm 14.5 Note that the source code is in context and works with copy/paste.

```
<script>

// MAIN SEQUENCE
var z = 'GCTACGGCCGCGCATTCGATCGCGATTCGTTTG';

// ORDER
var h = 'ACGT';

// TRAINING MODELS
var s0 = h+'ATGCGCGCGGCATTCGTACGATGCGTACGGTCTA';
var s1 = h+'ATTACTATATTATTTCGCGTAATACTGGTCTATATCGGCATT';

var q;

// GET THE TWO TRANSITION MATRICES
var p = [] = DPD(s0, '(+)');
var b = [] = DPD(s1, '(-)');

document.write('Model (+):'+s0+'<br>'+SMC(p)+'<hr>');
document.write('Model (-):'+s1+'<br>'+SMC(b)+'<hr>');

p[0][0] = '(L)';
for(var i=0; i<p.length; i++){
    for(var j=0; j<p[i].length; j++){
        if (i>0 && j>0) {

            if(p[i][j]==0){p[i][j]=0.000000000000000001;}
            if(b[i][j]==0){b[i][j]=0.000000000000000001;}

            q = p[i][j]/b[i][j];
```

```
            p[i][j] = Log(2,q).toFixed(2);
        }
    }
}

document.write('log likelihood:<br>'+SMC(p)+'<hr>');

// THE SCANNER
var signal = '';
var wl = 12;
var w = '';

var u = z.length - wl + 1;

for(var r=0; r<u; r++) {
    w = z.substr(r, wl);
    signal += w + ' = ' + SW(w).toFixed(2) + '<br>';
}

document.write('z-sequence:<br>'+z+'<hr>');
document.write('Signal:<br>'+signal+'<hr>');

function SW(s){

    var l = [];
    var f=0;

    l[0]="";
    l[1]="";

    for(var u=0; u<s.length; u++){

        l[0] = s.substr(u, 1);
        l[1] = s.substr(u + 1, 1);

        for(var i=0; i<p.length; i++){
            for(var j=0; j<p[i].length; j++){
                if(p[i][0]==l[0] && p[0][j]==l[1]){
                    f+=Number(p[i][j]);
                }
            }
        }
    }
    return f;
}
```

(Continued)

Additional algorithm 14.5 (Continued)

```javascript
function DPD(s, n){

    //-------------[ Phase one ]-------------/
    var a = [];
    var k = s.length;

    for(var i=0; i<=k; i++){
        var q = 1;
        for(var j=0; j<=a.length; j++){
            if (s[i] === a[j]) {q = 0;}
        }
        if (q === 1) {a.push(s[i]);}
    }

    //-------------[ Phase two ]-------------/
    var d = a.length-1;
    var m = [];
    var e = [];
    var l = [];

    for(var i=0; i<=d; i++){
        m[i]=[];
        e[i]=[];
        for(var j=0; j<=d; j++){

            m[i][j]=0;

            if (j === 0) {
                e[i][0]=a[i];
                e[i][1]=0;
            }
        }
    }

    //-------------[ Phase three ]-------------/
    l[0]="";
    l[1]="";

    for(var i=0; i<s.length-1; i++){

        l[0] = s.substr(i, 1);
        l[1] = s.substr(i + 1, 1);

        for(var j=0; j<=d; j++){
```

```javascript
            if (l[0] === e[j][0]) {
                e[j][1] = e[j][1] + 1;
                r = j;
            }
            if (l[1] === e[j][0]) {c=j;}
        }
        m[r][c] = m[r][c] + 1;
    }

    //-------------[ Phase four ]---------------/

    for(var i=0; i<=d; i++){
        for(var j=0; j<=d; j++){
            if (e[i][1] > 0) {
                m[i][j]=(m[i][j]/e[i][1]).toFixed(2);
            }
        }
    }

    //-----------------[ END ]------------------/
    var t = [];

    for(var i=0; i<=d+1; i++){

        t[i]=[];

        for(var j=0; j<=d+1; j++){

            if (j == 0 && i == 0) {t[i][j] = n;}
            if (i>0 && j>0) {t[i][j] = m[i-1][j-1];}
            if (i == 0 && j>0 && j<=d+1) {t[i][j]=a[j-1];}
            if (j == 0 && i>0 && i<=d+1) {t[i][j]=a[i-1];}
        }
    }

    return t;
}

function Log(n, v) {
  return Math.log(v) / Math.log(n);
}

// SHOW MATRIX CONTENT
function SMC(m) {
    var r = "<table border=1>";
```

(Continued)

Additional algorithm 14.5 (Continued)

```
        for(var i=0; i<m.length; i++) {
            r += "<tr>";
            for(var j=0; j<m[i].length; j++){
                r += "<td>"+m[i][j]+"</td>";
            }
            r += "</tr>";
        }
        r += "</table>";

        return r;
}

</script>
```

Output:

Model (+):ACGTATGCGCGCGGCATTCGTACGATGCGTACGGTCTA

(+)	A	C	G	T
A	0.00	0.50	0.00	0.50
C	0.10	0.00	0.80	0.10
G	0.08	0.42	0.17	0.33
T	0.44	0.22	0.22	0.11

Model (-):ACGTATTACTATATTATTTCGCGTAATACTGGTCTATATCGGCATT

(-)	A	C	G	T
A	0.08	0.25	0.00	0.67
C	0.13	0.00	0.50	0.38
G	0.00	0.29	0.29	0.43
T	0.50	0.17	0.06	0.28

log likelihood:

(L)	A	C	G	T
A	-56.15	1.00	0.00	-0.42
C	-0.38	0.00	0.68	-1.93
G	56.15	0.53	-0.77	-0.38
T	-0.18	0.37	1.87	-1.35

z-sequence:
GCTACGGCCGCGCATTCGATCGCGATTCGTTTG

Signal:
GCTACGGCCGCG = 1.75
CTACGGCCGCGC = 1.75
TACGGCCGCGCA = 3.30

```
ACGGCCGCGCAT = 3.06
CGGCCGCGCATT = 0.71
GGCCGCGCATTC = 0.40
GCCGCGCATTCG = 1.85
CCGCGCATTCGA = 57.47
CGCGCATTCGAT = 57.05
GCGCATTCGATC = 56.74
CGCATTCGATCG = 56.89
GCATTCGATCGC = 56.74
CATTCGATCGCG = 56.89
ATTCGATCGCGA = 113.42
TTCGATCGCGAT = 113.42
TCGATCGCGATT = 113.42
CGATCGCGATTC = 113.42
GATCGCGATTCG = 113.42
ATCGCGATTCGT = 56.89
TCGCGATTCGTT = 55.96
CGCGATTCGTTT = 54.24
GCGATTCGTTTG = 55.43
```

The above implementation shows both the content of the p and b transition matrices (labeled "+" and "−") and the content of the LLM (labeled "(L)"). For exemplification, the output also shows the contents of each sliding window and the associated score. The succession of these scores represents the signal above the z-sequence and indicates the regions that more or less resemble the "+" model. Of course, for a more direct use, the implementation of Additional algorithm 14.6 must show only the main signal. Thus, to obtain only the main signal in the output, the previous implementation can be cleaned and rewritten as (Additional algorithm 14.6):

Additional algorithm 14.6 Note that the source code is out of context and is intended for explanation of the method.

```
// MAIN SEQUENCE
var z = 'GCTACGGCCGCGCATTCGATCGCGAT';

// ORDER
var h = 'ACGT';

// TRAINING MODELS
var s0 = h+'ATGCGCGCGGCATTCGTACGATGCGTACGGTCTA';
```

(Continued)

Additional algorithm 14.6 (Continued)

```
var s1 = h+'ATTACTATATTATTTCGCGTAATACTGGTCTATATCGGCATT';

var q;

// GET THE TWO TRANSITION MATRICES
var p = [] = DPD(s0, '(+)');
var b = [] = DPD(s1, '(-)');

p[0][0] = '(L)';
for(var i=0; i<p.length; i++){
    for(var j=0; j<p[i].length; j++){
        if (i>0 && j>0) {

            if(p[i][j]==0){p[i][j]=0.000000000000000001;}
            if(b[i][j]==0){b[i][j]=0.000000000000000001;}

            q = p[i][j]/b[i][j];
            p[i][j] = Log(2,q).toFixed(2);
        }
    }
}

// THE SCANNER
var signal = '';
var wl = 12;
var w = '';

var u = z.length - wl + 1;

for(var r=0; r<u; r++) {
    w = z.substr(r, wl);
    signal += SW(w).toFixed(2) + ',';
}

document.write('Signal:<br>'+signal);

Output:

Signal:
1.75,1.75,3.30,3.06,0.71,0.40,1.85,57.47,57.05,56.74,
56.89,56.74,56.89,113.42,113.42,
```

Note that in the above implementation the *DPD, SW,* and *SMC* functions are no longer shown to avoid redundancy. Also, the previous *z*-sequence has been slightly shortened so that the result from the output does not exceed the space allocated on the page. Additional algorithm 14.5 and/or Additional algorithm 14.6 includes all the methods that have been discussed in this chapter and concludes the use of LLM to determine the similarity between a model and an uncharacterized *z*-sequence.

14.5   A Scanner That Uses a Known LLM

Once the LLM has been built, it can be stored in text format and it may be used directly by a scanner. Such a scanner can use two inputs. The first input would consist of the *z*-sequence and the second input would be the LLM data. The implementation below shows two parts. The part that loads the LLM and the part of the scanner that reads and successively generates a score from the sliding windows to form a signal (Additional algorithm 14.7).

Additional algorithm 14.7   Note that the source code is in context and works with copy/paste.

```
<script>

// MAIN SEQUENCE
var z = 'GCTACGGCCGCGCATTCGATCGCGAT';

var c = '|(L)\tA\tC\tG\tT' +
        '|A\t-56.15\t1\t0\t-0.42' +
        '|C\t-0.38\t0\t0.68\t-1.93' +
        '|G\t56.15\t0.53\t-0.77\t-0.38' +
        '|T\t-0.18\t0.37\t1.87\t-1.35';

var n = [];
var m = [];
var p = [];

m = c.split('|');

// LOAD LLM
for(var i=1; i<m.length; i++) {
```

(Continued)

Additional algorithm 14.7 (Continued)

```
     p[i-1]=[];
     n = m[i].split('\t');
     for(var j=0; j<n.length; j++){
         p[i-1][j]=n[j]
     }
}

// SCANNER
var signal = '';
var wl = 12;
var w = '';

var u = z.length - wl + 1;

for(var r=0; r<u; r++) {
    w = z.substr(r, wl);
    signal += SW(w).toFixed(2) + ',';
}

document.write('Signal:<br>'+signal);

function SW(s){

var l = [];
var f=0;

    l[0]="";
    l[1]="";

    for(var u=0; u<s.length; u++){

        l[0] = s.substr(u, 1);
        l[1] = s.substr(u + 1, 1);

        for(var i=0; i<p.length; i++){
            for(var j=0; j<p[i].length; j++){
                if(p[i][0]==l[0] && p[0][j]==l[1]){
```

```
                        f+=Number(p[i][j]);
                }
            }
        }
    }

    return f;
}

</script>

Output:

Signal:
1.75,1.75,3.30,3.06,0.71,0.40,1.85,57.47,57.05,56.74,
56.89,56.74,56.89,113.42,113.42,
```

The result of the above implementation is the same as the result presented in Additional algorithm 14.6. In this version, the complexity of the algorithm is reduced because the construction of a LLM from the two transition matrices is no longer necessary. Here, variable c contains the LLM data, separated by the tab character (in java script the tab character is expressed by "\t"). Moreover, a simple artifice is introduced here to differentiate between the LLM rows. This differentiation is made by using a delimiting character introduced at the beginning of each line in c. The vertical line character was chosen as the delimiter ("|"). Nevertheless, any delimiter can be used as long as it has a low probability of occurring in the LLM data and is aesthetically intuitive for the programmer. First, the content of the c variable is split, and the resulting segments are stored in the array variable m. Thus, the m variable holds each LLM row as a continuous string. Note: the increment of variable i begins from 1 because the splitting of c into m pieces leaves $m[0]$ empty. Consequently, in the same cycle, the completion of matrix p elements is done from i-1. Secondly, each row from m is split using the tab character ("\t") and the individual values in the row are kept temporarily in the array variable n. At the same time, the values extracted from variable n are loaded successively on the rows of matrix p (the LLM). Once loaded, the p matrix is a LLM and can be used directly as such by a scanner.

14.6 The Meaning of Scores

Each score value over z was calculated by adding the relevant values from each position in the LLM. A series of scores above the z-sequence can indicate which regions are similar to the $s0$ sequence ("+" model) or to the $s1$ sequence ("−" model). A score value above zero indicates that a sliding window contains a region from the z-sequence that is similar to the "+" model (Table 14.1). In contrast, a score value below zero indicates that a sliding window contains a region from the z-sequence that is similar to the "−" model. A score value of zero indicates that a sliding window contains a region from the z-sequence that is similar to both models. Special experimental conditions can be created to highlight the significance of the score values (Figure 14.3). To force

Table 14.1   The meaning of scores.

Score	Meaning
Score = 0	Equally similar to both models
Score > 0	Similar to the structure of the "+" model
Score < 0	Similar to the structure of the "−" model

The table describes the significance of the observed values. A score value above zero indicates that a sliding window contains a region from the z-sequence that is more likely to resemble the "+" model. A score value of zero indicates that a sliding window contains a region from the z-sequence that is equally likely to resemble the "+" or the "−" model. A score value below zero indicates that a sliding window contains a region from the z-sequence that is more likely to be closer in structure with the "−" model.

Figure 14.3   An experiment for understanding scores. (a) Shows the contents of the sliding windows above the z-sequence; and (b) the score values calculated from the LLM for these sliding windows. (c) Shows the plot of the score values on a bar chart whose axis contains both positive and negative values. The first half (top) of the chart shows positive values that indicate a resemblance of the region from the z-sequence to the $s0$ sequence of the "+" model. The second half (bottom) of the chart shows negative values that indicate the regions in the z-sequence that resemble the $s1$ sequence of the "−" model. Notice the columns of the chart correlate with the positions shown by (a) and (b) regions of the panel. For a reference system, the same chart is represented by a line chart in a small window on the upper right corner. The left side of the figure indicates the meaning of the elements on the right side of the figure. The z-sequence is placed on a relative position below the chart and there is no direct correlation between the two, because the number of sliding windows is equal to the number of letters in the z-sequence minus the total number of positions in a sliding window.

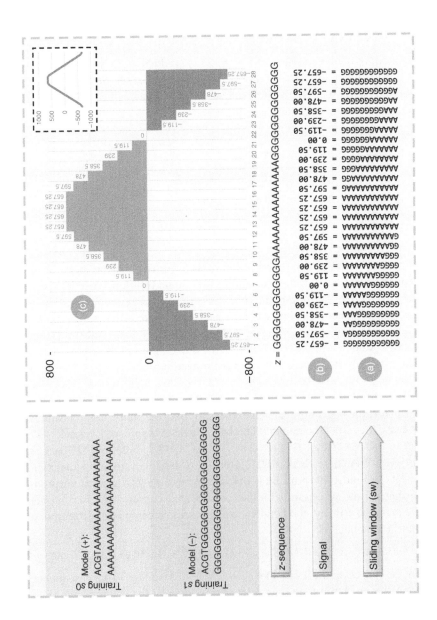

the algorithm into providing a limited number of results (that can be easily studied), two special training sequences (i.e. *s0* and *s1*) can be considered. For example, a first sequence *s0* can be composed of single repetitions of "A" for model "+":

s0 = "AAAAAAAAAAAAAAAAAAAAAAAAAAAAAAAAAAAAAAAAAAAA"

The content of the *h* variable (i.e. *h* = "ATGC") is added to the *s0* sequence on the left side to obtain a preferred order of the states on the columns and rows of the transition matrix. The composite sequence (i.e. *h* + *s0*) generates the following transition matrix using the DPD algorithm:

(+)	A	C	G	T
A	0.97	0.03	0.00	0.00
C	0.00	0.00	1.00	0.00
G	0.00	0.00	0.00	1.00
T	1.00	0.00	0.00	0.00

The impact of variable *h* on the values from the transition matrix can be easily observed. Therefore, a short discussion that is separate from the current experiment, is now in order. The question that arises in the case of an ordered transition matrix would be: What is the impact of the h variable on an MCG? As previously discussed, the content of the *h* variable slightly affects the values of the transition probabilities. Note that in the matrix above, the transition from "A" to "A" shows a transition probability value of 0.97, and the transition from "A" to "C" shows a transition probability of 0.03. In the case of a Markov chains generator (MCG), the transition probability for "AC" is negligible and if by chance a transition from "A" to "C" is made, the MCG immediately returns to the "AA" transition in just four steps because the rest of the transition probabilities show values of 1, namely: "A" to "C" ($Tr_{AC} = 0.03$), "C" to "G" ($Tr_{CG} = 1$), "G" to "T" ($Tr_{GT} = 1$), and "T" back to "A" ($Tr_{TA} = 1$). From here, there is a 0.97 probability that the system will transition back to state "A" and a 0.03 probability of transition to state "C." A transition probability value of 1 means certainty for the transition to a certain state. To continue with the main experiment, a second *s1* sequence can be composed of "G" mono repetitions for model "−":

s1 = "GGGGGGGGGGGGGGGGGGGGGGGGGGGGGGGGGGGGGGGGGGGG"

Again, the content of the *h* variable (i.e. *h* = "ATGC") is added to the *s1* sequence on the left side to obtain the same order of the states on the columns and rows of the transition matrix for the "–" model. The composite sequence (i.e. $h + s1$) generates the following transition matrix using the DPD algorithm:

(–)	A	C	G	T
A	0.00	1.00	0.00	0.00
C	0.00	0.00	1.00	0.00
G	0.00	0.00	0.97	0.03
T	0.00	0.00	1.00	0.00

In the transition matrix of the "–" model from above, we can see the same variation as in the matrix of the "+" model, namely the probability of transition from "G" to "G" is 0.97 and the probability of transition from "G" to "T" shows a value of 0.03. Once a transition is made by chance from "G" to "T," the return to the "G" state is made in four steps in the same manner as in the "+" model.

Nevertheless, the meaning of these transitions and their behavior in an MCG was discussed besides the main point, namely the current experiment. Here, the DPD is used to produce the transition matrices of the two models. In turn, the transition matrices of the two models yield the following LLM:

(L)	A	C	G	T
A	59.75	-5.06	0.00	0.00
C	0.00	0.00	0.00	0.00
G	0.00	0.00	-59.75	5.06
T	59.79	0.00	-59.79	0.00

Thus, the result of an analysis based on the LLM would be easier to comprehend if the sequence that is analyzed (the z-sequence) were composed of pieces from the two models (i.e. $s0$ and $s1$). But what can be expected from the result of the analysis of a mixed z-sequence? Below, a z-sequence that contains such regions is presented:

$$z = \text{"GGGGGGGGGGGGGAAAAAAAAAAAAAAAAGGGGGGGGGGGGGG"}$$

The first and the last region of the z-sequence show negative score values, indicating a resemblance to the "–" model, whereas the middle area of the z-sequence generates positive score values that point out a resemblance to the properties of the $s0$ sequence from the "+" model (Figure 14.3a–c). In other words, regions in the z-sequence that are the same as the $s0$ sequence will generate the highest positive score values. On the other hand, regions in the z-sequence that are the same as the *s1* sequence will generate the lowest negative score values. To repeat the experiment, please use Additional algorithm 14.6 from above and the three sequences shown in this subchapter (i.e *s0*, *s1*, and *z*). This concludes the meaning of the score values.

14.7 Beyond Bioinformatics

Markov chains has applications in all areas of science. But what could be another example of application for Markov chains? An intuitive example that comes to mind can be given in the case of literature. Let us imagine two very good poets, both with their own style of writing lyrics. The question is: Could we capture their style and meaning of lyrics with a memory unit such as a transition matrix? Once the transition matrix is constructed from the poetry written by the poet (the training sequence), the transition matrix can be viewed as "the memory of the poet." Moreover, if a Markov chain generator (MCG; described in the previous chapter) can mimic the style/essence of a poet (it can "bring to life" that poet after his death by generating similar poems), the method from this chapter is able to detect to which poet an anonymous poem belongs. Obviously, to make a more accurate detection, high-order Markov chains can be used and the transition matrices would be much larger (more rows than columns). However, a detection for more complex cases can also be done with the first-order Markov chains. The states can be represented not by individual letters but by the unique words found in the poem. In this way, the first-order Markov chains can be used further with no advanced modification to the implementations described so far. Then, the total number of states would be represented by the number of types of words existing in the text/poetry and the transitions between words/states would form a transition matrix that can be used by an MCG to generate a text in the style and meaning of the poet. Thus, writings can make us immortal in a manner of speaking because they transpose some of the important cerebral patterns of each individual on a piece of paper. Here, one can already observe a universality of the processes, whether the system is represented by us humans or the weather or the DNA/RNA information. Moreover, another question would be: Could we find out mathematically what differentiates the two poets? Here, the valuable LLM comes into use, which highlights the contrast between the transition matrix that comes from the writings of one poet and the transition matrix constructed from the writings of the other poet. Moreover, if a poem is presented with an unknown author, can we find out which of the two poets wrote the poem? Again, this is the very purpose of the method presented in this chapter. Once the transition matrices (e.g. *p* and *b*) have been computed from the writings (e.g. *s0* and *s1*) of the two poets, the LLM is further constructed. Consequently, the background model ("−") can be represented by one poet and the "+" model by the other poet. By scanning the anonymous poem (the z-sequence) with the help of a LLM adapted to words, a signal can be obtained whose score values can indicate whether the text of the poem resembles the style of one or the other poet. Moreover, the signal can indicate which parts of the text (the z-sequence) have been written by one poet or the other poet. Another use that comes to mind for Markov chains is related to the information security area. In information security, the method (as in the case of PSSM) can be used to heuristically detect malware files. A model (i.e. "+") can represent sections of malware files (i.e. $s0$) from which a transition matrix can be created and the background model ("−") can be represented by sections (i.e. $s1$) from genuine files from which a transition matrix can also be made. A LLM built from the transition matrices of the two models can be used by an antivirus scanner to detect new malware files. A note worth mentioning here, which is not usually discussed in the popularization of science, would be that Markov chains can be a more powerful and reliable method of prediction when compared to a regular neural network. However, in the era when the term "artificial intelligence" is distorted, misused, and mentioned in the wrong context, such indications may be useful for the reader.

14.8 Conclusions

The current chapter explored a detection method based on transition matrices. The DPD algorithm was transformed into a function by which two models were generated. A model "+" that was represented by a $s0$ sequence with a known function/meaning and a $s1$ sequence that represented a background model. The transition matrices of the two models were compared in a LLM to emphasize the main differences. These differences were materialized by a LLM. Once calculated, the LLM was further used to find regions similar to the "+" model within a z-sequence. Toward the end of the chapter, the meaning of the score values was discussed and some alternative ideas were explored through worldly examples. This chapter and the previous chapter encompassed all the important notions behind the Markov chains principles.

15

Spectral Forecast (I)

15.1 Introduction

Spectral forecast is a novel general-purpose prediction model recently published in the journal CHAOS [250]. This chapter explains how spectral forecast works in detail and suggests some interesting possible applications of the method in bioinformatics. Here, this novel prediction method is described, implemented, and tested. The model revolves around three known states: two extreme outcomes (A and B) and one measurement (P). These states are represented by either vectors or matrices that include sets of homologous parameters. An information spectrum is described as a series of predicted states (M_1, M_2, M_3, ..., M_d) generated between the two extreme outcomes (A and B). These states are successively calculated using the spectral forecast equation. The predicted states are compared with a known state (P) from the measurements to generate a series of similarity index values. The trend generated by the values of the similarity index shows how a system may behave against these two extreme outcomes (A and B). Note that the implementations described in this chapter are also present online in more advanced versions that can intuitively explain the usefulness of the method (see the online webpage of the book).

15.2 The Spectral Forecast Model

A general-purpose prediction method is presented, called the spectral forecast model. The spectral forecast model requires three matrices (or vectors, depending on the case) of equal size and uses two equations to determine the behavior against two possible outcomes. In this approach, three *known* matrices are used: matrix A, matrix B, and matrix P. However, the first equation in the model may be of great value for novel prediction approaches. That is, a matrix M can be further used to formulate the entire spectrum of unknown information between a matrix

Algorithms in Bioinformatics: Theory and Implementation, First Edition. Paul A. Gagniuc.
© 2021 John Wiley & Sons, Inc. Published 2021 by John Wiley & Sons, Inc.
Companion website: www.wiley.com/go/gagniuc/algorithmsinbioinformatics

A and a matrix *B*. For this calculation, a novel equation called "spectral forecast" is devised, namely:

$$M_{ijd} = \left[\left(\frac{d}{\text{Max}(A_{ij})} \right) \times A_{ij} \right] + \left[\left(\frac{(\text{Max}(d) - d)}{\text{Max}(B_{ij})} \right) \times B_{ij} \right]$$

where M_{ij} represents the predicted matrix at every discrete step (d). A_{ij} and B_{ij} represent two *known* matrices. Also, d stands for distance and represents a specific discrete step between matrix *A* and matrix *B*. Max(d) represents the total number of discrete steps that can be taken from matrix *A* to matrix *B*. The value of Max(d) is the maximum value found above the elements of the two matrices *A* and *B* (thus, it can be either Max(A_{ij}) or Max(B_{ij}); namely Max(Max(A_{ij}), Max(B_{ij})). Therefore, for all values of d (from 0 to Max(d)), M_{ijd} can be considered a 3D structure (Figure 15.1a). In a second step, the spectral forecast model involves a successive comparison between a matrix *M* at each distance/step d and a matrix *P* representing a sample (Figure 15.1b,c). Such a comparison can be made using a second equation, as follows:

$$S(d) = \frac{(M_{ijd} \times P_{ij})^2}{\left(\sum (M_{ijd})^2 \times \sum (P_{ij})^2 \right)}$$

where *S* is a similarity index and it represents the normalized dot-product of M_{ij} and P_{ij}. The M_{ij} stands for the predicted matrix at every discrete step and P_{ij} is the matrix originated from a sample of the same size/type with matrix *A* and matrix *B*. The similarity index can take values between 0 and 1. As the similarity between the corresponding *i,j* elements of matrix *M* and *P* increases, the similarity index *S* tends closer toward 1. As the differences between the values of the corresponding *i,j* elements of matrix *M* and *P* are more frequent, the similarity index *S* tends closer toward 0. Such a successive comparison between matrix *M* and *P* generates a number of similarity index values equal to Max(d). Thus, the trend shown by the similarity index values would indicate the tendency of the data from matrix *P* toward one of the two models (matrix *A* or matrix *B*) [250]. In such trends, the important cases are those that show a maximum similarity index between the two matrices *M* and *P* over distance d. These peak values may be a direct indication of the state represented by the data from matrix *P*. Thus, matrices *A* and *B* act as benchmarks for matrix *P*. Note that benchmarks are standard points of reference against which an object in the same category may be compared with. Thus, in bioinformatics, spectral forecast can be used as in the previous examples, where a matrix *A* can be the sample model and a matrix *B* can be a background model. Similarly, matrix *M* can take the place of the LLM (for a comparison, see the previous chapters). Next, a matrix *P* can be obtained based on the content of a sliding window over a sequence *z*. A comparison between matrix *P* and all matrices *M* may indicate the tendency for the next steps taken by the sliding window. Of course, the

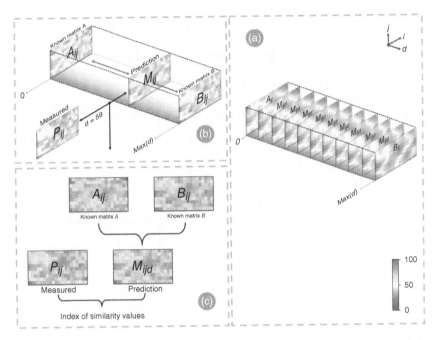

Figure 15.1 **A graphical representation of the spectral forecast equation and model.** An arbitrary example uses matrices of 10 × 20 in size. The matrices involved in the spectral forecast model are represented by a series of heatmaps. The values in the matrices are represented by colors (or shades of gray for the printed version) whose significance is described in the lower right corner of the figure. (a) Shows the M_{ijd} object in a three-dimensional shape. The panel graphically presents the effect of the spectral forecast equation and points to the notion of distance between the two matrices A and B. (b) Shows the spectral forecast model and the dynamics involved in performing calculations with this model. (c) Shows the steps by which the similarity score values are obtained.

model can be adapted to different experimental needs; thus, other implementation configurations can be found.

15.3 The Spectral Forecast Equation

The spectral forecast model and the spectral forecast equation can have different purposes. The spectral forecast equation may have multiple uses on a wide range of values. One of these uses would be a normalization between two unrelated matrices with the same dimension. For instance, elements of matrix A may contain integers in any range and the elements of matrix B may contain probability values. In this case, to provide a proportionality between the two matrices, the spectral forecast equation will mix the two matrices (A and B) in M according to the value set for distance d. In the case of two probability matrices A and B, the spectral forecast equation performs a normalization in favor of one of the matrices based on distance d (however, the equation as described here does not produce a probability matrix in M).

In other words, as the value of d decreases, the homologous elements of matrix M will be more similar to the values found in matrix A and less similar to the values from the elements of matrix B. In contrast, as the value of d will be higher, the values from the elements of matrix M will be more similar to the values from the elements of matrix B. Consequently, when $d = 50$ (presuming that $\text{Max}(d) = 100$), matrix M will represent a mix equally similar to matrix A and matrix B. The spectral forecast equation can be used in several conjunctures. A first possible use is related to vectors, where a vector M_i represents a combination between two vectors A_i and B_i at distance/index d:

$$M_{id} = \left[\left(\frac{d}{\text{Max}(A_i)} \right) \times A_i \right] + \left[\frac{(\text{Max}(d) - d)}{\text{Max}(B_i)} \times B_i \right]$$

where $\text{Max}(A_i)$ and $\text{Max}(B_i)$ represent the maximum values above the components of the two vectors. A second possibility is the computation of a matrix M_{ij} with values in between the counterpart elements of the two matrices A_{ij} and B_{ij}:

$$M_{ijd} = \left[\left(\frac{d}{\text{Max}(A_{ij})} \right) \times A_{ij} \right] + \left[\frac{(\text{Max}(d) - d)}{\text{Max}(B_{ij})} \times B_{ij} \right]$$

But what are counterpart elements over two or more than two matrices? The counterpart elements are the elements with the same i and j coordinates over any number of matrices of the same size. For instance, element $A_{1,1}$ and element $B_{1,1}$ are counterpart elements. On the same rationale, the element $A_{1,1}$, element $B_{1,1}$, and the element $M_{1,1,64}$ are counterpart elements (where 64 is an arbitrary value for d). Multidimensional mathematical objects can also be used (vectors – 1D; matrices – 2D; tensors – 3D, and so on).

15.4 The Spectral Forecast Inner Workings

The first question that arises is: how does the main equation work? and why it works? Let us consider two matrices: matrix A and matrix B. The maximum values over the elements of each matrix are known. For instance, consider that for matrix A the maximum value among the elements is 78 ($\text{Max}(A_{ij}) = 78$). In contrast, matrix B also shows a maximum value, namely 84 ($\text{Max}(B_{ij}) = 84$). In order to highlight how the equation works, a gradual experiment can be considered. The first step in deciphering the mode of operation, considers the two parts of the equation independently by using a matrix A. In the second step, two separate matrices are considered. A matrix A for the left side of the equation and a matrix B for the right side of the equation. These two approaches are meant to present the behavior of the equation in different setups.

15.4.1 Each Part on a Single Matrix

To understand how the equation works, several experiments can be performed separately on the two sides of the equation. Consider the two parts of the spectral forecast equation: the first part on the left and the second part on the right (Table 15.1). For example, what happens if the index d varies between extremes?

Table 15.1 Results of the two halves of the spectral forecast equation.

Left side	Right side
$M_{ijd} = \left[\left(\dfrac{d}{\text{Max}(A_{ij})} \right) \times A_{ij} \right]$	$M_{ijd} = \left[\dfrac{(\text{Max}(d) - d)}{\text{Max}(B_{ij})} \times A_{ij} \right]$
$M_{ijd} = \left[\left(\dfrac{78}{78} \right) \times A_{ij} \right]$	$M_{ijd} = \left[\dfrac{(78 - 78)}{78} \times A_{ij} \right]$
$M_{ijd} = [1 \times A_{ij}]$	$M_{ijd} = \left[\frac{0}{78} \times A_{ij} \right]$
$M_{ijd} = A_{ij}$	$M_{ijd} = [0 \times A_{ij}]$
	$M_{ijd} = 0$
$M_{ijd} = \left[\left(\dfrac{d}{\text{Max}(A_{ij})} \right) \times A_{ij} \right]$	$M_{ijd} = \left[\dfrac{(\text{Max}(d) - d)}{\text{Max}(A_{ij})} \times A_{ij} \right]$
$M_{ijd} = \left[\left(\frac{0}{78} \right) \times A_{ij} \right]$	$M_{ijd} = \left[\dfrac{(78 - 0)}{78} \times A_{ij} \right]$
$M_{ijd} = [0 \times A_{ij}]$	$M_{ijd} = \left[\frac{78}{78} \times A_{ij} \right]$
$M_{ijd} = 0$	$M_{ijd} = [1 \times A_{ij}]$
	$M_{ijd} = A_{ij}$

Shows the separate behavior of the two parts of the spectral forecast equation applied for a single matrix A. The first column of the table shows the behavior of the first part of the equation for $d = \text{Max}(d)$ and $d = 0$. The second column shows the behavior of the second part of the equation for $d = \text{Max}(d)$ and $d = 0$. For $d = \text{Max}(d)$ the first part of the equation will provide the original value in M_{ij}, whereas $d = \text{Max}(d)$ for the second part of the equation will provide a value of zero, and vice versa. Here it can be seen that the two parts of the equation have complementary results that can be exploited for different purposes.

Namely between 0 and Max(d)? Above, are the calculations for each of the two parts, for d taken as zero and d taken as Max(d).

When $d = $Max($A_{ij}$), the first part of the equation will provide the original value in M_{ij}. In contrast, when $d = 0$ (where Max(d) = Max(A_{ij})), the second part of the equation provides the original value in M_{ij} (Table 15.1 – diagonal on the table). Since they are complementary, $d = 0$ for the first part or $d = $Max($A_{ij}$) for the second part (where Max(d) = Max(A_{ij})), will lead to a value of zero (Table 15.1 – antidiagonal on the table). In other words, $d = $Max($A_{ij}$) on the left side of the equation returns back the value from A_{ij}, and on the right side, it returns zero. A $d = 0$ on the right side of the equation, it returns back the same value A_{ij}, and on the left side it returns zero.

15.4.2 Both Parts on a Single Matrix

To observe the relationship between the two parts of the equation, a new approach can be considered. To extend the example, consider both parts by using a single matrix, namely matrix A (Table 15.2).

Notice that for $d = 0$ or $d = $Max($A_{ij}$), the final result of the equation shows the initial value (A_{ij}) for any of the elements. In this case, *the two parts* of the equation *are complementary* to each other for any value of d. This complementarity allows for a competition between the two parts of the equation, namely between the left side and the right side. On the left side, the parenthesis of the fraction is not necessary because the division operation is first, then the multiplication operation follows. Thus, the parentheses were introduced for easy eye tracking.

Table 15.2 Experiments on the spectral forecast equation by using a single matrix.

If $d = 0$, Max(A_{ij}) = Max(d)	If $d = $Max($d$) = Max($A_{ij}$)
$M_{ijd} = \left[\left(\dfrac{78}{78}\right) \times A_{ij}\right] + \left[\dfrac{(78-78)}{78} \times A_{ij}\right]$	$M_{ijd} = \left[\left(\dfrac{0}{78}\right) \times A_{ij}\right] + \left[\dfrac{(78-0)}{78} \times A_{ij}\right]$
$M_{ijd} = [1 \times A_{ij}] + [0 \times A_{ij}]$	$M_{ijd} = [0 \times A_{ij}] + [1 \times A_{ij}]$
$M_{ijd} = A_{ij}$	$M_{ijd} = A_{ij}$

Here the complementarity between the two parts of the equation is highlighted. The table shows the behavior of the spectral forecast equation applied for a single matrix A. The first column of the table shows the behavior of the equation for $d = 0$. The second column shows the behavior of the equation for $d = $Max($d$). Since the equation uses the same matrix on both sides, for both $d = 0$ and $d = $Max($d$) the equation shows the same result, namely A_{ij}.

15.4.3 Both Parts on Separate Matrices

Up to this point, the experiment proved that the two parts of the equation show results that are complementary to each other. Here, two different matrices are introduced, namely matrix A and matrix B. The maximum value among the elements of matrix B is chosen arbitrarily at 84, and the maximum value above matrix A is chosen at 78. Thus, the experiment can be continued for two separate matrices. The introduction of a matrix B is made for the second part (the right side) of the equation (Table 15.3).

Table 15.3 Experiments on the spectral forecast equation by using two different matrices.

If $d = \text{Max}(A_{ij})$, $\text{Max}(d) = \text{Max}(A_{ij})$	If $d = 0$, $\text{Max}(d) = \text{Max}(B_{ij})$
$M_{ijd} = \left[\left(\dfrac{78}{78}\right) \times A_{ij}\right] + \left[\dfrac{(78-78)}{84} \times B_{ij}\right]$	$M_{ijd} = \left[\left(\dfrac{0}{78}\right) \times A_{ij}\right] + \left[\dfrac{(84-0)}{84} \times B_{ij}\right]$
$M_{ijd} = [1 \times A_{ij}] + \left[\dfrac{0}{84} \times B_{ij}\right]$	$M_{ijd} = [0 \times A_{ij}] + \left[\dfrac{84}{84} \times B_{ij}\right]$
$M_{ijd} = [A_{ij}] + [0 \times B_{ij}]$	$M_{ijd} = [0 \times A_{ij}] + [1 \times B_{ij}]$
$M_{ijd} = [A_{ij}] + [0]$	$M_{ijd} = [0] + [B_{ij}]$
$M_{ijd} = A_{ij}$	$M_{ijd} = B_{ij}$
$M_{ijd} = \left[\left(\dfrac{0}{78}\right) \times A_{ij}\right] + \left[\dfrac{(78-0)}{84} \times B_{ij}\right]$	$M_{ijd} = \left[\left(\dfrac{84}{78}\right) \times A_{ij}\right] + \left[\dfrac{(84-84)}{84} \times B_{ij}\right]$
$M_{ijd} = [0 \times A_{ij}] + \left[\dfrac{78}{84} \times B_{ij}\right]$	$M_{ijd} = \left[\dfrac{84}{78} \times A_{ij}\right] + \left[\dfrac{0}{84} \times B_{ij}\right]$
$M_{ijd} = [0] + \left[\dfrac{78}{84} \times B_{ij}\right]$	$M_{ijd} = \left[\dfrac{84}{78} \times A_{ij}\right] + [0 \times B_{ij}]$
$M_{ijd} = \left[\dfrac{78}{84} \times B_{ij}\right]$	$M_{ijd} = \left[\dfrac{84}{78} \times A_{ij}\right] + [0]$
$M_{ijd} = \left[\dfrac{\text{Max}(A_{ij})}{\text{Max}(B_{ij})} \times B_{ij}\right]$	$M_{ijd} = \left[\dfrac{84}{78} \times A_{ij}\right]$
	$M_{ijd} = \left[\dfrac{\text{Max}(B_{ij})}{\text{Max}(A_{ij})} \times A_{ij}\right]$

Here the complementarity between the two parts of the equation works the same as above and is highlighted by using two matrices, namely matrix A and matrix B. Thus, the table shows the behavior of the spectral forecast equation applied for matrix A on the left side and for matrix B on the right side. The first column of the table shows the behavior of the equation for $d = 0$. The second column shows the behavior of the equation for $d = \text{Max}(d)$. Since the equation uses the same matrix on both sides, for $d = 0$ and $d = \text{Max}(d)$, it shows the same result, namely A_{ij}.

The table above shows that extreme values for index d, favor either the value in element A_{ij} or the value in element B_{ij}. Nevertheless, three examples of manual calculation can shed more light on the behavior of the spectral forecast equation. Of course, a manual calculation is impractical for all elements of matrix M. Therefore, a calculation is presented here only for a single element of matrix M at different values taken by d, whereas the other elements (denoted by "?") can be calculated independently in the same manner. In a first example, the two matrices A and B contain the same maximum values and different values between the elements $A_{1,1}$ and $B_{1,1}$, where the element in $B_{1,1}$ contains the maximum value. In a second example, the maximum values remain the same as in the first case; however, the values in elements $A_{1,1}$ and $B_{1,1}$ are narrowed, where the value in $B_{1,1}$ is less than the maximum value. In a final example, the maximum values above the matrices differ and the values in the $A_{1,1}$ and $B_{1,1}$ elements also differ. Thus, the last example will show a realistic situation that contains a variability in data often encountered in practice.

15.4.4 Concrete Example 1

The spectral forecast equation is stated below for a recapitulation of those discussed up to this point. Two matrices can be considered: matrix A and matrix B (Table 15.4):

$$M_{ijd} = \left[\left(\frac{d}{\text{Max}(A_{ij})} \right) \times A_{ij} \right] + \left[\frac{(\text{Max}(d) - d)}{\text{Max}(B_{ij})} \times B_{ij} \right]$$

where i and j represent the coordinates inside the M matrix and d represents the distance/index. Therefore, for all possible values of d, M_{ijd} represents a three-dimensional structure. However, probably the best example is the calculation method for one element in the M matrix for various values of d (Table 15.5).

Table 15.4 Set up for example 1.

Matrix A			Matrix M			Matrix B		
1	33	100	$M[1,1]$	$M[1,2]$	$M[1,3]$	100	75	70
10	12	31	$M[2,1]$	$M[2,2]$	$M[2,3]$	48	55	61
36	43	10	$M[3,1]$	$M[3,2]$	$M[3,3]$	54	62	59

The table shows the values in the two matrices A and B and the element in the matrix M for which the calculation is made using the spectral forecast equation. The current example shows equal maximum values over both matrices (Max(d) = 100).

Table 15.5 Step-by-step calculation for example 1.

M[1,1]	Matrix *M*

For $d = 0$:

$$M_{1,1,0} = \left[\left(\frac{0}{100}\right) \times A_{ij}\right] + \left[\frac{(100-0)}{100} \times B_{ij}\right]$$

$$M_{1,1,0} = [0 \times A_{1,1}] + \left[\frac{100}{100} \times B_{1,1}\right]$$

$$M_{1,1,0} = [0 \times 1] + [1 \times 100]$$

$$M_{1,1,0} = [0] + [100]$$

$$M_{1,1,0} = 100$$

100	?	?
?	?	?
?	?	?

For $d = 1$:

$$M_{1,1,1} = \left[\left(\frac{1}{100}\right) \times A_{ij}\right] + \left[\frac{(100-1)}{100} \times B_{ij}\right]$$

$$M_{1,1,1} = \left[\frac{1}{100} \times A_{1,1}\right] + \left[\frac{99}{100} \times B_{1,1}\right]$$

$$M_{1,1,1} = [0.01 \times 1] + [0.99 \times 100]$$

$$M_{1,1,1} = [0.01] + [99]$$

$$M_{1,1,1} = 99.01$$

99.01	?	?
?	?	?
?	?	?

For $d = 50$:

$$M_{1,1,50} = \left[\left(\frac{50}{100}\right) \times A_{ij}\right] + \left[\frac{(100-50)}{100} \times B_{ij}\right]$$

$$M_{1,1,50} = \left[\frac{50}{100} \times A_{1,1}\right] + \left[\frac{50}{100} \times B_{1,1}\right]$$

$$M_{1,1,50} = [0.5 \times 1] + [0.5 \times 100]$$

$$M_{1,1,50} = [0.5] + [50]$$

$$M_{1,1,50} = 50.5$$

50.5	?	?
?	?	?
?	?	?

(Continued)

Table 15.5 (Continued)

M[1,1]	Matrix M

For $d = 70$:

$$M_{1,1,70} = \left[\left(\frac{70}{100}\right) \times A_{ij}\right] + \left[\frac{(100 - 70)}{100} \times B_{ij}\right]$$

$$M_{1,1,70} = \left[\frac{70}{100} \times A_{1,1}\right] + \left[\frac{30}{100} \times B_{1,1}\right]$$

$$M_{1,1,70} = [0.7 \times 1] + [0.3 \times 100]$$

$$M_{1,1,70} = [0.7] + [30]$$

$$M_{1,1,70} = 30.7$$

30.7	?	?
?	?	?
?	?	?

For $d = 100$:

$$M_{1,1,100} = \left[\left(\frac{100}{100}\right) \times A_{1,1}\right] + \left[\frac{(100 - 100)}{100} \times B_{1,1}\right]$$

$$M_{1,1,100} = \left[\frac{100}{100} \times A_{1,1}\right] + \left[\frac{0}{100} \times B_{1,1}\right]$$

$$M_{1,1,100} = [1 \times 1] + [0 \times 100]$$

$$M_{1,1,100} = [1] + [0]$$

$$M_{1,1,100} = 1$$

1	?	?
?	?	?
?	?	?

The values for the element M_{11} are calculated using the spectral forecast equation and different distances d, namely $d = 0$, $d = 1$, $d = 50$, $d = 70$, $d = 100$. The left side of the table shows the calculation method for each distance d and the right side of the table shows the final value for the element M_{11}. The behavior of the equation can be observed by following the events in the two complementary parts (the left side and the right side of the equation). The table indicates the proportionality of the values in M_{11}. The element M_{11} shows the same value as the value from the element B_{11} when $d = 0$ or the same value as the value from the element A_{11} when $d = 100$. Other distances d in the range 0 and 100 provide values in M_{11} that are proportionally dependent on the maximum values above the two matrices A and B, namely Max(d) = 100 for both.

The maximum values above each matrix are chosen for ease. In matrix A, the maximum value above the elements was set to 100 (Max(A_{ij}) = 100). Also, the maximum value above matrix B was also set to 100 (Max(B_{ij}) = 100). To calculate the value of an element in a matrix M at different distances d, we consider only element $M_{1,1}$ between $A_{1,1}$ and $B_{1,1}$. For an easy understanding of the complementarity provided by the equation, the element $A_{1,1}$ stores the value 1 and the element $B_{1,1}$ stores the value 100, which coincides with the maximum value over the B matrix ($A_{1,1} = 1$; $B_{1,1} = 100$). Table 15.5 shows the values for $M_{1,1}$ from distance $d = 0$ to distance $d = 100$.

Note that for $d = 0$, the element $M_{1,1}$ shows a value similar to the value in $B_{1,1}$, while for $d = 100$, the element $M_{1,1}$ shows a value of 1, which is similar to the value from element $A_{1,1}$. For intermediate values of d, the values in $M_{1,1}$ show a form of interpolation between $A_{1,1}$ and $B_{1,1}$.

15.4.5 Concrete Example 2

In a second example, we can consider the same maximum value for the two matrices and different values for the elements $B_{1,1}$ and $A_{1,1}$ (Table 15.6). Thus, the maximum values are positioned on other elements of the two matrices and the values from elements $A_{1,1}$ and $B_{1,1}$ are taken into account in the following example (Table 15.7).

In matrix A, the maximum value above the elements was set to 100 ($\text{Max}(A_{ij}) = 100$), whereas the maximum value above matrix B was also set to 100 ($\text{Max}(B_{ij}) = 100$). As in the previous example, to calculate the value of an element in a matrix M at different distances d, we consider only element $M_{1,1}$, between $A_{1,1}$ and $B_{1,1}$. This time, the element $A_{1,1}$ stores the value 10 and the element $B_{1,1}$ stores the value 20, which is different from the maximum value over the B matrix ($A_{1,1} = 10$; $B_{1,1} = 20$; $\text{Max}(A_{ij}) = \text{Max}(B_{ij}) = 100$).

Note that for $d = 0$, the element $M_{1,1}$ shows a value of 20, which is similar to the value shown in $B_{1,1}$. On the other hand, for $d = 100$, the element $M_{1,1}$ shows a value of 10, which is similar to the value from element $A_{1,1}$. Since the maximum values on both matrices are equal, the intermediate values of d allow for a form of interpolation between $A_{1,1}$ and $B_{1,1}$.

Table 15.6 Set up for example 2.

Matrix A			Matrix M			Matrix B		
10	33	49	$M[1,1]$	$M[1,2]$	$M[1,3]$	**20**	75	70
10	12	31	$M[2,1]$	$M[2,2]$	$M[2,3]$	48	55	61
36	43	100	$M[3,1]$	$M[3,2]$	$M[3,3]$	54	62	100

The table indicates the values in the two matrices A and B and the element in the matrix M for which the calculation is made using the spectral forecast equation. The current example shows equal maximum values over both matrices ($\text{Max}(d) = 100$).

Table 15.7 Step-by-step calculation for example 2.

M[1,1]	Matrix M

For $d = 0$:

$$M_{1,1,0} = \left[\left(\frac{0}{100}\right) \times A_{ij}\right] + \left[\frac{(100-0)}{100} \times B_{ij}\right]$$

$$M_{1,1,0} = [0 \times A_{1,1}] + \left[\frac{100}{100} \times B_{1,1}\right]$$

$$M_{1,1,0} = [0 \times 10] + [1 \times 20]$$

$$M_{1,1,0} = [0] + [20]$$

$$M_{1,1,0} = 20$$

20	?	?
?	?	?
?	?	?

For $d = 1$:

$$M_{1,1,1} = \left[\left(\frac{1}{100}\right) \times A_{ij}\right] + \left[\frac{(100-1)}{100} \times B_{ij}\right]$$

$$M_{1,1,1} = \left[\frac{1}{100} \times A_{1,1}\right] + \left[\frac{99}{100} \times B_{1,1}\right]$$

$$M_{1,1,1} = [0.01 \times 10] + [0.99 \times 20]$$

$$M_{1,1,1} = [0.1] + [19.8]$$

$$M_{1,1,1} = 19.9$$

19.9	?	?
?	?	?
?	?	?

For $d = 50$:

$$M_{1,1,50} = \left[\left(\frac{50}{100}\right) \times A_{ij}\right] + \left[\frac{(100-50)}{100} \times B_{ij}\right]$$

$$M_{1,1,50} = \left[\frac{50}{100} \times A_{1,1}\right] + \left[\frac{50}{100} \times B_{1,1}\right]$$

$$M_{1,1,50} = [0.5 \times 10] + [0.5 \times 20]$$

$$M_{1,1,50} = [5] + [10]$$

$$M_{1,1,50} = 15$$

15	?	?
?	?	?
?	?	?

For $d = 70$:

$$M_{1,1,70} = \left[\left(\frac{70}{100}\right) \times A_{ij}\right] + \left[\frac{(100-70)}{100} \times B_{ij}\right]$$

$$M_{1,1,70} = \left[\frac{70}{100} \times A_{1,1}\right] + \left[\frac{30}{100} \times B_{1,1}\right]$$

$$M_{1,1,70} = [0.7 \times 10] + [0.3 \times 20]$$

$$M_{1,1,70} = [7] + [6]$$

$$M_{1,1,70} = 13$$

13	?	?
?	?	?
?	?	?

Table 15.7 (Continued)

M[1,1]	Matrix M

For $d = 100$:

$$M_{1,1,100} = \left[\left(\tfrac{100}{100}\right) \times A_{1,1}\right] + \left[\tfrac{(100-100)}{100} \times B_{1,1}\right]$$

$$M_{1,1,100} = \left[\tfrac{100}{100} \times A_{1,1}\right] + \left[\tfrac{0}{100} \times B_{1,1}\right]$$

$$M_{1,1,100} = [1 \times 10] + [0 \times 20]$$

$$M_{1,1,100} = [10] + [0]$$

$$M_{1,1,100} = 10$$

10	?	?
?	?	?
?	?	?

The values for the element M_{11} are calculated using the spectral forecast equation and different distances d, namely: $d = 0$, $d = 1$, $d = 50$, $d = 70$, $d = 100$. The left side of the table shows the calculation method for each distance d, and the right side of the table shows the final value for element M_{11}. The behavior of the equation can be observed by following the events in the two complementary parts (the left side and the right side of the equation). The table indicates the proportionality of the values in M_{11}. The element M_{11} shows the same value as the value from the element B_{11} when $d = 0$ or the same value as the value from the element A_{11} when $d = 100$. Other distances d in the range 0 and 100 provide values in M_{11} that are proportionally dependent on the maximum values above the two matrices A and B. What differentiates the present example from the previous one is the narrowing of the difference between the values shown in A_{11} and B_{11}.

15.4.6 Concrete Example 3

Up to this point, the maximum values on both matrices were equal, namely 100. This last example shows different maximum values over the two matrices A and B, and different values in the elements $A_{1,1}$ and $B_{1,1}$ (Table 15.8).

In matrix A, the maximum value above the elements was set to 78 ($\text{Max}(A_{ij}) = 78$), whereas the maximum value above matrix B was also set to 84 ($\text{Max}(B_{ij}) = 84$). Again, to calculate the value of an element in a matrix M at different distances d, we consider only element $M_{1,1}$ between $A_{1,1}$ and $B_{1,1}$. To point out the complementarity provided by the equation, the element $A_{1,1}$ stores the value 10 and the element $B_{1,1}$ stores the value 20. Both values shown in the elements of $A_{1,1}$ and $B_{1,1}$ are different from the maximum values of their respective matrices ($A_{1,1} = 10$; $B_{1,1} = 20$; $\text{Max}(A_{ij}) = 78$; $\text{Max}(B_{ij}) = 84$). Note that $\text{Max}(d)$ represents the total number of discrete steps that can be taken from matrix A to matrix B. The value of $\text{Max}(d)$ is the maximum value found above the elements of the two matrices A and B (thus, it can be either $\text{Max}(A_{ij})$ or $\text{Max}(B_{ij})$; namely $\text{Max}(\text{Max}(A_{ij}), \text{Max}(B_{ij}))$. In the other two examples, both

Table 15.8 Set up for example 3.

Matrix A			Matrix M			Matrix B		
10	33	49	$M[1,1]$	$M[1,2]$	$M[1,3]$	**20**	75	70
10	12	31	$M[2,1]$	$M[2,2]$	$M[2,3]$	48	84	61
36	43	78	$M[3,1]$	$M[3,2]$	$M[3,3]$	54	62	9

The table indicates the values in the two matrices A and B and the element in the matrix M for which the calculation is made using the spectral forecast equation. The current example shows different maximum values for matrix A and matrix B.

matrices contained 100 as the maximum value and, of course, Max(d) was equal to 100. Here, the maximum value above the two matrices is 84. Thus, in this example, Max(d) = 84. Note that only in this example is the spectral forecast utility explored in depth (Table 15.9). Here, the value of the spectral forecast equation begins to be apparent. Matrix M will contain proportional values between matrix A and matrix B, function of the maximum values of matrix A and matrix B and the value shown by d. In other words, this proportion of values above the elements of matrix M shows how much the M matrix resembles matrix A and how much it resembles matrix B. This proportionality above the M matrix provides the possibility to use the spectral forecast equation as a mediator between two models. For example, a background model can be stored in matrix A and a sample model can be stored in matrix B. Thus, the M matrix can take the place of a more dynamic mediation matrix as in the case of the LLM, which was described in the previous chapters. This concludes the explanations regarding the functionality of the spectral forecast equation. In the case of vectors the equation works the same, namely the maximum values above the components of the vectors are taken into account.

15.5 Implementations

Up to this point, the theoretical parts of spectral forecast have been discussed. These theoretical explanations are less exciting without an implementation. Thus, in this subchapter, both the spectral forecast equation and the spectral forecast model are used. Note that the equation and the model intersect but may have different applicability. As described above, the spectral forecast equation can be used for both signals and matrices.

Table 15.9 Step-by-step calculation for example 3.

For $d = 0$:

$$M_{1,1,0} = \left[\left(\frac{0}{78}\right) \times A_{ij}\right] + \left[\frac{(84 - 0)}{84} \times B_{ij}\right]$$

$$M_{1,1,0} = [0 \times A_{1,1}] + \left[\frac{84}{84} \times B_{1,1}\right]$$

$$M_{1,1,0} = [0 \times 10] + [1 \times 20]$$

$$M_{1,1,0} = [0] + [20]$$

$$M_{1,1,0} = 20$$

For $d = 1$:

$$M_{1,1,1} = \left[\left(\frac{1}{78}\right) \times A_{ij}\right] + \left[\frac{(84 - 1)}{84} \times B_{ij}\right]$$

$$M_{1,1,1} = \left[\frac{1}{78} \times A_{1,1}\right] + \left[\frac{83}{84} \times B_{1,1}\right]$$

$$M_{1,1,1} = [0.012820513 \times 10] + [0.98809523809 \times 20]$$

$$M_{1,1,1} = [0.128205128] + [19.7619047619]$$

$$M_{1,1,1} = 19.8901098899$$

For $d = 50$:

$$M_{1,1,50} = \left[\left(\frac{50}{78}\right) \times A_{ij}\right] + \left[\frac{(84 - 50)}{84} \times B_{ij}\right]$$

$$M_{1,1,50} = \left[\frac{50}{78} \times A_{1,1}\right] + \left[\frac{34}{84} \times B_{1,1}\right]$$

$$M_{1,1,50} = [0.641025641 \times 10] + [0.40476190476 \times 20]$$

$$M_{1,1,50} = [6.41025641] + [8.09523809524]$$

$$M_{1,1,50} = 14.5054945052$$

For $d = 70$:

$$M_{1,1,70} = \left[\left(\frac{70}{78}\right) \times A_{ij}\right] + \left[\frac{(84 - 70)}{84} \times B_{ij}\right]$$

$$M_{1,1,70} = \left[\frac{70}{78} \times A_{1,1}\right] + \left[\frac{14}{84} \times B_{1,1}\right]$$

$$M_{1,1,70} = [0.897435897 \times 10] + [0.16666666666 \times 20]$$

$$M_{1,1,70} = [8.974358974] + [3.33333333333]$$

$$M_{1,1,70} = 12.3076923073$$

(Continued)

Table 15.9 (Continued)

For $d = 84$:

$$M_{1,1,84} = \left[\left(\frac{84}{78} \right) \times A_{ij} \right] + \left[\frac{(84 - 84)}{84} \times B_{ij} \right]$$

$$M_{1,1,84} = \left[\frac{84}{78} \times A_{1,1} \right] + \left[\frac{0}{84} \times B_{1,1} \right]$$

$$M_{1,1,84} = [1.07692307692 \times 10] + [0 \times 20]$$

$$M_{1,1,84} = [10.7692307692] + [0]$$

$$M_{1,1,84} = 10.7692307692$$

The values for element M_{11} are calculated using the spectral forecast equation and different distances d, namely: $d = 0$, $d = 1$, $d = 50$, $d = 70$, $d = 84$. The left side of the table shows the calculation method for each distance d and the right side of the table shows the final value for the element M_{11}. The behavior of the equation can be observed by following the events in the two complementary parts (the left side and the right side of the equation). The table indicates the proportionality of the values in M_{11}. The element M_{11} shows the same value as the value from the element B_{11} when $d = 0$ or the same value as the value from the element A_{11} when $d = 84$. Other distances d in the range 0 and 84 provide values in M_{11} that are proportionally dependent on the maximum values above the two matrices A and B. Matrix A shows a maximum value of 78 and matrix B shows a maximum value of 84. Note that the present example contains both different values for A_{11} and B_{11} and different maximum values.

15.5.1 Spectral Forecast for Signals

One of the easiest applications to understand the spectral forecast equation is related to signals/vectors. A question that can be asked would be: given two signals A and B, what would a third signal M look like if it must resemble both signals in a certain proportion? For instance, a signal M may be required to resemble signal A in a proportion of 23% and signal B in a remaining proportion of 77%. To begin the example, two different number sequences may represent the A and B signals. Each number in the sequence can be the signal strength and the position of the terms will represent time/steps. Thus, two signals are considered, a signal A:

$$A = 10.3, 23.4, 44.8, 63.2, 44.1, 35.1, 46.5, 62.6, 50.4, 28.9, 24.7$$

and a second signal B:

$$B = 18.8, 43.1, 52.2, 45.5, 46.8, 46.6, 67.9, 66.3, 70.4, 62, 39.7$$

Note that the above signals are only some random numbers as the meaning of the signals is not important. In Additional algorithm 15.1, the spectral forecast equation is used in a simple implementation that constructs an intermediate signal (M) between signal A and signal B based on distance $d = 33$. Note that value 33 was arbitrarily chosen for this demonstration.

Additional algorithm 15.1 Note that the source code is in context and works with copy/paste.

```
<script>

//Spectral forecast for signals
var tA = [];
var tB = [];

var A ='10.3,23.4,44.8,63.2,44.1,35.1,46.5,62.6,50.4,28.9,24.7';
var M ='';
var B ='18.8,43.1,52.2,45.5,46.8,46.6,67.9,66.3,70.4,62,39.7';

var tA = A.split(',');
var maxA = Math.max.apply(null, tA);

var tB = B.split(',');
var maxB = Math.max.apply(null, tB);

var d = 33;

var max = Math.max(maxA, maxB)

for(var i=0; i<tA.length; i++) {
    var tmp = ((d/maxA)*tA[i])+(((max-d)/maxB)*tB[i]);
    M += tmp.toFixed(2);
    if(i<tA.length-1){M += ','}
}

document.write('Signal A:<br>'+A+'<br>');
document.write('Max(A[i]):'+maxA+'<hr>');

document.write('Signal M:<br>'+M+'<hr>');

document.write('Signal B:<br>'+B+'<br>');
document.write('Max(B[i]):'+maxB+'<hr>');

</script>

Output:

Signal A:
10.3,23.4,44.8,63.2,44.1,35.1,46.5,62.6,50.4,28.9,24.7
Max(A[i]):63.2

Signal M:
15.37,35.12,51.12,57.17,47.89,43.08,60.35,67.91,63.72,48.03,33.99
```

(Continued)

Additional algorithm 15.1 (Continued)

```
Signal B:
18.8,43.1,52.2,45.5,46.8,46.6,67.9,66.3,70.4,62,39.7
Max(B[i]):70.4
```

In Additional algorithm 15.1, both signals A and B are stored as string variables. The numeric values in these string variables are initially extracted as before, by using the native Javascript *split* function and are then inserted into array variables (i.e. tA, tB). To find the maximum values in these signals (i.e. maxA, maxB), the native Javascript *max* function is used above the array variables. Then, the value for variable d is declared and the maximum value for variable d is further calculated. Notice that the maximum value above the two signals becomes the maximum value for d (Math.max(maxA, maxB)). In the last step, the M signal is built using the spectral forecast equation and a loop that iterates over the positions of both signals A and B:

$$M_{id} = \left[\left(\frac{d}{\mathrm{Max}(A_i)} \right) \times A_i \right] + \left[\frac{(\mathrm{Max}(d) - d)}{\mathrm{Max}(B_i)} \times B_i \right]$$

Which in the above implementation supports the following translation:

$$tmp = \left[\left(\frac{d}{\max\ A} \right) \times tA_i \right] + \left[\frac{(\max\ -d)}{\max\ B} \times tB_i \right]$$

Inside the loop, a variable *tmp* holds the value calculated by the spectral forecast equation, then, this value (*tmp*) is added to variable M in which the intermediate signal is constructed. At the end of these steps, the contents of variables A, M, B, and the maximum values per signal (i.e. maxA, maxB) are displayed (Additional algorithm 15.1).

15.5.2 What Does the Value of d Mean?

Here the value for d is set manually, but the number can be misleading. The output above shows three signals: Signal A, signal B as well as signal M, which was built for a distance $d = 33$. In the above case, signal A contains a maximum value of 63.2 and signal B contains a maximum value of 70.4. A second maximization between the two maximum values (i.e. maxA, maxB) indicates the upper bound (maximum value for d) for distance d (Max(d) = 70.4). Thus, the maximum value for distance d (Max(d)) represents 100%. Consequently, a value for d of 33 indicates the following proportions:

$$G\% = \frac{100}{\max} \times d = \frac{100}{70.4} \times 33 = 46.87\%$$

where $G\%$ expresses the similarity of signal M with one of the two reference signals, namely A and B. One way of interpreting the value of $G\%$ is that 46.87%

expresses the similarity of signal M with signal A, whereas the remaining 53.13% (100 – 46.87%) expresses how similar signal M is to the signal found in B. But what happens if the value of d exceeds the maximum value over the two signals A and B? For instance, suppose that its value is well above Max = 70.4, namely 1000. Thus, for $d = 1000$, the following results are true:

Signal A: 10.3, 23.4, 44.8, 63.2, 44.1, 35. 1, 46.5, 62.6, 50.4, 28.9, 24.7

Max(A_i): 63.2

Signal M: −85.27, −198.86, 19.58, 399.19, 79.81, −59.95, −160.83, 115.04, −132.13, −361.40, −133.40

Signal B: 18.8, 43.1, 52.2, 45.5, 46.8, 46.6, 67.9, 66.3, 70.4, 62, 39.7

Max(B_i): 70.4

Note that the M signal becomes very contrasted for values of d above the maximum value found above the two signals A and B. The larger d becomes ($d >$ max), or the smaller d becomes (below zero; $d < 0$), the more extreme the characteristics of M are in favor of one of the signals (A or B). Nevertheless, the output from Additional algorithm 15.1, that shows the signal from variable M, is not very easy to digest in this current state. To make a simple graphical display of the relationship between these three signals (A, B, and M), a second implementation shown in Additional algorithm 15.2 brings only a few additions to what was discussed in Additional algorithm 15.1, as follows:

Additional algorithm 15.2 Note that the source code is in context and works with copy/paste.

```
<canvas id="bio" height="300" width="1100"></canvas>

<script>

//Spectral forecast for signals
var tA = [];
var tB = [];

var A ='10.3,23.4,44.8,63.2,44.1,35.1,46.5,62.6,50.4,28.9,24.7';
var M ='';
var B ='18.8,43.1,52.2,45.5,46.8,46.6,67.9,66.3,70.4,62,39.7';

var tA = A.split(',');
var maxA = Math.max.apply(null, tA);

var tB = B.split(',');
var maxB = Math.max.apply(null, tB);
```

(Continued)

Additional algorithm 15.2 (Continued)

```javascript
var d = 60;

var max = Math.max(maxA, maxB);

for(var i=0; i<tA.length; i++) {
    var tmp=((d/maxA)*tA[i])+(((max-d)/maxB)*tB[i]);
    M+=tmp.toFixed(2);
    if(i<tA.length-1){M+=','}
}

Chart(A, '#ff0000', 'y')
Chart(M, '#000000', 'n')
Chart(B, '#ff0000', 'n')

function Chart(q,c,e) {

    var s = q.split(",");
    var mx = Math.max.apply(null, s);

    var canvas = document.getElementById('bio');

    var w = canvas.width;
    var h = canvas.height;

    if (canvas.getContext) {

        var ctx = canvas.getContext('2d');

        if(e=='y'){ctx.clearRect(0, 0, w, h);}

        ctx.moveTo(0, 0);
        ctx.beginPath();

        for (var i=0; i<=s.length-1; i++)
        {
            var y = h-((h / mx) * s[i]);
            var x = (w / s.length) * i;

            ctx.lineTo(x, y);
        }
        ctx.lineWidth = 2;
        ctx.strokeStyle = c;
        ctx.stroke();
    }
}

</script>
```

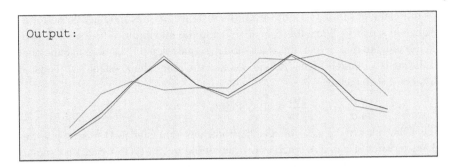

```
Output:
```

A function *Chart* that draws signals on a canvas object is added to Additional algorithm 15.2. This function is called for each signal separately, while the previous plots can be maintained. The *Chart* function receives three parameters. A parameter (q) containing the signal (in this case 11 values – each value separated by a delimiter, namely ", "). A second parameter is the color of the line identifying each signal (c). A third parameter (e) that indicates to the function whether the canvas object should be cleared or not ('y' for clear/erase). Inside the *Chart* function, the maximum value above the sequence of numbers (s) is detected and this value is stored in the *mx* variable. The *Chart* function contains a loop that makes a number of iterations (i) equal to the number of terms present in the number sequence (s). Inside the main loop, the coordinates above the canvas object are calculated based on the maximum value, namely according to the value found in the *mx* variable. Thus, the *y*-axis is represented by the height (h) of the canvas object divided by the value in the *mx* variable (h/mx), and the result is multiplied by the current value in the number sequence (s[i]). To position the zero values at the bottom of the chart, the *y*-axis is reversed by subtracting the result (the *y* value) from the height (h) of the canvas object:

$$y = h - \left(\frac{h}{mx} \times s_i \right) \qquad .$$

In contrast, the *x*-axis is calculated by dividing the length of the canvas object by the total number of terms in the sequence of numbers (w/s.length), and the result is multiplied by the iteration number (i):

$$x = \left(\frac{w}{|s|} \times i \right)$$

Once the two values are computed, the line is drawn from the previous coordinates to the current *x* and *y* coordinates (ctx.lineTo(x, y);). This concludes the discussions related to the *Chart* function. The output shown by Additional algorithm 15.2 already indicates the relationship of signal *M* with the other two reference signals. Signals *A* and *B* remained the same as in the previous example

(Additional algorithm 15.1); however, the distance (d) was modified to 60 for a change in scenario. Since this time $d = 60$ and the maximum value over the two signals is again max $= 70.4$, the black plotted signal (signal P) is more like the one of the signals, namely signal A, and less like the other signal (signal B). Thus, a distance $d = 60$ means:

$$G\% = \frac{100}{max} \times d = \frac{100}{70.4} \times 60 = 85.22\%$$

The value shown by $G\%$, of 85.22%, indicates that signal M resembles the characteristics of signal A, whereas the remaining 14.78% ($100 - 85.22\%$) of signal M resembles the signal found in B. Several experiments can be performed using Additional algorithm 15.2 by changing the value for variable d. For ease, a more sophisticated implementation of Additional algorithm 15.2 can be found online on the main webpage of this book. These two implementations conclude the spectral forecast utility for signals.

15.5.3 Spectral Forecast for Matrices

A more complex example that uses the spectral forecast equation is related to matrices. A question that can be asked this time would be: given two matrices A and B, what would a third matrix (M) look like if it must resemble both A and B in a certain proportion? For instance, a matrix M may be required to resemble matrix A in a proportion of 60% and matrix B for the remaining proportion of 40%. To begin the example, different matrices may represent the A matrix and the B matrix. Thus, the two matrices from Table 15.8 are considered here, a matrix A:

$$A = \begin{pmatrix} 10 & 33 & 49 \\ 10 & 12 & 31 \\ 36 & 43 & 78 \end{pmatrix}$$

and a second matrix B:

$$B = \begin{pmatrix} 20 & 75 & 70 \\ 48 & 84 & 61 \\ 54 & 62 & 9 \end{pmatrix}$$

A more complex implementation related to matrices and the spectral forecast equation is shown in Additional algorithm 15.3. In this case, the spectral forecast equation provides an intermediate matrix (M) based on a specific distance ($d = 42$), whose corresponding values are found between matrix A and matrix B. Note that value 42 was arbitrarily chosen for this demonstration. In Additional algorithm 15.3, both matrices A and B are stored as string variables (tA and tB). The numeric values in these string variables are initially extracted as before, by using the native Javascript *split* function and are then inserted into array variables (i.e. A and B). To find the element containing the maximum value above a matrix (i.e. `maxA`, `maxB`), the native Javascript *max* function is used above the array variables (i.e. `A` and `B`). Then, the value for variable *d* is declared (`d=42`) and the maximum value for variable *d* is further calculated. Notice that the maximum value above the two matrices *A* and *B* becomes the maximum value for *d* (`Math.max(maxA, maxB)`). In the last step, the *M* matrix is built using the spectral forecast equation, and a loop that iterates over the elements of both matrices *A* and *B*:

$$M_{ijd} = \left[\left(\frac{d}{\text{Max}(A_{ij})} \right) \times A_{ij} \right] + \left[\frac{(\text{Max}(d) - d)}{\text{Max}(B_{ij})} \times B_{ij} \right]$$

which in the above implementation supports the following translation:

$$M_{ijd} = \left[\left(\frac{d}{\max A} \right) \times A_{ij} \right] + \left[\frac{(\max - d)}{\max B} \times B_{ij} \right]$$

Inside the loop, a matrix *M* holds the values calculated by the spectral forecast equation. At the end of these steps, the contents of variables *A*, *M*, *B*, and the maximum values per matrix (i.e. `maxA`, `maxB`) are displayed (Additional algorithm 15.3).

Additional algorithm 15.3 Note that the source code is in context and works with copy/paste.

```
<script>

//Spectral forecast for matrices

var tA ='|10\t33\t49' +
        '|10\t12\t31' +
        '|36\t43\t78';

var tB ='|20\t75\t70' +
        '|48\t84\t61' +
        '|54\t62\t9';

var tmp;

tmp = load(tA);
var A = tmp[0];
var maxA = tmp[1];

tmp = load(tB);
var B = tmp[0];
var maxB = tmp[1];
```

(Continued)

Additional algorithm 15.3 (Continued)

```
var M = [];
var d = 42;

var max = Math.max(maxA, maxB)

for(var i=0; i<A.length; i++) {
    M[i]=[];
    for(var j=0; j<A[i].length; j++){
        M[i][j]=((d/maxA)*A[i][j])+(((max-d)/maxB)*B[i][j]);
        M[i][j]=M[i][j].toFixed(2);
    }
}

document.write('Matrix A:<br>'+SMC(A)+'<br>');
document.write('Max(A[i,j]):'+maxA+'<hr>');

document.write('Matrix M:<br>'+SMC(M)+'<hr>');

document.write('Matrix B:<br>'+SMC(B)+'<br>');
document.write('Max(B[i,j]):'+maxB+'<hr>');

//LOAD MATRICES FROM STRINGS
function load(t){

    var n = [];
    var m = [];
    var L = [];

    m = t.split('|');
    var max = 0;

    for(var i=1; i<m.length; i++) {
        L[i-1]=[];
        n = m[i].split('\t');
        for(var j=0; j<n.length; j++){
            L[i-1][j]=Number(n[j]);
            if(max<=L[i-1][j]){max=L[i-1][j];}
        }
    }

    return [L, max];
}

// SHOW MATRIX CONTENT
function SMC(m) {
    var r = "<table border=1>";
```

```
    for(var i=0; i<m.length; i++) {
        r += "<tr>";
        for(var j=0; j<m[i].length; j++){
            r += "<td>"+m[i][j]+"</td>";
        }
        r += "</tr>";
    }
    r += "</table>";

    return r;
}

</script>

Output:

Matrix A:
 10 33 49
 10 12 31
 36 43 78

Max(A[i,j]):78

Matrix M:
 15.38 55.27 61.38
 29.38 48.46 47.19
 46.38 54.15 46.50

Matrix B:
 20 75 70
 48 84 61
 54 62 9

Max(B[i,j]):84
```

Note that the value for *d* is set manually, but the value can be misleading just as explained in the previous subchapter. The output above shows three matrices: Matrix *A*, matrix *B*, and matrix *M*, which was built for a distance $d = 42$. In the above case, matrix *A* contains a maximum value of 78 and matrix *B* contains a maximum value of 84. A second maximization between the two maximum values (i.e. maxA, maxB) indicates the upper bound (maximum value for *d*) for distance $d = 84$. Thus, the maximum value for distance *d* represents 100%, just as explained in the previous subchapter. Thus, a value for d of 42 indicates the following proportions:

$$G\% = \frac{100}{\text{max}} \times d = \frac{100}{84} \times 42 = 50\%$$

where $G\%$ indicates that the values in matrix M are equally similar to the values in matrices A and B. Namely, that overall, matrix M resembles matrix A in a proportion of 50%, and matrix B in a proportion of 50% (100 – 50%). This can also be directly observed in the output of Additional algorithm 15.3. But what happens if the value of d exceeds the maximum value over the two matrices (max)? This topic was covered in the previous subchapter, but the conclusion was that the larger d becomes ($d >$ max), or the smaller d becomes (below zero; $d < 0$), the more extreme the characteristics of M are in favor of one of the matrices (A or B). A more sophisticated implementation for Additional algorithm 15.3 can be found online on the main webpage of this book. This implementation concludes the spectral forecast utility for matrices.

15.6   The Spectral Forecast Model for Predictions

Up to this point, the utility of the spectral forecast equation for vectors and matrices has been described. But the equation is somewhat separate in utility from the model itself. To make a prediction, the spectral forecast equation is used in conjunction with a similarity equation. The prediction process consists in a successive comparison between a vector or matrix M and a similar object P of the same size, named here "the measurement."

15.6.1   The Spectral Forecast Model for Signals

The implementation in Additional algorithm 15.4 shows a version of the model applied to signals. First, three signals are declared. The two reference signals A and B, and signal P, which in theory should come from a measurement and about which a prediction of the evolution in time is desired. However, the three signals are in fact represented by a series of random numbers which have no special meaning:

$A = 10.3, 23.4, 44.8, 63.2, 44.1, 35.1, 46.5$

$B = 18.8, 43.1, 52.2, 45.5, 46.8, 46.6, 67.9$

$P = 18.8, 43.1, 52.2, 45.5, 46.8, 46.6, 67.9$

Note that for verification purposes, signal P is the same as signal B. Initially, the maximum values present in each signal are detected (i.e. `maxA` and `maxB`). The maximum value for distance d is set by a new maximization between the maximum values above the two signals A and B (i.e. `Math.max(maxA, maxB)`).

Then the *forecast* function is called, which iterates a variable *d* from zero to Max(*d*) (i.e. max). At each iteration, the *spectral* function is called. The value of *d* is injected into the *spectral* function to calculate the values that make up the *M* signal:

$$M_{id} = \left[\left(\frac{d}{\max A} \right) \times A_i \right] + \left[\frac{(\max - d)}{\max B} \times B_i \right]$$

Once the *M* signal is calculated, the *spectral* function passes the signal as a string to the *index* function, which compares signal *M* to signal *P*, by using the equation described at the beginning of the chapter, namely:

$$S(d) = \frac{(M_{id} \times P_i)^2}{\left(\sum (M_{id})^2 \times \sum (P_i)^2 \right)}$$

The *index* function returns the value of the similarity score (which is situated between zero and one) back to the *spectral* function, which in turn returns the similarity score to the *forecast* function. The *forecast* function adds the returned value to a string variable (t) and continues to the next iteration of *d*. The process continues until the iteration reaches Max(*d*), after which the result is printed in the output.

Additional algorithm 15.4 Note that the source code is in context and works with copy/paste.

```
<script>

//Spectral forecast model for signals
var t = '';

var tA = [];
var tB = [];

var A ='10.3,23.4,44.8,63.2,44.1,35.1,46.5';
var B ='18.8,43.1,52.2,45.5,46.8,46.6,67.9';

var P ='18.8,43.1,52.2,45.5,46.8,46.6,67.9';

var tA = A.split(',');
var tB = B.split(',');

var maxA = Math.max.apply(null, tA);
var maxB = Math.max.apply(null, tB);

var tP = P.split(',');

var max = Math.max(maxA, maxB)
```

(Continued)

Additional algorithm 15.4 (Continued)

```javascript
document.write(forecast()+'<hr>');

function forecast(){
    var t = '';
    for(var d=0; d<max; d++) {
        var tmp = spectral(d);
        t+='For <i>d</i>>='+d+', the index <i>s</i>['+d+'] = '+tmp;
        if(d<max-1){t+='<br>';}
    }
    return t;
}

function spectral(d){
    var M='';
    for(var i=0; i<tA.length; i++) {
        var tmp=((d/maxA)*tA[i])+(((max-d)/maxB)*tB[i]);
        M+=tmp.toFixed(2);
        if(i<tA.length-1){M+=','}
    }
    return index(M);
}

function index(M){
    var u=0;
    var k=0;
    var h=0;

    var tM = M.split(',');

    for(var i=0; i<tM.length; i++) {
        u+=(Number(tP[i])*Number(tM[i]));    //(Pi*Mi)
        k+=Number(tP[i])**2;                 //(Pi^2)
        h+=Number(tM[i])**2;                 //(Mi^2)
    }
    //s[d] = (Pi*Mi)^2 / ((Pi^2) * (Mi^2))
    return u**2/(k*h);
}

</script>

Output:

For d=0, the index s[0] = 1
For d=1, the index s[1] = 0.9999850076555136
For d=2, the index s[2] = 0.9999387427857133
For d=3, the index s[3] = 0.9998625844708812
For d=4, the index s[4] = 0.9997538595896938
```

```
For d=5, the index s[5] = 0.9996155238016333
For d=6, the index s[6] = 0.9994426889240177
For d=7, the index s[7] = 0.9992402483561409
For d=8, the index s[8] = 0.9990046202642195
For d=9, the index s[9] = 0.9987358369579232
For d=10, the index s[10] = 0.9984347350782586
For d=11, the index s[11] = 0.9981021166386677
For d=12, the index s[12] = 0.9977365440429127
For d=13, the index s[13] = 0.9973328716473806
For d=14, the index s[14] = 0.9969017570641806
For d=15, the index s[15] = 0.99643140611362
For d=16, the index s[16] = 0.9959290801798455
For d=17, the index s[17] = 0.9953880581456316
For d=18, the index s[18] = 0.9948180212832275
For d=19, the index s[19] = 0.9942107403546273
For d=20, the index s[20] = 0.9935714681355112
For d=21, the index s[21] = 0.9928947325515898
For d=22, the index s[22] = 0.9921789747583631
For d=23, the index s[23] = 0.9914309146149758
For d=24, the index s[24] = 0.9906480976644273
For d=25, the index s[25] = 0.9898251220295177
For d=26, the index s[26] = 0.988967817423824
For d=27, the index s[27] = 0.9880784807152327
For d=28, the index s[28] = 0.9871384222209092
For d=29, the index s[29] = 0.986179454056766
For d=30, the index s[30] = 0.98516419395104
For d=31, the index s[31] = 0.9841343077509911
For d=32, the index s[32] = 0.9830572708198193
For d=33, the index s[33] = 0.981937647309787
For d=34, the index s[34] = 0.9807946892709519
For d=35, the index s[35] = 0.9795908670430284
For d=36, the index s[36] = 0.9783733389854155
For d=37, the index s[37] = 0.9770978776621292
For d=38, the index s[38] = 0.9757990283415907
For d=39, the index s[39] = 0.9744469929981138
For d=40, the index s[40] = 0.9730741054818983
For d=41, the index s[41] = 0.9716689767735215
For d=42, the index s[42] = 0.9702021650263695
For d=43, the index s[43] = 0.9687159662012149
For d=44, the index s[44] = 0.967172114599994
For d=45, the index s[45] = 0.9656128531974031
For d=46, the index s[46] = 0.9639918675267266
For d=47, the index s[47] = 0.9623598914075397
For d=48, the index s[48] = 0.9606690916627004
For d=49, the index s[49] = 0.9589656557914885
For d=50, the index s[50] = 0.9571983617356936
For d=51, the index s[51] = 0.9554008395448116
For d=52, the index s[52] = 0.9535833375181383
```

(Continued)

Additional algorithm 15.4 (Continued)

```
For d=53, the index s[53] = 0.9517015962081136
For d=54, the index s[54] = 0.9497902124712339
For d=55, the index s[55] = 0.9478396617738604
For d=56, the index s[56] = 0.945844869165415
For d=57, the index s[57] = 0.9438186139592625
For d=58, the index s[58] = 0.9417479533657269
For d=59, the index s[59] = 0.9396492733245702
For d=60, the index s[60] = 0.9375211980712375
For d=61, the index s[61] = 0.9353402867772599
For d=62, the index s[62] = 0.9331269614492081
For d=63, the index s[63] = 0.9308689716146904
For d=64, the index s[64] = 0.9285816216405253
For d=65, the index s[65] = 0.9262613598536346
For d=66, the index s[66] = 0.923891509787129
For d=67, the index s[67] = 0.9215162769343779
```

The output from Additional algorithm 15.1 shows a range of similarity scores between 1 and 0.9215. Again, a value of zero means a total dissimilarity between signal P and signal M, whereas a value of 1 means that the two signals are identical. As mentioned earlier, signal P and signal B are identical specifically for testing purposes. Thus, between signal M and signal P, the expectation at distance zero will be a similarity score equal to 1. On the other hand, the range of scores should decrease indicating a smaller and smaller similarity between signal P and signal M at higher values of d. What can be seen in the Additional algorithm 15.4 output is in line with the expectations mentioned above. Namely, for $s[d = 0]$ the value is exactly 1. The values of the similarity score decrease as d increases, and reach a final value of 0.9215 on the last step of d ($s[d = 67]$). Nevertheless, the output from Additional algorithm 15.4 is not easy to digest in its current numerical state. To make sense of the evolution of these similarity scores, a simple graphical display may point out a useful trend. Thus, a second implementation shown in Additional algorithm 15.5 brings only a simple chart, which is able to graphically express even small differences in the similarity score values:

Additional algorithm 15.5 Note that the source code is in context and works with copy/paste.

```
<canvas id="bio" height="300" width="1100"></canvas>

<script>

//Spectral forecast model for signals
var t = '';
```

```
var tA = [];
var tB = [];

var A ='10.3,23.4,44.8,63.2,44.1,35.1,46.5';
var B ='18.8,43.1,52.2,45.5,46.8,46.6,67.9';
var P ='18.8,43.1,52.2,45.5,46.8,46.6,67.9';

var tA = A.split(',');
var tB = B.split(',');

var maxA = Math.max.apply(null, tA);
var maxB = Math.max.apply(null, tB);

var tP = P.split(',');

var max = Math.max(maxA, maxB)

Chart(forecast(), '#000000', 'y');

function forecast(){
    var t = '';
    for(var d=0; d<max; d++) {
        var tmp = spectral(d);
        t+=tmp;
        if(d<max-1){t+=',';}
    }
    return t;
}

function spectral(d){
    var M='';
    for(var i=0; i<tA.length; i++) {
        var tmp=((d/maxA)*tA[i])+(((max-d)/maxB)*tB[i]);
        M+=tmp.toFixed(2);
        if(i<tA.length-1){M+=','}
    }
    return index(M);
}

function index(M){
    var u=0;
    var k=0;
    var h=0;

    var tM = M.split(',');
```

(Continued)

Additional algorithm 15.5 (Continued)

```
    for(var i=0; i<tM.length; i++) {
        u+=(Number(tP[i])*Number(tM[i]));   //(Pi*Mi)
        k+=Number(tP[i])**2;                //(Pi^2)
        h+=Number(tM[i])**2;                //(Mi^2)
    }
    //s[d] = (Pi*Mi)^2 / ((Pi^2) * (Mi^2))
    return u**2/(k*h);
}


// just a chart
function Chart(q,c,e) {

    var s = q.split(",");
    var mx = Math.max.apply(null, s);
    var mn = Math.min.apply(null, s);

    var canvas = document.getElementById('bio');

    var w = canvas.width;
    var h = canvas.height;

    if (canvas.getContext) {

        var ctx = canvas.getContext('2d');

        if(e=='y'){ctx.clearRect(0, 0, w, h);}

        ctx.moveTo(0, 0);
        ctx.beginPath();

        for (var i=0; i<=s.length-1; i++)
        {
            var y = (h-80) - ((h-80) / (mx-mn)) * (s[i]-mn);
            var x = ((w-80) / s.length) * i;

            ctx.lineTo(x+40, y+40);

            ctx.font = "20px Arial";

            if(i==0){
                ctx.textAlign = 'left';
                ctx.fillText(Number(s[i]).toFixed(5), x+45, y+60);
            }

            if(i==s.length-1){
                ctx.textAlign = 'right';
```

```
                    ctx.fillText(Number(s[i]).toFixed(5), x+35, y+60);
        }
}

ctx.lineWidth = 2;
ctx.strokeStyle = c;
ctx.stroke();

//left vertical line
ctx.moveTo(x+40, 0);
ctx.lineTo(x+40, h);
ctx.stroke();

//right vertical line
ctx.moveTo(40, 0);
ctx.lineTo(40, h);
ctx.stroke();

//top line
ctx.moveTo(40, 1);
ctx.lineTo(x+40, 1);
ctx.stroke();

//bottom line
ctx.moveTo(40, h-1);
ctx.lineTo(x+40, h-1);
ctx.stroke();

//text
var text = 'Model B';
var dim = ctx.measureText(text).width;
ctx.save();
ctx.translate((40/2)+10,(h/2)+(dim/2)+14);
ctx.rotate(Math.PI + (Math.PI/2));
ctx.font = "30px Arial";
ctx.fillStyle = "#000000";
ctx.textAlign = "left";
ctx.fillText(text, 0, 0);
ctx.restore();

text = 'Model A';
dim = ctx.measureText(text).width
ctx.save();
ctx.translate(x+40+10,(h/2)-(dim/2)-14);
ctx.rotate(Math.PI / 2);
ctx.font = "30px Arial";
ctx.fillStyle = "#000000";
ctx.textAlign = "left";
```

(Continued)

Additional algorithm 15.5 (Continued)

```
            ctx.fillText(text, 0, 0);
            ctx.restore();

            text = 'Model P vs M';
            dim = ctx.measureText(text).width;
            ctx.save();
            ctx.translate((w/2)-(dim/2)-30,h-10);
            ctx.font = "30px Arial";
            ctx.fillStyle = "#000000";
            ctx.textAlign = "left";
            ctx.fillText(text, 0, 0);
            ctx.restore();
        }
    }

</script>

Output:
```

Note that in the above output (Additional algorithm 15.5) the *y*-axis starts from zero and ends at 1. However, the values shown by the similarity scores can be particularly close. For a better visualization, the implementation of the chart narrows the *y*-axis and shows only the region between the two values. To obtain this relative reduction, the minimum similarity score value was taken into account. Thus, the following change was made to the *Chart* function:

$$y = h - \left(\frac{h}{(mx - mn)} \times (s_i - mn) \right)$$

where *mn* is the minimum value and *mx* is the maximum value found in the list of similarity scores, *h* is the canvas height, and s_i is the current value of the similarity index. Note that the inner workings of the *Chart* function were fully described for the previous implementations (see Additional algorithm 15.2). This concludes the changes related to the *Chart* function.

15.6.2 Experiments on the Similarity Index Values

For a more detailed exploration of the trend provided by the similarity score values, a few additional experiments can be set. For instance, in a first experiment, signal P can be the same as signal A:

$$A = 10.3, 23.4, 44.8, 63.2, 44.1, 35.1, 46.5$$
$$B = 18.8, 43.1, 52.2, 45.5, 46.8, 46.6, 67.9$$
$$P = 10.3, 23.4, 44.8, 63.2, 44.1, 35.1, 46.5$$

The point would be that a repeated comparison between all M signals and the P signal would indicate a trend of the similarity index values that favors the prediction towards signal A. Namely, it will show small values of the similarity index for small values of d, and large values of the similarity index for large values of d. The M signal for $d = 0$ is similar to signal B. In other words, the trend must indicate a first minimum value of the similarity index for $d = 0$, which denotes that signal P is different from signal M (the first signal M is the same as B for $d = 0$). As the value of d increases ($d > 0$, $d \leq \mathrm{Max}(d)$), the values of the similarity index are increasing, indicating that signal P increasingly resembles signal M (signal M is the same as A for $d = \mathrm{Max}(d)$). For the last value of d ($d = \mathrm{Max}(d)$), the similarity index shows a value of about 1, which indicates that signal P is the same as signal M and therefore the same as signal A (Figure 15.2a). In a second experiment, an inverse situation can be set, namely a signal P can be the same as signal B:

$$A = 10.3, 23.4, 44.8, 63.2, 44.1, 35.1, 46.5$$
$$B = 18.8, 43.1, 52.2, 45.5, 46.8, 46.6, 67.9$$
$$P = 18.8, 43.1, 52.2, 45.5, 46.8, 46.6, 67.9$$

Again, signal M for $d = 0$ is similar to signal B. By running the experiment, the trend indicates a first value of the similarity index, which is equal to 1 (for $d = 0$), which denotes that signal P is the same as signal M, and therefore the same as signal B. As the value of d increases ($d > 0$, $d \leq \mathrm{Max}(d)$), the similarity index values are decreasing, indicating that signal P less and less resembles the M signals (Figure 15.2b). For the last value of d ($d = \mathrm{Max}(d)$), the similarity index shows a minimum value, which indicates that signal P is radically different from signal M, and therefore totally different from signal A. In a third experiment, signal P represents a combination between signal A and signal B. Namely, the first half of signal P is represented by the first half of signal B and the second half of signal P

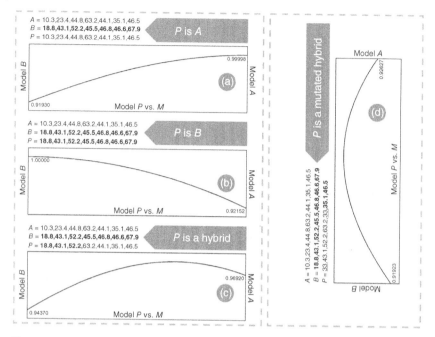

Figure 15.2 Experimental predictions using the spectral forecast model. (a) Signal P is a copy of signal A. Thus, a successive comparison between signal M and signal P shows a list of similarity scores that start from a lower value (0.91930) and end up to approximately 1 (0.99998). Since the similarity score values start from a value lower than 1 and end up as ~1, as expected, this trend shows that P is A. (b) This time signal P is a copy of signal B. Thus, a successive comparison between signal M and signal P shows similarity scores that start from 1 and end up to a lower value (0.92152). Since the similarity score values start from the maximum value of 1 and end up to a value lower than 1, this trend shows that P is B. (c) Signal P represents a combination between signal A and signal B. Namely, the first half of signal P is represented by the first half of signal A and the second half of signal P is represented by the second half of signal B. Thus, signal P is a hybrid. The panel indicates, as expected, a maximum similarity with an M signal located somewhere between the two signals A and B. This successive comparison between signal M and signal P indicates that signal P tends over time to signal A. (d) Signal P represents again a combination between signal A and signal B; however, two mutations are added to signal P. The chart shows an equilibrium in the sense that signal P does not tend to signal A, nor to signal B. In this case the interpretation may indicate a tipping point, namely an uncertainty about the evolution of signal P.

is represented by the second half of signal A. Thus, signal P is a hybrid:

$$A = 10.3, 23.4, 44.8, \underline{63.2, 44.1, 35.1, 46.5}$$

$$B = \underline{18.8, 43.1, 52.2}, 45.5, 46.8, 46.6, 67.9$$

$$P = \underline{18.8, 43.1, 52.2}, \underline{63.2, 44.1, 35.1, 46.5}$$

As expected, signal P shows a maximum similarity with an M signal located somewhere between the two signals A and B. This successive comparison between signal M and signal P indicates that signal P tends over time toward signal A (the reason for this directionality has some explanations below). In a complementary way, a new construction of signal P, by using the first half of signal A and the second half of signal B, leads to the exact same trend, namely:

$$A = \underline{10.3, 23.4, 44.8, 63.2}, 44.1, 35.1, 46.5$$

$$B = 18.8, 43.1, 52.2, 45.5, \underline{46.8, 46.6, 67.9}$$

$$P = \underline{10.3, 23.4, 44.8, 63.2, 46.8, 46.6, 67.9}$$

Note that the trend for the above case is not shown, however, changes to variables A, B, and P from Additional algorithm 15.5 will show a similar trend to that present in Figure 15.2c.

In a fourth experiment, signal P represents a combination between signal A and signal B and two mutations are added to signal P, namely:

$$A = 10.3, 23.4, 44.8, \underline{63.2}, 44.1, \underline{35.1, 46.5}$$

$$B = 18.8, \underline{43.1, 52.2}, 45.5, 46.8, 46.6, 67.9$$

$$P = \mathbf{33.0}, \underline{43.1, 52.2}, 63.2, \mathbf{33.0}, \underline{35.1, 46.5}$$

The trend shown by the similarity index values indicate an equilibrium point in which signal P resembles signal A and signal B in equal measure (Figure 15.2d). In this case, the equilibrium may be interpreted in two ways. If the process represented by A and B is reversible, the equilibrium point may be seen as a tipping point that indicates an uncertainty about the evolution of signal P (signal P does not tend to signal A, nor to signal B). However, if the process represented by A and B is irreversible, then the equilibrium point may be interpreted just as the middle of the transition to A (if B is the starting point). For instance, in the third experiment, it was argued that signal P tends over time toward signal A. An obvious question would be: Why can't the evolution of P be in the other direction in time? How do we know that the evolution is toward A? This question requires a little more consideration regarding the interpretation of what A and B mean. Many physical or biological processes have only one direction. For example, in an experiment published in the journal *Chaos*, the spectral forecast model was described and it was used to predict the onset of a disease, namely type II diabetes. In that experiment, a matrix A represented a normal group, a matrix B represented a diabetic group, and a matrix P represented a previously uncharacterized individual. Thus, the model underlined the future evolution of the undiagnosed individual (P). The point here is that diabetes is an irreversible disease. Thus, a maximum peak in between A and B indicates an evolution over time toward diabetes (called prediabetes). The reasoning suggests a timeline between A and B for the onset. Here, the interpretation has a particularly important role. Thus, there are many irreversible processes in which the model can be interpreted as directional (from *A* to *B*). Also, there are reversible processes in which the model can be interpreted as a balance between *A* and *B*.

15.6.3 The Spectral Forecast Model for Matrices

Of course, the spectral forecast method can also be applied to matrices. Note that in the case of matrices and other multidimensional objects, the spectral forecast equation and the similarity index equation work the same without any adaptation. For example, the implementation below uses three matrices, *A*, *B*, and *P* (Additional algorithm 15.6). Like in the case of signals from above (Additional algorithm 15.5), the current implementation successively computes one matrix *M* for each distance *d*:

$$M_{ijd} = \left[\left(\frac{d}{\text{Max}(A_{ij})} \right) \times A_{ij} \right] + \left[\frac{(\text{Max}(d) - d)}{\text{Max}(B_{ij})} \times B_{ij} \right]$$

Simultaneously, each matrix *M* is compared to matrix *P* to calculate a value for the similarity index:

$$S(d) = \frac{(M_{ijd} \times P_{ij})^2}{\left(\sum (M_{ijd})^2 \times \sum (P_{ij})^2 \right)}$$

Thus, a series of similarity values are obtained, which can be interpreted as discussed in the case of signals. The implementation below shows those discussed above (Additional algorithm 15.6).

Additional algorithm 15.6 Note that the source code is in context and works with copy/paste.

```
<canvas id="bio" height="300" width="1100"></canvas>

<script>

//Spectral forecast model for matrices

var tA ='|10\t33\t49' +
        '|10\t12\t31' +
        '|36\t43\t78';

var tB ='|20\t75\t70' +
        '|48\t84\t61' +
        '|54\t62\t9';

var tP ='|10\t33\t49' +
```

```
         '|10\t12\t31' +
         '|36\t43\t78';

var tmp;

tmp = load(tA);
var A = tmp[0];
var maxA = tmp[1];

tmp = load(tB);
var B = tmp[0];
var maxB = tmp[1];

tmp = load(tP);
var P = tmp[0];

var M = [];

var max = Math.max(maxA, maxB)

Chart(forecast(), '#000000', 'y');

function forecast(){
    var t = '';
    for(var d=0; d<max; d++) {
        var tmp = spectral(d);
        t+=tmp;
        if(d<max-1){t+=',';}
    }
    return t;
}

function spectral(d){
    for(var i=0; i<A.length; i++) {
        M[i]=[];
        for(var j=0; j<A[i].length; j++){
            M[i][j]=((d/maxA)*A[i][j])+(((max-d)/maxB)*B[i][j]);
            M[i][j]= M[i][j].toFixed(2);
        }
    }
    return index(M);
}

function index(M){
    var u=0;
    var k=0;
    var h=0;

    for(var i=0; i<M.length; i++) {
```

(Continued)

Additional algorithm 15.6 (Continued)

```
        for(var j=0; j<M[i].length; j++){
            u += (Number(P[i][j])*Number(M[i][j]));  //(Pij*Mij)
            k += Number(P[i][j])**2;                 //(Pij^2)
            h += Number(M[i][j])**2;                 //(Mij^2)
        }
    }
    //s[d] = (Pij*Mij)^2 / ((Pij^2) * (Mij^2))
    return u**2/(k*h);
}


//LOAD MATRICES FROM STRINGS
function load(t){
    var n = [];
    var m = [];
    var L = [];

    m = t.split('|');
    var max = 0;

    for(var i=1; i<m.length; i++) {
        L[i-1]=[];
        n = m[i].split('\t');
        for(var j=0; j<n.length; j++){
            L[i-1][j]=Number(n[j]);
            if(max<=L[i-1][j]){max=L[i-1][j];}
        }
    }
    return [L, max];
}

// just a chart
function Chart(q,c,e) {

    var s = q.split(",");
    var mx = Math.max.apply(null, s);
    var mn = Math.min.apply(null, s);

    var canvas = document.getElementById('bio');

    var w = canvas.width;
    var h = canvas.height;

    if (canvas.getContext) {

        var ctx = canvas.getContext('2d');
```

```
        if(e=='y'){ctx.clearRect(0, 0, w, h);}

        ctx.moveTo(0, 0);
        ctx.beginPath();

        for (var i=0; i<=s.length-1; i++)
        {
            var y = (h-80) - ((h-80) / (mx-mn)) * (s[i]-mn);
            var x = ((w-80) / s.length) * i;

            ctx.lineTo(x+40, y+40);
            ctx.font = "20px Arial";

            if(i==0){
                ctx.textAlign = 'left';
                ctx.fillText(Number(s[i]).toFixed(5), x+45, y+60);
            }

            if(i==s.length-1){
                ctx.textAlign = 'right';
                ctx.fillText(Number(s[i]).toFixed(5), x+35, y+60);
            }
        }

        ctx.lineWidth = 2;
        ctx.strokeStyle = c;
        ctx.stroke();

        //left vertical line
        ctx.moveTo(x+40, 0);
        ctx.lineTo(x+40, h);
        ctx.stroke();

        //right vertical line
        ctx.moveTo(40, 0);
        ctx.lineTo(40, h);
        ctx.stroke();

        //top line
        ctx.moveTo(40, 1);
        ctx.lineTo(x+40, 1);
        ctx.stroke();

        //bottom line
        ctx.moveTo(40, h-1);
        ctx.lineTo(x+40, h-1);
        ctx.stroke();
```

(Continued)

Additional algorithm 15.6 (Continued)

```
        //text
        write(ctx, w, h, x, y, 'Model A', 'R');
        write(ctx, w, h, x, y, 'Model B', 'L');
        write(ctx, w, h, x, y, 'Model P vs M', 'B');
    }
}

// write text on chart
function write(ctx, w, h, x, y, text, l){

    ctx.save();
    var dim = ctx.measureText(text).width;

    if(l=='B'){ctx.translate((w/2)-(dim/2)-30,h-10);}

    if(l=='R'){
        ctx.translate(x+40+10,(h/2)-(dim/2)-14);
        ctx.rotate(Math.PI / 2);
    }

    if(l=='L'){
        ctx.translate((40/2)+10,(h/2)+(dim/2)+14);
        ctx.rotate(Math.PI + (Math.PI/2));
    }

    ctx.font = "30px Arial";
    ctx.fillStyle = "#000000";
    ctx.textAlign = "left";
    ctx.fillText(text, 0, 0);
    ctx.restore();
}

</script>


Output:
```

In principle, a detailed description of the above implementation is not necessary since Additional algorithm 15.6 represents an adaptation of Additional algorithm 15.3 to the main format of Additional algorithm 15.5. The above output shows an expected trend because matrix P was set intentionally with the same values as matrix A. Note that the experiments on the similarity index values described for signals are also valid for matrices and other multidimensional objects. But how can this model be adapted to the needs of bioinformatics? Bioinformatics generally works with discrete objects (symbols) placed in the form of sequences (i.e. ordinary text). Any analysis of such sequences leads to data structures that can be easily analyzed using spectral forecast, either the model as a whole or only by using the spectral forecast equation.

15.7 Conclusions

This chapter described the spectral forecast model as well as some possible practical uses with implications in different approaches for prediction. In the first instance, the spectral forecast model was presented as an alternative to different analysis. In a second stage, the spectral forecast equation was presented independent of the main model and two approaches were introduced, namely the use of the equation on vectors and matrices. In a second stage, the equation was explained in detail with the help of experiments and concrete cases that showed the mode of operation. Toward the end of the chapter, a series of implementations were presented for both vectors and matrices. Moreover, in the case of vectors, spectral forecast was also presented as a possible mediator between two signals. Future uses for the spectral forecast model may include the field of meteorology, medical diagnostics, forensics, economic forecasts, or in the field of genetics for establishing the relationship between species. In biology, the spectral forecast equation can be used for tissue structure prediction based on two groups of histological slides.

16

Entropy vs. Content (I)

16.1 Introduction

Information entropy (IE) and Information content (IC) are two methods that quantitatively measure information. This chapter emphasizes the importance of IC through a very detailed comparison with IE. Information entropy provides a measure of the average amount of information that is needed to represent an event drawn from a probability distribution for a random variable. Therefore, this chapter first describes the information entropy and highlights the theoretical notions behind the concept. It details these notions directly in connection with sequences of symbols, bypassing the simpler examples that can be given as an introduction. Furthermore, these principles are reinforced with the help of some concrete step-by-step examples. Then, in a second part, these theoretical notions related to information entropy are implemented and the bridge between the theory and the implementation is described in detail. In the last part of the chapter, the results provided by information entropy are compared in parallel with the results provided by the information content model (self-sequence alignment), described previously in chapter 5. This comparison is made to highlight the qualitative differences between information entropy and the new method of information content described as a primary source in this work.

16.2 Information Entropy

Before making a comparison between the information content and information entropy, the information entropy must be described in the context of symbol sequences. The information entropy equation was introduced by

Algorithms in Bioinformatics: Theory and Implementation, First Edition. Paul A. Gagniuc.
© 2021 John Wiley & Sons, Inc. Published 2021 by John Wiley & Sons, Inc.
Companion website: www.wiley.com/go/gagniuc/algorithmsinbioinformatics

Dr. Claude Elwood Shannon, and it is particularly well known in the scientific community, namely:

$$\text{Entropy} = -\sum_{i=1}^{n} p_i \times \log_2 p_i$$

which is equivalent to:

$$\text{Entropy} = \sum_{i=1}^{n} p_i \times \log_2 \left(\frac{1}{p_i} \right)$$

where n represents the total number of symbols in the alphabet of a sequence and p_i represents the probability of occurrence of a symbol i found in the alphabet. The above formula will be used further, as it is a more "positive" expression. Note that the first version of the formula is written as a comment in the implementation (Additional algorithm 16.1). Nevertheless, what does information entropy mean in bioinformatics? Bioinformatics as a field works mainly with sequences of symbols. Thus, calculating the entropy of a sequence or calculating the entropy of a region from a sequence is very important. But how to calculate the information entropy? To continue with some concrete examples, consider the following sequences of symbols:

Sequence 1 = "GGGGGGGG"

Sequence 2 = "TTAAGGGG"

Sequence 3 = "TTAAGGCC"

Sequence 4 = "TTTAAGCC"

One question that can be asked is: what is the information entropy value for each of these symbol sequences? Please consider the first sequence of symbols:

Sequence 1 = "GGGGGGGG"

The alphabet of this sequence contains a single character, namely "G." The probability of randomly choosing a letter "G" from this sequence, is the number of "G" characters counted in the sequence, divided by the total length of the sequence:

$$p(G) = \frac{\text{Count}(G)}{\text{Sequence length}} = \frac{8}{8} = 1$$

Thus, by plugging the probability value into the equation, the value of entropy in this case would be:

$$\text{Entropy} = \sum_{i=1}^{n} p_i \times \log_2 \left(\frac{1}{p_i} \right) = \left(1 \times \log_2 \left(\frac{1}{1} \right) \right) = \left(p(G) \times \log_2 \left(\frac{1}{p(G)} \right) \right)$$
$$= \left(1 \times \log_2 (1) \right) = (1 \times 0) = 0$$

Note that $\log_2(1)$ is zero. Since the alphabet of the sequence is represented by a single symbol, the value for n will be 1 ($n = 1$). Thus, the entropy of the sequence "GGGGGGGG" is zero. In a second case, consider the following sequence:

Sequence 2 = "TTAAGGGG"

The alphabet of this sequence contains three characters, namely "T," "A," and "G." The probability of randomly choosing each letter of the alphabet from this sequence will be:

$$p(A) = \frac{\text{Count}(A)}{\text{Sequence length}} = \frac{2}{8} = \frac{1}{4} = 0.25$$

$$p(T) = \frac{\text{Count}(T)}{\text{Sequence length}} = \frac{2}{8} = \frac{1}{4} = 0.25$$

$$p(G) = \frac{\text{Count}(G)}{\text{Sequence length}} = \frac{4}{8} = \frac{2}{4} = \frac{1}{2} = 0.5$$

Note that the sum of the probabilities is 1:

$$p(A) + p(T) + p(G) = 1$$

Since the alphabet of the sequence is represented by three symbols, the value for n in the information entropy formula will be 3 ($n = 3$). Thus, the entropy of the sequence in this case would be:

$$
\begin{aligned}
\text{Entropy} &= \sum_{i=1}^{n} p_i \times \log_2\left(\frac{1}{p_i}\right) \\
&= \left(p(A) \times \log_2\left(\frac{1}{p(A)}\right)\right) + \left(p(T) \times \log_2\left(\frac{1}{p(T)}\right)\right) \\
&\quad + \left(p(G) \times \log_2\left(\frac{1}{p(G)}\right)\right) \\
&= \left(0.25 \times \log_2\left(\frac{1}{0.25}\right)\right) + \left(0.25 \times \log_2\left(\frac{1}{0.25}\right)\right) + \left(0.5 \times \log_2\left(\frac{1}{0.5}\right)\right) \\
&= \left(0.25 \times \log_2(4)\right) + \left(0.25 \times \log_2(4)\right) + \left(0.5 \times \log_2(2)\right) \\
&= (0.25 \times 2) + (0.25 \times 2) + (0.5 \times 1) = (0.5) + (0.5) + (0.5) = 1.5
\end{aligned}
$$

The entropy of the "TTAAGGGG" sequence shows a value of 1.5. For a third case, consider the following sequence:

Sequence 3 = "TTAAGGCC"

This time, the alphabet of this sequence contains four characters, namely "T," "A," "G," and "C." The probability of randomly choosing each letter of the alphabet from this sequence will be:

$$p(A) = \frac{\text{Count}(A)}{\text{Sequence length}} = \frac{2}{8} = \frac{1}{4} = 0.25$$

$$p(T) = \frac{\text{Count}(T)}{\text{Sequence length}} = \frac{2}{8} = \frac{1}{4} = 0.25$$

$$p(G) = \frac{\text{Count}(G)}{\text{Sequence length}} = \frac{2}{8} = \frac{1}{4} = 0.25$$

$$p(C) = \frac{\text{Count}(C)}{\text{Sequence length}} = \frac{2}{8} = \frac{1}{4} = 0.25$$

Note again that the sum of the probabilities is 1:

$$p(A) + p(T) + p(G) + p(C) = 1$$

Since the alphabet of the sequence is represented by four symbols ("T," "A," "G," and "C"), the value for n in the information entropy formula will be 4 ($n = 4$). Thus, the entropy of the sequence in this case would be:

$$\text{Entropy} = \sum_{i=1}^{n} p_i \times \log_2\left(\frac{1}{p_i}\right)$$

$$= \left(p(A) \times \log_2\left(\frac{1}{p(A)}\right)\right) + \left(p(T) \times \log_2\left(\frac{1}{p(T)}\right)\right)$$

$$+ \left(p(G) \times \log_2\left(\frac{1}{p(G)}\right)\right) + \left(p(C) \times \log_2\left(\frac{1}{p(C)}\right)\right)$$

$$= \left(0.25 \times \log_2\left(\frac{1}{0.25}\right)\right) + \left(0.25 \times \log_2\left(\frac{1}{0.25}\right)\right) + \left(0.25 \times \log_2\left(\frac{1}{0.25}\right)\right)$$

$$+ \left(0.25 \times \log_2\left(\frac{1}{0.25}\right)\right)$$

$$= \left(0.25 \times \log_2(4)\right) + \left(0.25 \times \log_2(4)\right) + \left(0.25 \times \log_2(4)\right)$$

$$+ \left(0.25 \times \log_2(4)\right)$$

$$= (0.25 \times 2) + (0.25 \times 2) + (0.25 \times 2) + (0.25 \times 2)$$

$$= (0.5) + (0.5) + (0.5) + (0.5) = 2$$

Based on the above, the entropy of the "TTAAGGCC" sequence is 2. In a fourth and final case, consider the following sequence:

Sequence 4 = "TTTAAGCC"

Again, the alphabet of this sequence contains four characters, namely "T," "A," "G," and "C." The probability of randomly choosing each letter of the alphabet from this sequence will be:

$$p(T) = \frac{\text{Count}(T)}{\text{Sequence length}} = \frac{3}{8} = 0.375$$

$$p(A) = \frac{\text{Count}(A)}{\text{Sequence length}} = \frac{2}{8} = \frac{1}{4} = 0.25$$

$$p(G) = \frac{\text{Count}(G)}{\text{Sequence length}} = \frac{1}{8} = 0.125$$

$$p(C) = \frac{\text{Count}(C)}{\text{Sequence length}} = \frac{2}{8} = \frac{1}{4} = 0.25$$

Thus, the information entropy is:

$$\text{Entropy} = \sum_{i=1}^{n} p_i \times \log_2\left(\frac{1}{p_i}\right)$$

$$= \left(p(T) \times \log_2\left(\frac{1}{p(T)}\right)\right) + \left(p(A) \times \log_2\left(\frac{1}{p(A)}\right)\right)$$

$$+ \left(p(G) \times \log_2\left(\frac{1}{p(G)}\right)\right) + \left(p(C) \times \log_2\left(\frac{1}{p(C)}\right)\right)$$

$$= \left(0.375 \times \log_2\left(\frac{1}{0.375}\right)\right) + \left(0.25 \times \log_2\left(\frac{1}{0.25}\right)\right)$$

$$+ \left(0.125 \times \log_2\left(\frac{1}{0.125}\right)\right) + \left(0.25 \times \log_2\left(\frac{1}{0.25}\right)\right)$$

$$= \left(0.375 \times \log_2(2.66666666667)\right) + \left(0.25 \times \log_2(4)\right)$$

$$+ \left(0.125 \times \log_2(8)\right) + \left(0.25 \times \log_2(4)\right)$$

$$= (0.375 \times 1.41503749928) + (0.25 \times 2) + (0.125 \times 3) + (0.25 \times 2)$$

$$= (0.53063906223) + (0.5) + (0.375) + (0.5) = 1.9$$

As the result from above indicates, the entropy of this sequence is 1.9. Note that the order of the symbols in the sequence does not matter when calculating the entropy and the symbols can be rearranged at random. As long as each letter is present in the same proportion in the sequence, the entropy will have the same value (i.e. "TTAAGCCT" or "TTTAAGCC," or "TATAGTCC," and so on, show an entropy of 1.9). On paper, the above calculations may look complicated; however, the implementation can be reasonably simple. But how should this formula be implemented for practical purposes?

16.3 Implementation

Initially, as in many of the implementations described throughout the chapters, a detection of the alphabet of the sequence is made by using a dedicated algorithm (see a detailed description in Additional algorithm 9.5). The alphabet detector extracts the unique letters from the input sequence and keeps these letters reliably inside an array variable (a). The length of variable a shows how many types of letters are present in the alphabet of the sequence. Once these letters are known, a loop iterates an index variable (i) from zero to a maximum value that is represented by the length of variable a. At each step of the loop, the letter stored in the i element of variable a ($a[i]$) is used by the native Javascript *replace* function.

The *replace* function temporarily deletes all instances of the current letter (a[i]) from the main sequence (c) and temporarily stores the length of the remaining sequence in variable r. Next, the current letter from the i element of variable a (a[i]) is temporarily stored in a variable l. At this point, variable k holds the total length of the sequence (c) and variable r holds the total length of the sequence from which the letters of type a(i) are missing. Thus, by subtracting the value found in variable r from the value stored in variable k, the number of letters of type a(i) present in sequence c is found. The result of this subtraction (k-r) is further divided by the total length of sequence c (k) to obtain the probability value for the a(i) letter ((k-r)/k). Next, the i element of variable a (a[i]) is repurposed. This probability value ((k-r)/k) replaces the letter in the i element of variable a (a[i]=(k-r)/k). Next, the probability value from a(i) is associated to the current letter found in variable l; thus, both are printed in the output. Since the i element of variable a actually contains the probability value for the current letter of the alphabet, it is directly used for the computation of entropy. Namely, the value of a(i) is multiplied by the logarithm in base two of unity divided by the value of a(i) (Log(2, (1/a[i]))). The result is added to variable e at each step of the loop, which at the end of loop (i<=a.length-1) will store the final value of information entropy for the input sequence (c). Note that the function for calculating the logarithm in any base is reused from Additional algorithm 9.11. The *SMC* (Show Matrix Content) function is also reused from Additional algorithm 3.2.

Additional algorithm 16.1 Note that the source code is in context and works with copy/paste.

```
<script>

// INFORMATION ENTROPY

var c = "TTTAAGCC";

//ALPHABET DETECTION
var a = [];
var t = c.split('');
var k = t.length;

for(var i=0; i<=k; i++){
    var q = 1;
    for(var j=0; j<=a.length; j++){
        if (t[i] === a[j]) {q = 0;}
    }
```

```
    if (q === 1) {a.push(t[i]);}
}

document.write('Sequence: '+c+ '<br>');
document.write('Alphabet of the sequence:'+SMC(a));

var e = 0;
var r = '';
var l = '';

for(var i=0; i<=a.length-1; i++){
    r = c.replace(new RegExp(a[i], 'g'),'').length;

    l = a[i];
    a[i]=(k-r)/k;

    document.write('p('+l+') = '+a[i]+ '<br>');

    //e += -(a[i]*Log(2,a[i]));
    e += (a[i]*Log(2,(1/a[i])));
}

document.write('Entropy = '+e+ '<br>');

function Log(n, v) {
  return Math.log(v) / Math.log(n);
}

// SHOW MATRIX CONTENT
function SMC(m) {
    var r = "<table border=1>";
    for(var i=0; i<m.length; i++) {
        r += "<tr>";
        for(var j=0; j<m[i].length; j++){
            r += "<td>"+m[i][j]+"</td>";
        }
        r += "</tr>";
    }
    r += "</table>";
```

(Continued)

Additional algorithm 16.1 (Continued)

```
    return r;
}

</script>

Output:

Sequence: TTTAAGCC
Alphabet of the sequence:
```

| T |
| A |
| G |
| C |

```
p(T) = 0.375
p(A) = 0.25
p(G) = 0.125
p(C) = 0.25
Entropy = 1.9056390622295662
```

The above example can be greatly reduced and transformed into a dedicated function that returns the entropy value for any sequence of symbols. Thus, the implementation below compresses everything that was discussed so far about information entropy (Additional algorithm 16.3).

Additional algorithm 16.2 Note that the source code is in context and works with copy/paste.

```
<script>

// INFORMATION ENTROPY

var c = "GAGAGAGA";

document.write(entropy(c) + '<br>');
```

```
//ENTROPY
function entropy(c){

    //ALPHABET DETECTION
    var a = [];
    var t = c.split('');
    var k = t.length;

    for(var i=0; i<=k; i++){
        var q = 1;
        for(var j=0; j<=a.length; j++){
            if (t[i] === a[j]) {q = 0;}
        }
        if (q === 1) {a.push(t[i]);}
    }

    var e = 0;
    var r = '';

    for(var i=0; i<=a.length-1; i++){
        r = c.replace(new RegExp(a[i], 'g'),'').length;
        a[i]=(k-r)/k;
        e += (a[i]*Log(2,(1/a[i])));
    }

    return e;
}


function Log(n, v) {
  return Math.log(v) / Math.log(n);
}

</script>

Output:

1
```

Notice that the alphabet detector remains a common part of the function. The example used this time to call the function is the "GAGAGAGA" sequence, which gives 1 as a result. The "GAGAGAGA" sequence is similar in structure to a binary sequence (i.e. "10101010" or "01010101"), and it was used here as a tribute to highlight the origin of information entropy (Dr. Claude Elwood Shannon, Bell Labs).

16.4 Information Content vs. Information Entropy

IE and IC are two methods that measure information. But what is the difference between the two? Moreover, what is the difference between their results? To test the two models against each other, a set of sequences are considered, namely:

Table 16.1 Information entropy vs. information content.

Information Entropy (IE)		
Sequence 1	"TTAAGCCT"	IE = 1.9
Sequence 2	"TTTAAGCC"	IE = 1.9
Sequence 3	"TATAGTCC"	IE = 1.9
Sequence 4	"CAGATTCT"	IE = 1.9

$$IE = \sum_{i=1}^{n} p_i \times \log_2\left(\frac{1}{p_i}\right)$$

Information Content (IC)		
Sequence 1	"TTAAGCCT"	IC = 72.45
Sequence 2	"TTTAAGCC"	IC = 89.46
Sequence 3	"TATAGTCC"	IC = 85.58
Sequence 4	"CAGATTCT"	IC = 83.20

$$s = \{x_1, \ldots, x_{|s|}\}$$

$$IC = 100 - \left(\frac{\sum_{u=1}^{|s|-1}\left(\frac{\sum_{i=1}^{|s|-u}f\left(x_i, x_{u+i}\right)}{(|s| - u) \times 100}\right)}{(|s| - 1)} \right)$$

$$f\left(x_i, x_{u+i}\right) = \begin{cases} +1, & x_i = x_{u+i} \\ 0, & x_i \neq x_{u+i} \end{cases}$$

The table shows the two models side by side. A set of four sequences are analyzed using the two models. The results show that information entropy (IE) provides the same value for all four sequences, whereas the information content (IC) shows different values for each sequence. Note that the IE uses probabilities and the IC model uses a matching function and classical statistics.

Sequence 1 = "TTAAGCCT"

Sequence 2 = "TTTAAGCC"

Sequence 3 = "TATAGTCC"

Sequence 4 = "CAGATTCT"

Note that each sequence from above has the same frequency of symbols/letters. Four experiments are presented for each of the two models (Table 16.1). The main observation on the results is that the order of the symbols in each sequence is considered by the IC model; however, this order is disregarded by the IE model. Thus, IE shows the same value for all four sequences, whereas the IC shows a different value for each sequence (Table 16.1).

To understand how these calculations were performed, an example for each model can be auspicious. Consider the more complex sequence of the four, namely the sequence: "CAGATTCT." From the IC perspective, the calculations will be performed as follows:

$$
IC = 100 - \left(\frac{\sum_{u=1}^{|s|-1} \left(\frac{\sum_{i=1}^{|s|-u} f\left(x_i, x_{u+i}\right)}{(|s| - u) \times 100} \right)}{(|s| - 1)} \right)
$$

First the numerator is solved:

$$
\sum_{u=1}^{|s|-1} \left(\frac{\sum_{i=1}^{|s|-u} f\left(x_i, x_{u+i}\right)}{(|s| - u) \times 100} \right)
$$

$$
= \left(\frac{f(C,A) + f(A,G) + f(G,A) + f(A,T) + f(T,T) + f(T,C) + f(C,T)}{(8-1) \times 100} \right)
$$

$$
+ \left(\frac{f(C,G) + f(A,A) + f(G,T) + f(A,T) + f(T,C) + f(T,T)}{(8-2) \times 100} \right)
$$

$$
+ \left(\frac{f(C,A) + f(A,T) + f(G,T) + f(A,C) + f(T,T)}{(8-3) \times 100} \right)
$$

$$
+ \left(\frac{f(C,T) + f(A,T) + f(G,C) + f(A,T)}{(8-4) \times 100} \right) + \left(\frac{f(C,T) + f(A,C) + f(G,T)}{(8-5) \times 100} \right)
$$

$$
+ \left(\frac{f(C,C) + f(A,T)}{(8-6) \times 100} \right) + \left(\frac{f(C,T)}{(8-7) \times 100} \right)
$$

$$
= \left(\frac{0+0+0+0+1+0+0}{(8-1) \times 100} \right) + \left(\frac{0+1+0+0+0+1}{(8-2) \times 100} \right)
$$

$$
+ \left(\frac{0+0+0+0+1}{(8-3) \times 100} \right) + \left(\frac{0+0+0+0}{(8-4) \times 100} \right) + \left(\frac{0+0+0}{(8-5) \times 100} \right)
$$

$$+ \left(\frac{(1+0)}{(8-6) \times 100} \right) + \left(\frac{(0)}{(8-7) \times 100} \right)$$

$$= 14.28571429 + 33.33333333 + 20 + 0 + 0 + 50 + 0 = 117.61904762$$

Therefore, the result from the numerator is 117.61904762. Note that the exact calculation method is graphically shown in Table 5.1. Secondly, the new result can be inserted in the main equation and the denominator can divide the numerator:

$$IC = 100 - \left(\frac{\sum_{u=1}^{|s|-1} \left(\frac{\sum_{i=1}^{|s|-u} f\left(x_i, x_{u+i}\right)}{(|s|-u) \times 100} \right)}{(|s|-1)} \right) = 100 - \left(\frac{117.61904762}{(8-1)} \right)$$

$$= 100 - \left(\frac{117.61904762}{7} \right) = 100 - 16.8027210886$$

$$= 83.1972789114\%$$

Thus, the information content for sequence "CAGATTCT" shows a value of 83.20%. From the IE perspective, the calculations for the sequence "CAGATTCT" will be performed as follows:

$$p(T) = \frac{\text{Count}(T)}{\text{Sequence length}} = \frac{3}{8} = 0.375$$

$$p(A) = \frac{\text{Count}(A)}{\text{Sequence length}} = \frac{2}{8} = \frac{1}{4} = 0.25$$

$$p(G) = \frac{\text{Count}(G)}{\text{Sequence length}} = \frac{1}{8} = 0.125$$

$$p(C) = \frac{\text{Count}(C)}{\text{Sequence length}} = \frac{2}{8} = \frac{1}{4} = 0.25$$

Thus, the information entropy is:

$$IE = \sum_{i=1}^{n} p_i \times \log_2 \left(\frac{1}{p_i} \right) = \left(p(T) \times \log_2 \left(\frac{1}{p(T)} \right) \right)$$

$$+ \left(p(A) \times \log_2 \left(\frac{1}{p(A)} \right) \right) + \left(p(G) \times \log_2 \left(\frac{1}{p(G)} \right) \right)$$

$$+ \left(p(C) \times \log_2 \left(\frac{1}{p(C)} \right) \right) = \left(0.375 \times \log_2 \left(\frac{1}{0.375} \right) \right)$$

$$+ \left(0.25 \times \log_2 \left(\frac{1}{0.25} \right) \right) + \left(0.125 \times \log_2 \left(\frac{1}{0.125} \right) \right)$$

$$+ \left(0.25 \times \log_2 \left(\frac{1}{0.25} \right) \right) = \left(0.375 \times \log_2 (2.66666666667) \right)$$

$$+ \left(0.25 \times \log_2 (4) \right) + \left(0.125 \times \log_2 (8) \right) + \left(0.25 \times \log_2 (4) \right)$$

$$= (0.375 \times 1.41503749928) + (0.25 \times 2) + (0.125 \times 3) + (0.25 \times 2)$$

$$= (0.53063906223) + (0.5) + (0.375) + (0.5)$$

$$= 1.9056390622295662$$

As the result from above indicates, the entropy of sequence "CAGATTCT" is 1.9. Notice that the order of the symbols in the sequence is irrelevant for IE, and the symbols can be arranged at random as long as their frequency is constant. In other words, as long as each letter from the alphabet of the sequence shows the same frequency, the entropy will show the same value. On the other hand, the model for IC considers the order of the letters in the sequence. A rearrangement of the letters in the sequence will generate another value for IC. Therefore, the main difference between information entropy and the information content is given by the consideration of the order of the symbols/letters in the sequence.

16.4.1 Implementation

The implementation below brings the two models in the same context (Additional algorithm 16.3). The implementation for measuring the IC is brought directly from the "*Self-Sequence alignment*" chapter (Additional algorithm 5.2) to build the *content* function, whereas the implementation for measuring the IE is brought from above (Additional algorithm 16.2) to build the *entropy* function. Furthermore, a scanner from Additional algorithm 5.4 is adapted to the current situation, which traverses a z sequence with a 12-position sliding window (wl=12). The contents of each sliding window are injected into both the *entropy* function and the *content* function. The results returned by these functions are stored in separate string variables, namely a string variable in which a signal is built for IC values (signal_s), and a string variable in which a signal is built for IE values (signal_e). The numerical values in both signals are spaced by a common delimiter, namely a comma character (' , '). At the end of the loop, the two string variables are printed in the output.

Additional algorithm 16.3 Note that the source code is in context and works with copy/paste.

```
<script>

// SCANNER
// THE INFORMATION ENTROPY
// THE INFORMATION CONTENT

var z = "AAAAAACAGGTGAGTAAAAAAAA";

var signal_s = '';
```

(Continued)

Additional algorithm 16.3 (Continued)

```javascript
var signal_e = '';

var wl = 12;
var w = '';

var u = z.length - wl + 1;

for(var l=0; l<u; l++) {
    w = z.substr(l, wl);
    signal_s += content(w) + ',';
    signal_e += entropy(w) + ',';
}

document.write('Sequence: '+z+'<br>');
document.write('Information CONTENT: '+signal_s+'<br>');
document.write('Information ENTROPY: '+signal_e+'<br>');

//INFORMATION CONTENT (SELF-SEQUENCE ALIGNMENT)

function content(s)
{
    var t = 0;
    var m = 0;
    var x = [];

    x = s.split('');

    for (var u=1; u<=(s.length - 1); u++)
    {
        for (var i=0; i<=(s.length-u); i++)
        {m += f(x[i], x[u+i]);}

        t += (m / (s.length-u) * 100);
        m = 0;
    }

    return (100 - (t / (s.length - 1))).toFixed(1);
}

function f(x1, x2){
    if (x1 == x2) {return 1;} else {return 0;}
}

//INFORMATION ENTROPY
function entropy(c){

    //ALPHABET DETECTION
    var a = [];
    var t = c.split('');
    var k = t.length;

    for(var i=0; i<=k; i++){
        var q = 1;
        for(var j=0; j<=a.length; j++){
```

```
            if (t[i] === a[j]) {q = 0;}
        }
        if (q === 1) {a.push(t[i]);}
    }

    var e = 0;
    var r = '';

    for(var i=0; i<=a.length-1; i++){
        r = c.replace(new RegExp(a[i], 'g'),'').length;
        a[i]=(k-r)/k;
        //e += -(a[i]*Log(2,a[i]));
        e += (a[i]*Log(2,(1/a[i])));
    }

    return e.toFixed(1);
}


function Log(n, v) {
    return Math.log(v) / Math.log(n);
}

</script>

Output:

Sequence: AAAAAACAGGTGAGTAAAAAAAA
Information CONTENT: 75.2,60.1,71.3,78.7,67.7,66.8,75.6,62.3,76.8,72.5,
                    64.8,52.9,
Information ENTROPY: 1.6,1.6,1.6,1.8,1.8,1.8,1.8,1.5,1.5,1.4,1.3,1.0,
```

The output from Additional algorithm 16.3 shows the successive values from the main signals provided by the two functions (*entropy* and *content*). The values in the two signals obviously have different scales. To observe the differences or the similarities, the two signals can be normalized. This is not too complicated, because their normalization is done by default in the *Chart* function presented in several lines of code for the implementation of other algorithms. Thus, since the information entropy (`signal_e`) signal and the information content (`signal_s`) signal are already in a string format, they can be directly injected into the *Chart* function, as follows:

Additional algorithm 16.4 Note that the source code is in context and works with copy/paste.

```
<canvas id="bio" height="300" width="1100"></canvas>
```

(Continued)

Additional algorithm 16.4 (Continued)

```
<script>

// SCANNER
// THE INFORMATION ENTROPY
// THE INFORMATION CONTENT

var z = "AAAAAACAGGTGAGTAAAAAAAA";

var signal_s = '';
var signal_e = '';

var wl = 12;
var w = '';

var u = z.length - wl + 1;

for(var l=0; l<u; l++) {
    w = z.substr(l, wl);
    signal_s += content(w);
    signal_e += entropy(w);

    if(l<u-1){
        signal_s += ',';
        signal_e += ',';
    }
}

document.write('<hr>Sequence: '+z+'<br>');
document.write('Information CONTENT: '+signal_s+'<br>');
document.write('Information ENTROPY: '+signal_e+'<br>');

Chart(signal_s, '#ff0000', 'y')
Chart(signal_e, '#000000', 'n')

//INFORMATION CONTENT (SELF-SEQUENCE ALIGNMENT)

function content(s)
{
    var t = 0;
    var m = 0;
    var x = [];

    x = s.split('');

    for (var u=1; u<=(s.length - 1); u++)
    {
        for (var i=0; i<=(s.length-u); i++)
        {m += f(x[i], x[u+i]);}
```

```
        t += (m / (s.length-u) * 100);
        m = 0;
    }

    return (100 - (t / (s.length - 1))).toFixed(1);
}

function f(x1, x2){
    if (x1 == x2) {return 1;} else {return 0;}
}

//INFORMATION ENTROPY
function entropy(c){

    //ALPHABET DETECTION
    var a = [];
    var t = c.split('');
    var k = t.length;

    for(var i=0; i<=k; i++){
        var q = 1;
        for(var j=0; j<=a.length; j++){
            if (t[i] === a[j]) {q = 0;}
        }
        if (q === 1) {a.push(t[i]);}
    }

    var e = 0;
    var r = '';

    for(var i=0; i<=a.length-1; i++){
        r = c.replace(new RegExp(a[i], 'g'),'').length;
        a[i]=(k-r)/k;
        //e += -(a[i]*Log(2,a[i]));
        e += (a[i]*Log(2,(1/a[i])));
    }

    return e.toFixed(1);
}


function Log(n, v) {
  return Math.log(v) / Math.log(n);
}


function Chart(q,c,e) {

    var s = q.split(",");
```

(Continued)

Additional algorithm 16.4 (Continued)

```
    var mx = Math.max.apply(null, s);

    var canvas = document.getElementById('bio');

    var w = canvas.width;
    var h = canvas.height;

    if (canvas.getContext) {

        var ctx = canvas.getContext('2d');

        if(e=='y'){ctx.clearRect(0, 0, w, h);}

        ctx.moveTo(0, 0);
        ctx.beginPath();

        for (var i=0; i<=s.length-1; i++)
        {
            var y = h-((h / mx) * s[i]);
            var x = (w / s.length) * i;

            ctx.lineTo(x, y+2);
        }
        ctx.lineWidth = 2;
        ctx.strokeStyle = c;
        ctx.stroke();
    }
}

</script>
```

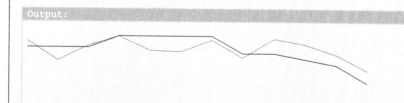

Output:

Sequence: AAAAAACAGGTGAGTAAAAAAAA
Information CONTENT: 75.2,60.1,71.3,78.7,67.7,66.8,75.6,62.3,76.8,72.5,64.8,52.9
Information ENTROPY: 1.6,1.6,1.6,1.8,1.8,1.8,1.8,1.5,1.5,1.4,1.3,1.0

Again, the *Chart* function in Additional algorithm 16.4 normalizes the two signals by using a constraint on the vertical size of the canvas object (h). In other words, the *h* variable in the *Chart* function means 100% for both signals. Thus, the two signals would overlap perfectly only if their proportions were the same. However, the output of Additional algorithm 16.4 suggests that the two methods behave differently. For longer and more complex signals, IC follows IE in a general trend; however, the results of the two models are different and may bear different meanings in practice. Although the two models essentially measure information, they behave differently in detail and have different discrimination qualities. The IC model accounts for the order inside a sequence of symbols and shows superior information details compared to IE in which the order of the symbols is irrelevant. Note that a more complex implementation for the comparison between the two models can be found online on the main website of this book.

16.4.2 Additional Considerations

Although the entropy formula seems very simple compared to the information content model, the entropy equation still requires a slightly more complex implementation because it uses the alphabet detector, which is strictly necessary. On the other hand, the model for information content seems sophisticated as an equation, but it is very direct on the implementation side. Note that the main difference between the two approaches is that the information entropy is calculated using probabilities, while the model for detecting the information content is mechanistic in nature and is based more on classical statistics. The implementation of information entropy is in this case computationally faster, but this is largely achieved through the *replace* function, which is native to javascript. On the other hand, the implementation of the information content model does not use native functions that speed up processing (although good solutions could perhaps be easily found).

16.5 Conclusions

This chapter was concerned with the importance and the correct use of two models that are able to measure information in biological sequences. In a first phase, this chapter described the IE from a theoretical point of view, by using some examples involving sequences of symbols. Subsequently, the theory was used for the implementation process, which eventually led to a dedicated *entropy* function that received a single parameter (namely any sequence of symbols) and returned the value of Information Entropy. In a second phase, the entropy function was further used in parallel with a function from another model described in this book, called *self-sequence alignment* (information content).

Therefore, this chapter highlighted the importance of the IC model through a decent comparison with information entropy. A sequence z was analyzed with the help of the two models by using the sliding window approach. Thus, two signals were obtained whose values were plotted on a chart for comparison. In terms of results, additional experiments showed a difference between the two models, both quantitatively and qualitatively. The main point of the discussion was that self-sequence alignment is a model that should not be confused with the results provided by the information entropy model. Please note that this chapter was an addition to the self-sequence alignment model.

17

Philosophical Transactions

17.1 Introduction

Bioinformatics as a science is a multidisciplinary field and the methods used for different analyzes can be seen from a broader framework, connected with several fields, including philosophy. In particular, all scientific methods are strongly connected in one way or another by the meaning of randomness. Thus, the correct understanding of the role of independent random variables in the universe in which we live in, is of particular importance for prediction methods used in either bioinformatics or other fields of study. The subjects discussed here revolve around information entropy, random numbers, pseudo-random numbers, chaos theory, and noise. Thus, this chapter introduces some general discussions and perspectives on the physical world and touches with some abstract conclusions the field of biology. It explores the concept of complexity and frame of reference that is associated with the human condition. It further discusses determinism and deals with some important existential questions: Why in our frame of reference reality allows for independent random variables? Why do predictions even work in the real world? Why a computer program is fully deterministic? How the two relate? And why? Please note that all philosophical discussions represent an act of rebellion against the human condition, and their topics are somewhere between genius, clarity, and total nonsense, just as genius as a concept is on the verge of madness. These are difficult and complicated issues whose solutions have been sought since ancient times and the oldest known writings. Thus, these issues will definitely not be solved here, but rather discussed.

17.2 The Frame of Reference

Biological organisms are hunters of low entropy. This trend can be observed from the microscale to the macroscale. To reduce complexity, our own behavior

Algorithms in Bioinformatics: Theory and Implementation, First Edition. Paul A. Gagniuc.
© 2021 John Wiley & Sons, Inc. Published 2021 by John Wiley & Sons, Inc.
Companion website: www.wiley.com/go/gagniuc/algorithmsinbioinformatics

consists of detections and associations between patterns. The word "dynamic," "disorganized," or "indefinite" appears in our vocabulary when an apparent irreducibility occurs; when a lack of repetition/cyclicity or pattern is noticed, namely a deviation from the organization as expected from our reference frame. Mental compartmentalization is the method by which entropy is lowered to synthesize different solutions to partially understand complex problems that reach beyond our biological condition. Of course, such an awareness is not new in the scientific community. The categorization tendency can be seen from Aristotle's writings to the present day. This is the reason why for several hundred years we were (and still are) interested in mathematical models that have given meaning to these processes. For instance, good examples of such mathematical models are the more recent Navier–Stokes equations that describe the motion of fluids or mathematical techniques like Fourier transform, which help break down signals in their constitutive frequencies and many other mathematical models used today in cutting-edge research. But what about processes that show a degree of uncertainty and cannot be easily/intuitively broken down into categories and subcategories? Of course, the whole field of probability theory deals with these matters, and a good example is the Markov chains approach described in Chapters 13 and 14. But what makes many of the physical processes predictable only in the short term? For instance, why is the onset of a disease difficult to predict? Of course, we can consider the large number of variables that are involved in complex systems. But is that the only reason? What fundamental forces make our universe only partially predictable? It can be argued that the importance of noise is often underestimated in biology, and this is one of the central topics discussed in this chapter. What is noise? In an abstract way, noise can be seen as an intersection/addition of patterns at different scales and offsets that may be represented in many forms (i.e. the interaction of mechanical waves or/and the interaction of electromagnetic waves). Among the electromagnetic noise that make their mark on different events would be heat. Thermal agitation underlies the basic properties of matter, from the properties of crystalline structures to different chemical reactions, and last but not least, it underlies the "independent" thoughts and ideas that we have, as thermal agitation brings that drop of randomization that puts any phenomenon under the umbrella of chaos theory.

17.2.1   The Fundamental Layer of Complexity

But, how noise should look like in a concrete example? Moreover, how might the information appear? High information content implies high entropy, and this it is the natural state of the universe in which we live in (please see what information content means in Chapter 5). This default aspect is further discussed. Oscillations are the main fundamental property in our universe from which everything emerges. Thus, the simplest example that comes to mind can be given by using waveforms of different signals. A signal is a function that conveys discretized information over time about a particular phenomenon (i.e. weather, fog, thunder, chemical reactions, wave propagation). The term "signal" is highly versatile, and it means the variation of something over time. The concept of "time," however, is often versatile also. For instance, in bioinformatics, time can be represented by the length of a sequence of numbers or letters (usually DNA, RNA, or proteins). Signals can be sampled and analyzed in real time or recorded as data (a series of numbers, each number representing a value over a period of time) and analyzed later on. A waveform of a signal is the shape of its graph as a function of time. Sinusoidal waveforms are probably the most familiar type of waveforms. The addition of several sinusoidal waveforms of different frequencies, leads to the formation of a single, more complex waveform. To increase the complexity, the addition of these sinusoidal waveforms can be shifted in time relative to each other. As more arbitrary sinusoidal waveforms are added, the resulting waveform will show more and more irregularity (higher entropy; lack of pattern – more and more nondeterministic from our frame of reference). Moreover, several resultant waveforms can merge and produce another more complex waveform with even higher entropy. But is this irregular waveform truly nondeterministic? Since the constitutive parts and their conditions are known, it is safe to say that the new waveform is fully deterministic. The point of this example is that at their core, all nondeterministic signals are composites of deterministic waveforms. Real signals taken from different nonlinear physical phenomena show highly irregular waveforms, whose constitutive parts have unknown meaning in our frame of reference. At first glance, such signals may seem random and quickly can lead to the notion of noise, which is of the highest complexity. Thus, the unknown meaning of noise represents the irregular boundary of our frame of reference. Following the reasoning from above, it can be concluded that noise is the most refined piece of information. In an abstract manner, information appears to be emergent of fundamental waveforms. Space–time determines the offset between constitutive waveforms of a signal (of any type of physical signal). In other words, the constitutive waves of a signal can have different source locations in space and can meet in different periods of time until the formation of the signal that reaches a detector (whatever that may be). One of the biggest achievements in science has been the decomposition of more complex signals back into their basic waveforms. Today, this mathematical approach it is known as "Fourier transform." Based on this mathematical model, a complex signal can be decomposed into a set of sinusoidal waves of different frequencies and different offsets relative to each other. In turn, the addition of these sinusoidal waves with respect to their offset positions, reconstructs the original signal. Nevertheless, even the Fourier transform has limitations in decoding complexity. Again, the tendency of the universe in our frame of reference is to increase entropy and consequently the information content. For instance, a mechanical shock to a perfectly insulated solid material causes a mechanical wave. This mechanical wave propagates and reflects through the solid material until it is completely absorbed by conversion into a range of electromagnetic waves (photons) in the terahertz domain (heat). These electromagnetic waves scatter in all directions until thermal equilibrium is reached. Thermal equilibrium occurs in this solid when there is an equal probability of receiving or releasing heat (photons). Thus, thermal equilibrium in this case can be interpreted/perceived as noise. In other words, the mechanical shock leads to absorption of mechanical work into the noise. Moreover, in the current example, the solid behaves as an information amplifier because the mechanical wave (an action with low entropy relative to the result) is converted into electromagnetic waves that contain much more information (high entropy) than the original. Thus, the information content is interpreted here as entropy or noise.

17.2.2 On the Complexity of Life

A connection between noise and life is a good exercise for understanding the evolution of living organisms. Noise is ultimately responsible for all mutations. In turn, mutations are the engine of evolution. By association, noise drives the evolutionary process. One question that has been asked since ancient times is: how did life come about? There are partial hypotheses and experiments that indicate how these events could have taken place, but no one can be sure since no one was there to witness the events. The concept of "primordial soup" leads the mind to a visualization of high entropy and last but not least, it leads the mind to the concept of noise. But high entropy compared to what? When such a "primordial soup" is imagined we are actually visualizing a closed universe, and this is one of the main issues in understanding how life began (self-assembled). As reasoned above, maximum entropy can be interpreted as the highest level of complexity. On the other hand, if just enough energy is injected, chemical processes become reductionist in nature (bubble patterns appear from complexity). Here, the term "reductionism" refers to a decrease in entropy through chemical reactions that lead to more "sophisticated" molecules. Thus, biological organisms constantly decrease the information content (entropy) to exist in this frame of reference. Interestingly, from this "information content" perspective, life *is not complex* – thus, under this idea lies the low entropy of life forms. In an abstract way, low entropy is obtained by the formation of patterns from noise (Table 17.1). In contrast, the interaction of patterns (whatever embodiment those patterns may take) at different scales leads to high entropy.

In other words, obvious patterns lead to low entropy (low information) in our frame of reference, which further leads to the perception of dependent variables (Table 17.1). On the other hand, the tendency of the universe is to break these low entropy patterns and raise them to the level of noise. Thus, high entropy

Table 17.1 A link between entropy, information, patterns, and the perception of types of variables.

Entropy	Patterns	Variables	Information
Low entropy	Low period patterns	Dependent variables	Low information content
High entropy	High period patterns	Independent variables	High information content

The table indicates how in our frame of reference the low entropy in information is obtained by the appearance of redundant structures (patterns with periods that fit in our reference system) that easily lead to the concept of dependent variables. In contrast, the high entropy of information is obtained by the intersection of patterns with periods that exceed our frame of reference and that easily leads to the concept of independent random variables.

manifested through the interaction of patterns with different periods leads to a lack of patterns in our frame of reference, which further raises the information content and creates the intractable impression of independence (Table 17.1). This is true from information to physical processes, as a valid universality exists. For example, in a number sequence of a certain size, if the beginning and the end of a pattern of numbers is unknown, then the numbers over the sequence are unpredictable (independent) while if the beginning and end of a pattern is quantifiable/visible in that frame of reference (the length of the sequence), then it can be predicted. Note that independence usually refers to a uniform distribution; however, that is not always the case.

17.3 Random vs. Pseudo-random

An exercise of the mind can express the principles of determinism and free will in connection with different prediction methods. For instance, if we consider two parallel universes that start from exactly the same conditions in t_0, these two universes will evolve exactly the same until t_1. If one of the two universes is paused in t_1, the other universe will continue unabated. Resuming the paused universe in t_2 would produce a gap between the events in the two universes ($g = t_2 - t_1$). However, by simplistic reasoning, the two universes will evolve exactly the same, but in different periods with a gap g. Therefore, the same starting conditions lead to the same result. However, in our universe, the repetition of events by using the same initial conditions is not possible for slightly more complex systems. The problem with any experiment we undertake is that it will take place inside the universe we are currently in, and there will always be external and internal noise influences that will interact with the atoms of the system being tested. On computers, any simulation is a deterministic universe. For instance, the lack of true noise may limit many simulations related to protein interactions. Thus, these simulations arrive to the same result at each repetition. An additional layer for simulation of noise (random numbers) in these protein–protein interactions obviously lead to different results each time. However, if the noise is not external, from our universe, the simulation can exhibit a limited and predictable number of results. We might wonder: why is that? In computers, there are specialized mathematical functions (deterministic algorithms) for simulation of "random" numbers. The numbers generated by these functions are called pseudo-random numbers. This name comes from the fact that the random numbers are generated based on mathematical formulas, which explode an initial "seed" (usually, a number extracted from the computer clock) into a sequence of numbers that follow a pattern. Thus, each new number in the sequence will be dependent on the previous number provided by the same formula. An example of a pseudo-random generator is the linear congruential generator:

$$X_{n+1} = (a \times X_n + c) \bmod m$$

where X_1 is an integer seed, n indicates the total number of terms in the sequence, m is the modulus, a is the multiplier, and c is the increment. The pattern of such a function shows a characteristic distribution for long numerical sequences that exceed the period (the cycle of repetition). The range of numbers of the above function is between 0 and $m - 1$, with a uniform distribution of integers. The length of a period can be controlled by setting the value of m. However, a narrow portion over this sequence shows a small piece of the wider pattern, thus simulating randomness without showing the real distribution derived from the initial seed (X_1). The implementation from Additional algorithm 17.1 includes those discussed so far on the linear congruential generator:

Additional algorithm 17.1 Note that the source code is in context and works with copy/paste.

```
<script>

// A LINEAR CONGRUENTIAL GENERATOR

var x=3;

document.write(prandom(x));

function prandom(x){
var a = 11;
var m = 25;
var c = 17;
```

```
var r="";

    for (i = 0; i < 20; i++) {
      x = (a*x+c) % m;
      r += x + ",";
    }


    return r;
}

</script>

Output:
0,17,4,11,13,10,2,14,21,23,20,12,24,6,8,5,22,9,16,18
```

where x is the seed, a is the multiplier, m is the modulus, c is the increment, and r is the variable that stores the number sequence. A new seed value ($x=157$) accompanied by changes in the values of variables a (the multiplier) and c (the increment) can lead to much shorter and noticeable periods (Additional algorithm 17.2). For example, the implementation below produces a period of four different numbers (i.e. 22,2,12,7).

Additional algorithm 17.2 Note that the source code is in context and works with copy/paste.

```
<script>

// A LINEAR CONGRUENTIAL GENERATOR

var x=157;

document.write(prandom(x));

function prandom(x){
var a = 17;
var m = 25;
var c = 3;
```

(Continued)

Additional algorithm 17.2 (Continued)

```
var r="";

    for (i = 0; i < 20; i++) {
      x = (a*x+c) % m;
      r += x + ",";
    }


    return r;
}

</script>

Output:
22,2,12,7,22,2,12,7,22,2,12,7,22,2,12,7,22,2,12,7
```

Nevertheless, a sequence of pseudo-random numbers is deterministic from our frame of reference. A given seed will always determine the same sequence of pseudo-random numbers. Thus, the outcome of a simulation will be seed-dependent. This of course violates the concept of randomness, which implies unpredictability. However, it is these properties that make pseudo-random numbers particularly important for computer simulations in some situations. The importance is given by the fact that a "random" element is added, and the deterministic environment is preserved. Thus, the m value is directly proportional to the "independence" of the processes that take place inside the computer simulation. The point here is that any computer simulation that takes its input from our universe becomes as unpredictable as any other physical phenomenon. Thus, for a computer simulation to be accurate, it must insert the numerical equivalent of noise sampled directly from our universe. Such a source can be the electromagnetic noise converted to numbers inside a computer. Sampling of electric current fluctuations inside the computer can be another promising option. Numbers, which are produced with the help of noise, are known as true random numbers, and these can be sampled exclusively from physical phenomena in our universe.

17.4 Random Numbers and Noise

Random numbers are required to be independent and this property can be viewed only in the context of sequences and their distribution. Random numbers are especially important in experiments because these sequences represent a threshold, which indicates any phenomenon that is below the maximum entropy (please see the significance of self-sequence alignment in Chapter 5 or the background models from Chapter 9). If a device produces numbers that conform to a uniform distribution, then each number is considered truly random. Therefore, when such a generator is mentioned, it actually refers to its properties that provide an unpredictable/irreproducible and uniform distribution of a number sequence (a number sequence is a list of numbers). This property gives these devices the status of "true randomness," namely number sequences that cannot be predicted from our frame of reference. In computers, we take "independent" numerical values from various sources. Another method for acquisition of randomness ("independent" random variables) from physical phenomena is the electronic capture of atmospheric noise or the cosmic electromagnetic radiation and the thermal noise present in all electronic components. For instance, in video cameras, the thermal noise is manifested as video noise and can be seen in all video recordings made in low-visible-light environments. Thus, image frames taken from a video camera with a covered lens can be a source of random numbers. In televisions, the lack of a strong electromagnetic signal leads to the same video noise effect, and in radio receivers it is manifested as static sound. For this precise reason, all electronic sensitive instruments are cooled to decrease the thermal noise. For instance, digital image sensors in telescopes are cooled to a temperature as low as possible to reduce the effects of thermal noise. Otherwise, the internal noise and the photons coming from very distant galaxies would weigh the same for the detector/sensor. Another source used to generate random numbers are the detectors of radioactive decay. However, randomization samples taken on the basis of thermal noise can be similar in distribution with randomization samples taken from radioactive decay. All these methods are used to generate what it is called "genuine random numbers" or "true random numbers." But are these sources in their essence truly random? I will argue that what we call true randomization comes from the "deepest" layers of hidden variables, with causes and scales very far from our frame of reference. Thus, the deeper and more numerous the source layers are, the more independent these number distributions become in our frame of reference. In other words, relative to us, *noise can be perceived as the highest level of complexity.*

17.5 Determinism and Chaos

Determinism should be the short path in understanding the universe; however, it is far from simple. For a long time, our civilization associated determinism (sometimes even today) with linear phenomena, for obvious reasons (i.e. our behavior seeks low entropy even in the most complex situations). However, these views have begun to change in recent times. Seeded initially by Dr. Poincaré and others, a modern theory emerged, namely the chaos theory, which started with Dr. Edward Norton Lorenz and the evolution of computers in the 1960s. Chaos theory is concerned with the behavior of dynamical systems that are sensitive to the initial conditions. In physical processes, the conditions are subtly but continuously altered by the background noise. This is why many scientific experiments need special conditions (cooling, shielding, and so on). However, the continuous addition of randomization qualities from the background noise allows only for the possibility of short-term predictions for any physical process that can be cataloged as a dynamical system. This is the main reason why there are prediction limits in the case of physical processes, or simulations that use noise from different physical sources. One interesting notion is that the above observation makes our universe both deterministic and probabilistic at the same time. Above, many discussions have been related to noise and its impact on physical processes. But can these notions be demonstrated here?

17.5.1 Chaos Without Noise

A mathematical dynamical system can be represented by iterated functions. Thus, the best example regarding the importance of noise over the chaotic phenomena can begin with a nonlinear equation, such as:

$$X_{i+1} = 4 \times X_i \times (1 - X_i)$$

The above equation means the current value (X_{i+1}) is equal to four times the previous value $(4 \times X_i)$, further multiplied by the result of the unit minus the previous value $(1 - X_i)$. An initial X_0 value can be plugged into this equation, and the result can be inserted back into the equation for a second iteration to provide a new result and so on. For instance, consider a value for $X_0 = 0.62342375475$ and a series of iterations:

$$X_1 = 4 \times X_0 \times (1 - X_0)$$

$$X_1 = 4 \times 0.62342375475 \times (1 - 0.62342375475)$$

$$X_1 = 0.9390663070536474$$

Next iteration (iteration 2):

$$X_2 = 4 \times X_1 \times (1 - X_1)$$

$$X_2 = 4 \times 0.9390663070536474 \times (1 - 0.9390663070536474)$$

$$X_2 = 0.22888311204108905$$

Next iteration (iteration 3):

$$X_3 = 4 \times X_2 \times (1 - X_2)$$

$$X_3 = 4 \times 0.22888311204108905 \times (1 - 0.22888311204108905)$$

$$X_3 = 0.7059825322539013$$

Of course, these iterations can continue to infinity. Nonetheless, the results of several iterations are sufficient for the present purposes and can show the evolution of the values over time (steps/iterations). Moreover, a comparison can be made between the results of different iterations that start from slightly different X_0 values. For instance the current X_0 value can be considered as it is from the above, namely:

$$X_0 = 0.62342375475$$

Also, a second version can also be considered:

$$X_0 = 0.62342375474$$

Note that there is only a tiny difference between the two X_0 values on the last digit. Below, an implementation is shown by which a function *system* iterates the above equation for any initial value X_0 (Additional algorithm 17.3). To observe the evolution of the same system (the equation in this case), the function can be called twice for different X_0 values. For example, the *system* function is called once for the first value $X_0 = 0.62342375475$ and a second time for the next value $X_0 = 0.62342375474$, and the results of the function are printed after each call. The *system* function returns the result of the iterations in a sequence of numbers that are separated by a special delimiter (" | ", however, it can be any type of delimiter). Note that a variable *s* indicates to the *system* function how many times the iteration must be performed. To allow for easy eye tracking of the output values, the number of iterations has been set to five (s=5;).

Additional algorithm 17.3 Note that the source code is in context and works with copy/paste.

```
<script>

//CHAOS

var s = 5;

var x0 = 0.62342375475;
```

(Continued)

Additional algorithm 17.3 (Continued)

```
document.write(system(x0, s) + '<hr>');

var x0 = 0.62342375474;
document.write(system(x0, s) + '<hr>');

function system(x0, s){
    var x1 = 0;
    var r = '';

    for(var i=1; i<s; i++){

        x1 = 4 * x0 * (1 - x0);
        x0 = x1;

        r += x1 + "|";
    }
    return r;
}

</script>


Output:

0.9390663070536474|0.22888311204108905|0.7059825322539013|
0.8302847856250821|

0.9390663070635213|0.22888311200640635|0.7059825321786768|
0.8302847857490415|
```

Note that within the main cycle of the *system* function, immediately after the computation of the equation, the value in X_1 is copied into the variable X_0 (x0 = x1;), and the value in the variable X_1 is added in the number sequence from variable r (r += x1 + "|";). Then, in the next iteration, the value for X_1 is recalculated, and so on. However, the output from Additional algorithm 17.3 does not provide us with pertinent information on the evolution of the two X_0 values, and printing several values is not very useful without a graphical visualization method. Thus, the same implementation from Additional algorithm 17.3 can be extended for 100 iterations, and the results can be shown using a rudimentary chart (Additional algorithm 17.4). Such a rudimentary chart is constructed by using a simple function that draws lines above a canvas object (id="bio") based on the values returned by the main equation. For simplicity of implementation, the axes and numbers are not drawn, as the interest is related exclusively to the evolution of the values and not to exact measurements. Thus, for reference, the *y*-axis represents the values returned by the equation and the *x*-axis represents the iterations/steps.

Additional algorithm 17.4 Note that the source code is in context and works with copy/paste.

```
<canvas id="bio" height="300" width="1100"></canvas>

<script>

//CHAOS

var s = 100;

Chart(system(0.62342375475, s), '#ff0000','n');
Chart(system(0.62342375474, s), '#000000', 'n');

function system(x0, s){
    var x1 = 0;
    var r = 0;

    for(var i=1; i<s; i++){

        x1 = 4 * x0 * (1 - x0);
        x0 = x1;

        r+=x1 + ",";
    }
    return r;
}

//The following is just a rudimentary chart

function Chart(q,c,e) {

    var s = q.split(",");
    var mx = Math.max.apply(null, s);

    var oldy = 0;
    var oldn = 0;
```

(Continued)

Additional algorithm 17.4 (Continued)

```
var canvas = document.getElementById('bio');

var w = canvas.width;
var h = canvas.height;

if (canvas.getContext) {

    var ctx = canvas.getContext('2d');

    if(e=='y'){ctx.clearRect(0, 0, w, h);}

    ctx.moveTo(0, 0);
    ctx.beginPath();

    for (var i=0; i<=s.length-1; i++)
    {
        var y = h - ((h / mx) * s[i]);
        var x = (w / s.length) * i;

        ctx.moveTo(oldn, oldy);
        ctx.lineTo(x, y);

        oldy = y;
        oldn = x;
    }

    ctx.lineWidth = 2;
    ctx.strokeStyle = c;
    ctx.stroke();
    }
}

</script>
```

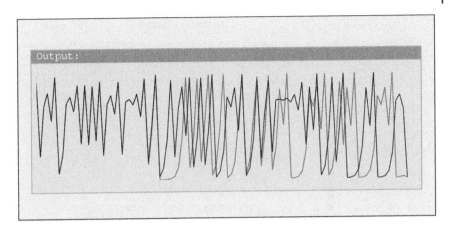

Note that the *Chart* function from Additional algorithm 17.4 receives three parameters. A parameter with all the values obtained above the 100 iterations (r), each value separated by a delimiter (" , "). A second parameter is the color of the line identifying each initial value (x0). A third parameter that indicates to the function whether the canvas object should be cleared or not ('y' for clear/erase). Inside the *Chart* function, the maximum value above the sequence of numbers (s) is detected, and this value is stored in the *mx* variable. The *Chart* function contains a loop that makes a number of iterations (i) equal to the number of terms present in the number sequence (s). Inside the main loop, the coordinates above the canvas object are calculated based on the maximum value, namely according to the value found in the *mx* variable. Thus, the *y*-axis is represented by the height of the canvas object divided by the value in the *mx* variable (h/mx), and the result is multiplied by the current value in the number sequence (s[i]). To position the zero values at the bottom of the chart, the *y*-axis is reversed by subtracting the result (the *y* value) from the height of the canvas object:

$$y = h - \left(\frac{h}{mx} \times s_i \right)$$

In contrast, the *x*-axis is calculated by dividing the length of the canvas object by the total number of terms in the sequence of numbers (w/s.length), and the result is multiplied by the iteration number (i):

$$x = \left(\frac{w}{|s|} \times i \right)$$

Once the two values are computed, the line is drawn from the previous coordinates to the current *x* and *y* coordinates (ctx.lineTo(x, y);). This concludes the discussions related to the *Chart* function. In the above output (Additional algorithm 17.4), the tipping point is observed when the two lines begin to separate

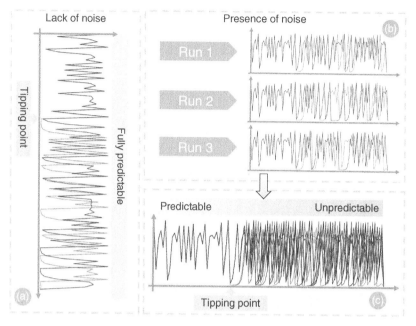

Figure 17.1 Limits of prediction, chaos, and noise. Each graph shows the evolution of the same nonlinear equation ($X_{n+1} = 4 \times X_n \times (1 - X_n)$) starting with two almost identical values ($X_0 = 0.62342375475$ and $X_0 = 0.62342375474$). (a) The panel on the left shows an experiment that observes the evolution of values (y-axis) for 100 iterations (x-axis) in the absence of noise. (b) The panel on the right shows an experiment that displays the evolution of values (y-axis) for 100 iterations (x-axis) in presence of noise. The noise in this case is a very small value (0.00000000000001), which is randomly added or subtracted from the main result (X_{n+1}) at each iteration in both evolutions. The place where the two evolutions in a chart begin to diverge is called the tipping point. Note that each run of the experiment is unique from the tipping point onward. (c) Limits of prediction. It shows an overlay of several experiments with the same parameters in the presence of noise. Note that up to the tipping point the experiments in the presence of noise do not differ from experiments in the absence of noise. However, after the tipping point, in the presence of noise, each experiment is unique because although it is imperceptible, noise leads to unique evolutions at each run. Thus, the presence of noise indicates the prediction limits for any nonlinear system in our frame of reference.

(Figure 17.1a). That is the point at which the two identical systems (the same equation started with slightly different X_0 values) begin to behave radically differently. The revelation here is that a tiny change in the initial conditions leads to a radically different behavior of the same system over time, and this can be seen in the output of Additional algorithm 17.4. With a little experimentation, a correlation can be observed between the tipping-point position on the chart and the digit position on which a modification is made. Namely, the farther the modified digit is in the value of X_0, the farther the tipping point will be on the chart. In other words, the smaller the modification in the initial conditions, the longer the two systems will be in synchronization (will behave the same). A careful inspection of the output from Additional algorithm 17.4 suggests that a noise-free chaotic system is deterministic and can be predicted with certainty, only if the initial conditions are known. Thus, the same initial values will always lead to the same distributions on the chart.

17.5.2 Chaos with Noise

Without noise, the nonlinear equation leads to the same result, namely the same input leads to the same output. But what happens to a dynamic system in the presence of noise? How can noise be simulated to understand the effects? In essence, noise is reduced to an imperceptible interaction with the system. For example, a very small value (n) added or subtracted randomly from the result of each iteration can be considered noise:

$$X_{i+1} = 4 \times X_i \times (1 - X_i) + n$$

where n is a constant value with either a positive or a negative sign, which is meant to simulate the background noise for this system. To implement this strategy, only one addition is made to Additional algorithm 17.4, as follows:

Additional algorithm 17.5 Note that the source code is in context and works with copy/paste.

```
<canvas id="bio" height="300" width="1100"></canvas>

<script>

// CHAOS

var s = 100;

Chart(system(0.62342375475, s), '#ff0000','n');
Chart(system(0.62342375474, s), '#000000', 'n');

function system(x0, s){
    var x1 = 0;
    var r = 0;

    for(var i=1; i<s; i++){

        x1 = 4 * x0 * (1 - x0);
        x0 = x1 + noise();

        r+=x1 + ",";
    }
```

(Continued)

Additional algorithm 17.5 (Continued)

```
      return r;
}

// insert noise

function noise(){
    var interact = [0.00000000000001,-0.0000000000001];
    var n = interact[Math.floor(Math.random()*interact.length)];
    return n;
}


// The same chart as before

function Chart(q,c,e) {

    var s = q.split(",");
    var mx = Math.max.apply(null, s);

    var canvas = document.getElementById('bio');

    var w = canvas.width;
    var h = canvas.height;

    if (canvas.getContext) {

        var ctx = canvas.getContext('2d');

        if(e=='y'){ctx.clearRect(0, 0, w, h);}

        ctx.moveTo(0, 0);
        ctx.beginPath();

        for (var i=0; i<=s.length-1; i++)
        {
            var y = h - ((h / mx) * s[i]);
            var x = (w / s.length) * i;

            ctx.lineTo(x, y);
        }
        ctx.lineWidth = 2;
        ctx.strokeStyle = c;
        ctx.stroke();
    }
}

</script>
```

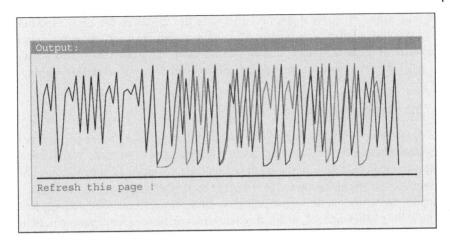

As can be seen above, the addition of noise is made by a new function called *noise* (Additional algorithm 17.5). The *noise* function is called at each iteration and it returns a positive or a negative value, namely the value can be randomly either 0.00000000000001 or −0.00000000000001. This simple approach leads to the observation of interesting effects when a chaotic system encounters noise (Figure 17.1b,c). Since the value for noise is particularly small compared to the calculated values, it fails to change the tipping point of the system. Moreover, for this particular setup of noise, only the region up to the tipping point remains constant, while the region after the tipping point evolves differently with each run (Figure 17.1b,c). The noise in the system can be raised or lowered depending on the values in the *noise* function. For example, a positive or negative value for a lower noise value, such as:

0.00000000000000001

will have a visible effect only after the tipping point. This indicates that the decrease in noise intensity leads to an increase of the prediction limit in a physical system (which is known in practice). The increase of the prediction limit by lowering the noise level is due to an expansion of the region of the distribution, which remains constant at each execution. Thus, a correlation exists between the noise value and the prediction limit, namely the limit at which the noise begins to have an effect over the evolution of the distribution. To repeat the experiments, equally delete or insert digits of zero inside the values of the *noise* function and run the Additional algorithm 17.5 (more zeros mean low-noise intensity; fewer zeros mean high-noise intensity). Note that here, noise means very subtle independent random variables that can interact with the system while is running.

17.5.3 Limits of Prediction

A dynamic system is anything that moves or changes over time (e.g. planetary motion, weather, the stock market, or chemical reactions). Can these dynamic systems be predicted? Some of them can be predicted, for example, the movement of planets or chemical reactions. On the other hand, the weather or the stock market are unpredictable for longer periods of time. The formula used above is just one of the mathematical dynamical systems that can be used for experimentation. A short example can be given for a parallel between Additional algorithm 17.4 and a concrete physical system. The above nonlinear equation can abstractly represent a container and the initial value X_0 can represent the initial position of a single molecule inside a liquid environment. Since this is an abstract example, the initial value X_0 may represent all the parameters, like position, translation, rotation, and vibration. Time is here represented by the number of steps taken by this system. Thus, consider this molecule tracked in a container for five minutes. It will start at the x, y, z coordinates relative to the container and end at the x', y', z' coordinates. If we had the possibility of a reload of the entire system five minutes before, we will certainly notice that the molecule we are following will reach the same point x', y', z' (much like Additional algorithm 17.4 does). In computer simulations, this is possible and repeatable. Computers are deterministic general-purpose machines defined by a strict set of instructions that are unaffected by independent random variables. On computers, 3D spaces are finite, without interference related to noise or other external factors that can influence the simulation. Thus, on a computer, the whole universe is the container, and for this reason the experiment is certainly deterministic and repeatable. In our universe, however, a container of liquid is influenced by imperceptible external and internal factors such as mechanical waves, electromagnetic waves (photons), and other particles that are imperceptible from our frame of reference (we cannot perceive them without specialized measurements). These influences are continuous and lead to a limitation in the measurements we can perform and also lead to unpredictability or short-term predictability of events (much like the results from Additional algorithm 17.5). The internal energy of matter is the main culprit for this limitation of prediction models. For this precise reason, in the real world, the repetition of a simple experiment cannot ensure exactly the same result. The two-arm pendulum experiment is a classic example in physics that demonstrates this very point.

17.5.4   On the Wings of Chaos

In a romantic interpretation, the universe can be seen as the most complex generator of high-entropy chaos. In contrast, we can be viewed as low-entropy chaotic bubbles floating on the ocean of noise of this machine. In other words, what it is perceived on a small scale as random walks within our frame of reference may be fully deterministic from a larger frame. Thus, the independent random variables of noise are themselves tiny parts of a collection of interacting chaotic systems. However, the weak interaction between chaotic systems is deterministic in nature. For example, the noise function that is based on computer randomization (i.e. the noise function from Additional algorithm 17.5) can be replaced by a chaotic system (a nonlinear equation that iterates particularly small values) from which samples at different offsets and equal intervals can be collected. Since the samples would only represent a slice through this chaotic system, the numerical values associated with these samples would seem as genuine independent random variables at first sight. However, these values are predictable if the chaotic system is fully known. Thus, we can conclude again that noise is a part of a chaotic system that is outside our frame of reference. Considering the results from noiseless chaos and noisy chaos, a pertinent question may arise, namely: Could future events be predicted without a deterministic universe? Well, the experiments from above show that absence of noise means certainty, whereas presence of different intensities of noise imposes different limits of prediction in the evolution of a chaotic system. By association, the fact that short-term events can be predicted for a physical process, it shows once again that we are living in a noisy, deterministic universe. Complementarily, the mimicry of physical phenomena on isolated universes, such as computer simulations, also points to hard determinism.

17.6   Free Will and Determinism

Without the background noise (whatever form the noise may take), all future events could perhaps be calculated with certainty, given the knowledge of the initial conditions (i.e. computer simulations). However, in the real world, noise is intrinsic to matter and continuously drives the chaotic phenomena (not just the initial conditions), which is fully deterministic, but impossible to predict on long term from our frame of reference. But how does noise over the chaotic phenomena relate to the notion of "free will"? Noise from real-time images/frames taken by video cameras can have a direct impact on the explanations regarding the lack of "free will." The best connection between noise, random numbers, and "free will" can be reasoned by imagining an experiment. For instance, consider a scenario involving a robot whose input is a video camera fixated on it. The frames provided by this camcorder would display a road showing a center line. While moving, the purpose of the robot is to maintain the center line in the middle of the framers provided by the video camera. In this thought experiment, the speed and the variations in the mechanical movements of the robot are considered constant and ideal. Each time the experiment is restarted with the same ideal initial conditions, the behavior of the robot will show subtle variations in the evolution of steering for maintaining direction on this central line (i.e. small variations in the input may lead to different tipping points and further to larger variations in the output). Thus, even if the speed of the robot is perfectly constant, with the ideal mechanics and identical initial conditions, each experiment may still be unique. The reason why it will position the line a little bit different in real time is related to the thermal noise produced by the sensor of the video camera. This noise effect manifests as individual pixels of different color shades, evenly distributed over the camcorder frames, and act as independent random variables for the algorithms used in the image analysis inside the robot. However, the same robot will take exactly the same "decisions" when the input is a video recording, thus, demonstrating the importance of noise in mimicking free will. A video recording is devoid of the random element (noise); thus, the same input will lead to the same output. In other words, once the random element (noise) is recorded and replayed, it becomes a constant in that frame of reference (i.e. the video recording).

17.6.1 The Greatest Disappointment

We humans are no different in this regard. In biological organisms, noise (e.g. thermal agitation, electromagnetic radiation, etc.) has a much deeper role because it is also intrinsic to cellular processes and makes the mirage of free will even more pronounced. As reasoned above, noise is not random in the cosmic frame; it may be random only for our frame of reference. Consequently, the "free will" of rationality is deterministic from a larger frame of reference. Thoughts could be defined in a simplistic manner as a sum of past events plus the background noise that can tilt the balance in favor or against a certain "decision" (a deterministic association). Moreover, the central nervous system of animals, even the simplest animals, can be seen as a chaotic processor and this provides the impression of "free will" (creativity and turbulence may have a commonality). In other words, our frame of reference allows us to escape the awareness regarding the lack of free will, because noise is unpredictable, which further leads to the impression of behavioral independence. But what about consciousness? No one knows exactly what consciousness is. There are different hypotheses, some that can only be partially tested and others that cannot be tested at all. However, consciousness may be a fragile phenomenon that always agrees with the result of information processing (similar to an "excuse" module for different deterministic actions). This is especially evident in the case of psychological manipulations in which an individual believes that the decision belongs to him. Or involuntary and natural psychological influences in cases where an individual becomes an "original" repeater (speaks phrases heard from others as if they originate from his own experience and thoughts). On the other hand, any action that is not "excused" (e.g. breastfeeding or involuntary movements), falls under the wing of "instinct," which further shows the limits and fragility of the conscious mind. These observations, of course, have ramifications in absolutely all aspects of consciousness. There are other everyday observable glitches that put into question consciousness and free will. For instance, an old Romanian saying is "vorbesti de lup si lupul la usa," which means "you're talking about the wolf and the wolf is at the door." Such popular phrases represent the everyday observable glitches and deepen the subtlety on which this "excuse" module (consciousness) can work with. Nevertheless, free will is a topic debated for millennia, but only in recent times, more and more do we understand as a civilization that this notion is false. Again, in the Romanian language, there are other very old popular sayings, which fit into the same narratives, namely "Omul potrivit la locul potrivit," which means "The right man in the right place," or "ce ti-e scris in frunte ti-e pus," which means "what is written is already on your forehead" (with reference to destiny), or "cine se aseamana se aduna," which means "whoever resembles gathers." These old European sayings and other similar ones that are present in different cultures all over the world are a hint of the recognition of determinism by the experience of previous generations. As a conclusion for the above discussions, *without noise there is no illusion of free will.*

17.6.2 The Most Powerful Processor in Existence

Often, the human brain is viewed as a powerful processor with a misleading disregard of the elementary unit that makes the brain development and the core functionality possible. The reason why the brain may be seen as a chaotic processor is related to a basic organelle present in all eukaryotic cells, namely the cell nucleus, which is fundamentally a chaotic processor. One claim that can be made here is that the cell nucleus is the most powerful and versatile processor in existence. The fact that this type of processor makes us and the other eukaryotic species, of course, cannot be a sufficiently credible argument to sustain this claim. Unlike our rigid electronic microprocessors, the cell nucleus is a dynamic 3D microprocessor driven by noise, which contains both memory and processing intertwined in the same cavity. Decoding a processor whose conformation changes continuously is a particularly difficult task and has been in research for decades without exact results on the inner dynamic rules that makes it alive. *In vivo*, the three-dimensional molecular dynamics of the cell nucleus makes it very difficult to study. For instance, fluorescently labeled molecules that bind to DNA or RNA short motifs are used today. However, once such an approach is made, the labeled molecules interfere with the internal processes of the nucleus and decrease the chances of complete and correct decoding regarding the mode of operation over time. *Post factum*, the study of the nucleus in a locked state brings only general information and nothing tangible about the dynamics and mode of operation (i.e. slices through the cell nucleus for electron microscopy). Currently, there are several hypotheses related to the structure of the nucleus and the chromosomal territories are known in general, but that's about it. Moreover, this biological processor does not work according to our reference system (i.e. the simple rules of logic implemented in classical electronics) and shows three crucial characteristics with which everyone can agree: (i) it is three dimensional, (ii) it is malleable, and (iii) it is in a continuous dynamics. Moreover, it is a processor that changes its state using the same code (i.e. different cell types). The general thinking about biological organisms when compared to computers is that biological organisms are analogous. But are biological organisms essentially analogous? It could be argued that the cell nucleus is neither analog nor binary, yet it is still a discrete processor. By nature, all molecules are discrete units, and this is fairly obvious with DNA or RNA polymers. More to the point, any DNA or RNA polymer shows a predisposition to a particular structural conformation when confined to the nuclear space (i.e. the local pH and ionic strength influences this 3D distribution to a certain degree). In practice, this predisposition becomes more obvious as the length of the polymer decreases (i.e. short RNA molecules). The Brownian motion dictated by thermal noise would in theory lead to all possible structural configurations of DNA or RNA molecules; however, this does not happen. Rather, random walks for DNA polymers in 3D exhibit a structural predisposition that is given by the order of the nucleotides in the sequence/polymer. For instance, a point mutation has the power to radically change the distribution (the preferred shape of the polymer in 3D), or, in most cases it may have no noticeable effect on this distribution (such experiments can be done using the objective digital stains described in Chapter 7). Of course, the preferential 3D conformation of DNA alone is not the only property responsible for the interworking of the cell nucleus. Dedicated proteins and nuclear mRNA also actively participate in the structural arrangements of DNA in the nucleus depending on different external signals that adjust the feedback of the gene network. Nonetheless, inside the cell nucleus, genes found along the DNA molecules/polymers are physically spaced apart through different folds according to these predispositions for spatial contortions. Furthermore, the physical distance between the genes dictates the fine adjustment of the gene network. Many RNA molecules or their protein products behave as signaling molecules (i.e. transcription factors) that have the ability to activate or inactivate the transcription of other genes. In other words, activation/inactivation of a gene leads to the presence or lack of an RNA product whose direct (the RNA molecule itself) or indirect (the protein derived from that RNA) structural properties can activate or inactivate other genes in the set and vice versa. Thus, the RNA feedback between genes makes the information processing circuit possible. Such a processing circuit is of course influenced by the background noise (e.g. electromagnetic noise), which manifests itself as Brownian motion. The Brownian motion subtly sets the activation/inactivation tipping points for any of these feedbacks by varying the physical distances between the folds that contain different genes (similar to a forest moved by the wind). Also, the Brownian motion and the concentration gradients dictate a specific stochastic distribution in the liquid medium of the nucleus that allows the diffusion of mRNA molecules from the origin of synthesis to other regions of a lower concentration (radial targeted random walks), thus, maintaining the feedback network of this processor (the cell nucleus). To add to the complexity, the rules concerning these feedbacks extend between different organelles in the same cell and can reach different cells within an organism. Of course, the laws behind this molecular dynamics represent the holy grail of bioinformatics. Moreover, these laws are without a doubt the ultimate goal of the field. To put things from a different perspective, even if we were to synthesize a simple cell nucleus following known examples from nature, we would not know exactly how and why it works, yet.

17.6.3 Certainty vs. Interpretation

Mechanistic views on genomics took root decades ago and were strongly influenced in parallel by the evolution of computers. The influence of the computer industry has been so great that the gene is today seen by many as a kind of biological "file." This is also in line with our human condition that requires clear patterns for understanding data (which is not necessarily that simple in genomics). Contrary to expectations, it is not yet clear what a gene really is. Of course, the alignment of sequenced mRNAs against sequenced genomes indicates the exact location of genes and their promoters. In turn, this provides a mechanistic clarity over the meaning of genes, and it provides a consensus in the scientific community. However, in the scientific literature, there are more and more exceptions in recent years that indicate huge variability/permissivity in gene locations (e.g. the arrays of closely located transcription start sites (TSSs) with different rates of initiation). This claim is also supported by enhancers of gene promoters that can vary considerably in distance from one gene to another, indicating a lack of a clear rule regarding the location of the gene. Although the above may seem outrageous, these considerations related to the definition of a gene are still relevant today. This further points out the crucial role that the random element can play in all biological processes. Following this rationale, a few questions can be asked: is a gene slowly constructed by directed mutations (i.e. the heterochromatin–euchromatin interplay; please see the recycle model [31]) across generations long before it gets activated? Are pseudogenes old dysfunctional genes or new ones in the making that may be activated in the future generations of a species? or both ? or does a gene simply evolve from other duplicated genes or pseudogenes in a mechanistic way? These are questions with serious implications that will need to be addressed in more detail in the future.

17.6.4 A Wisdom that Applies

The last subchapter is dedicated
to Col. Pavel Gagniuc, my father
(1949–2021)

A true story told by a retired Romanian army colonel (Col. Pavel Gagniuc) brings an interesting lesson on the certainty of scientific discoveries based on computer assistance. After World War II, in which the technology and organization of the German Nazi forces showed overwhelming superiority in combat, all future officers in the communist military school (which included Romania) were accepted according to one criterion, namely how good they were in the field of mathematics, and what was their innate talent for strategy (they usually go hand in hand). Later, in the early 1980s, within the Warsaw Pact, communist countries held annual military exercises together, for a better coordination in the event of an armed conflict with the "evil" in the West. Now, imagine that we are talking about an era when chess was a national sport and a time when Romania had the third or fourth largest army in the world under the "supreme leader" Nicolae Ceausescu. Today we see these events in the spirit of history; however, at that time they were particularly serious and with possible devastating effects on civilization as we know it. In these military exercises, one team from each communist country from the Warsaw Pact participated in this competition for ground-to-air missile launches (among many types of military exercises) and the winners would have brought the communist glory (aka the Klingonian empire) for one year to the communist country they came from. This particular exercise consisted of firing 10 ground-to-air missiles by each team into supersonic planes (a cheap commodity at that time for the Soviets; MIG 15, a small obsolete supersonic fighter jet similar in size to a modern military drone) piloted by remote control from the ground. Thus, observers throughout the communist bloc paid the utmost attention to these military exercises. At the beginning of the exercise, the Soviet Union come up with a computer-guided laser technology and many other new military developments. They targeted and hit 9 planes out of 10. The Germans come up with an automatic technology for guidance via video analysis of the target in real time. They targeted and hit 8 out of 10 planes. Poland come up with a sophisticated technology based on a gyroscopic system with computer corrections in flight. They targeted and hit 7 out of 10 planes. Of course, every country in the Warsaw Pact came up with something new at these meetings. Note that the military technologies described are fictional; however, the point is that the missiles were indeed automated and guided by the use of onboard computers. Now, it was Romania's turn to fire missiles in this exercise. The Romanian representative at that time was Col. Chiper ("Chiper" – means pepper in Romanian/Moldovan dialect – a "predestined" name for critical thinking), a silent old man, very good in implementing math logic in combat. His team included other Romanian officers, including, at that time, lieutenant Pavel Gagniuc from which I know of these events. All the young officers called him "father Chiper." Of course, there were no high hopes for the Romanian team at that time. However, Col. Chiper and his team, in the absence of the advanced technologies that their comrades in the other countries possessed, had to be particularly inventive and rely on what they had at their disposal, namely a topographic map of the region, 10 unguided ground-to-air missiles, and math, a lot of it (paper computing). Before the military exercise, the Romanian team had prepared a rudimentary manual system based exclusively on mathematical calculations made by hand prior to launch (so without guidance during the missile flight). The result of all that work was basically some precalculated bizarre divisions for prediction of interception written on a paper, glued to the military cap worn by the officer who was now aiming in real time. Thus, the direct overlapping relationship between these divisions on the cap, the optical aiming of the firing station and the moving aircraft, was the method used for interception. Cutting a lot from the story, the Romanian team hits all 10 planes, and wins the contest for this section of the exercise. Of course, there is also the possibility in which the Soviets wanted happy partners and let them have a chance through a predictable trajectory of the jets. Nonetheless, all the observers of the communist bloc, including the media of the Soviet Union, were shocked and eager to interview Col. Chiper. Thus, everyone hunted Col. Chiper and they finally caught with him out of a mess hall of a military unit, where he was accommodated during the exercises. With dozens of microphones thrown in his face, he listens to a few quick questions. Then, when the reporters were quiet, one of the reporters asked him: "Comrade Colonel, each country came up with extremely advanced military technologies, all based on state-of-the-art powerful computers. You didn't have any sophisticated technology and especially you didn't have computer-based guidance. How did you manage to win?" Being a quiet "master Yoda" type, he uttered one sentence that became the main headline in the Red Army newspaper the next day: "In calculator, prostie bagi, prostie iese!," translated later into Russian "В компьютер, глупо поставишь, глупость выходит!," which means "In a computer, stupidity you put in, stupidity you get out!" Of course, these words were said to "pepper" the pride of the Soviet Union; however, they brought to light a relentless truth. How much can discretization affect the prediction limit? This phrase also pointed to the subjectivity of the software and the interpretation of the incoming real-time data, and that, makes one think of today. This sentence also shows the long way from the idea to the actual implementation and reliability of any artificial system. Moreover, whether it is a bioinformatics software where the "complexity" can be turned into "complicated" fairly quickly, or it is about other software implementations in other fields, in the end every computation result should be viewed with a grain of salt as it is still partially under the subjective interpretation and beliefs of the scientist. For example, today, software libraries are a valuable shortcut in the software industry. From this point of view, software libraries should be embraced in the industry in those projects where accuracy is not critical. On the other hand, software libraries should be rejected by the scientific community in places where trust is not a commodity. A subjective interpretation in a software library (which no one really checks) can have serious consequences on the interpretation of data in many third-party scientific investigations and also it participates in the deprofessionalization of engineers working in research that may wake up helpless when these libraries are missing from "the list." In the spirit of the above, we can conclude that too much simplification or shortcuts for development "kills the cat," which in this case is the know-how of our global civilization. *On the finale of this story, I can only quote the deepest and most meaningful philosophical sentence that was ever spoken: "It is what it is.".*

17.7   Conclusions

A philosophical approach was presented here to motivate the prediction limits that exist in both bioinformatics and other fields. An obvious conclusion of this chapter first indicated that the deeper the level of complexity inside a system, the more difficult is the prediction of that system. Based on this observation, the discussions further suggested that pieces of chaotic systems reach our frame of reference as independent random variables. This was the basic idea behind the explanations in connection to the random number generators, namely pieces from complex systems lead to unpredictability which then led to the concept of independence. As a bridge between randomness and certainty, pseudo-random numbers have also been discussed to better point out what our frame of reference means. Thus, randomness was shown to be emergent of complexity. Noise has been described as a term with a rather subjective threshold that forms the boundary of our frame of reference. Moreover, it has been suggested here that long-term predictions are becoming less and less likely due to noise. Thus, the main point of the discussions was that noise-induced chaotic behavior gives rise to the phenomena of free will, but it also imposes the prediction limits. Nevertheless, the chaotic behavior is known to be fully deterministic, which may point out to the true nature of our universe. Many discussions were accompanied by simple experimental implementations whose advanced versions can be found online on the main webpage of this book.

Appendix A

Sequence logos are a visualization approach to graphically represent the sequence conservation. For this reason, their implementation was left to the end of this book. Note that the implementations described here are not limited to motif sequences for DNA, RNA, or proteins. Thus, these representations can be used for other fields as well. The implementations presented below show the methods by which sequence logos can be obtained both from matrices and directly from sets of motif sequences.

A.1 Association of Numerical Values with Letters

To start designing a general algorithm for displaying sequence logos, the plan must first include an already-known matrix. This matrix is stored as before in string format. The string format is loaded into a two-dimensional array (L) object by the *load* function. The *load* function returns both the array (L) and the maximum value above the array (max). Moreover, the *load* function associates to each value the corresponding letter from the first column of the matrix/array. By associating numerical values with the letters they belong to, the algorithm makes the labeling strategy possible and greatly reduces the size of the implementation (Additional algorithm A.1).

Additional algorithm A.1 Note that the source code is in context and works with copy/paste.

```
<script>

c = "|G\t-0.92\t-0.22\t0.69\t1.39\t-40.06\t0.69\t-0.92\t1.16\t-0.92" +
    "|A\t0.18\t0.47\t-0.22\t-40.06\t-40.06\t0.69\t1.03\t-0.92\t-40.06" +
```

(Continued)

Algorithms in Bioinformatics: Theory and Implementation, First Edition. Paul A. Gagniuc.
© 2021 John Wiley & Sons, Inc. Published 2021 by John Wiley & Sons, Inc.
Companion website: www.wiley.com/go/gagniuc/algorithmsinbioinformatics

Additional algorithm A.1 (Continued)

```
        "|T\t-0.22\t0.18\t-0.22\t-40.06\t1.39\t-40.06\t-40.06\t-0.92\t1.03" +
        "|C\t0.47\t-0.92\t-0.92\t-40.06\t-40.06\t-40.06\t-0.22\t-40.06\t-0.22";

var tmp;
tmp = load(c);

document.write(SMC(tmp[0])+ '<br>');

//LOAD MATRICES FROM STRINGS
function load(t){
    var n = [];
    var m = [];
    var L = [];

    m = t.split('|');
    var max = 0;

    for(var i=1; i<m.length; i++) {
        L[i-1]=[];
        n = m[i].split('\t');
        for(var j=0; j<n.length; j++){
            L[i-1][j]=n[j];
            if(j>0){L[i-1][j]+='|'+L[i-1][0];}
            if(max<=L[i-1][j]){max=L[i-1][j];}
        }
    }
    return [L, max];
}

// SHOW MATRIX CONTENT
function SMC(m) {
    var r = "<table border=1>";
    for(var i=0; i<m.length; i++) {
        r += "<tr>";
        for(var j=0; j<m[i].length; j++){
            r += "<td>"+m[i][j]+"</td>";
        }
        r += "</tr>";
    }
    r += "</table>";

    return r;
}

</script>

Output:

G -0.92|G -0.22|G  0.69|G   1.39|G -40.06|G   0.69|G -0.92|G  1.16|G  -0.92|G
A  0.18|A  0.47|A -0.22|A -40.06|A -40.06|A   0.69|A  1.03|A -0.92|A -40.06|A
T -0.22|T  0.18|T -0.22|T -40.06|T   1.39|T -40.06|T -40.06|T -0.92|T  1.03|T
C  0.47|C -0.92|C -0.92|C -40.06|C -40.06|C -40.06|C -0.22|C -40.06|C  -0.22|C
```

A.2 Sorting Values on Columns

The *load* function returns both the array and the maximum value above the array. Thus, once the matrix is loaded into memory, an additional strategy must be devised. Namely, the values on each column of the matrix must be sorted from the highest value to the lowest value or vice versa. However, this normal task is not that simple in practice. Each element in a column contains a string value composed of the numerical value, a delimiter (' | '), and the string value (the letter) associated with it. Thus, here the same *split* method based on a delimiter is used. By applying a nested loop, each column *j* is traversed and the string values on column *k* are extracted. On each step, the string value is split into a numeric value and a string value (the letter associated with the value). The numerical value and the letter are temporarily inserted into a $2 \times k$ array variable *a*. Next, the values of array variable *a* are sorted by a *iSort* function, which is able to sort on the first column (a[k][0]), where the numerical values are, and swap the letters concomitantly (a[k][1]). The *iSort* function returns the same variable *a* with numerical values sorted in ascending order (in *a*, the last element *k* contains the largest value). Once all the numerical values and their associated letters are sorted in variable *a*, the elements are passed back into a regular matrix *M*. Next, matrix *M* is printed in the output. Thus, up to this point, a matrix *M* contains sorted columns and the design of a logo can safely begin (Addtional Algorithm A.2).

Additional algorithm A.2 Note that the source code is in context and works with copy/paste.

```
<script>

//ORDER VALUES ON EACH COLUMN

c = "|G\t-0.92\t-0.22\t0.69\t1.39\t-40.06\t0.69\t-0.92\t1.16\t-0.92" +
    "|A\t0.18\t0.47\t-0.22\t-40.06\t-40.06\t0.69\t1.03\t-0.92\t-40.06" +
    "|T\t-0.22\t0.18\t-0.22\t-40.06\t1.39\t-40.06\t-40.06\t-0.92\t1.03" +
    "|C\t0.47\t-0.92\t-0.92\t-40.06\t-40.06\t-40.06\t-0.22\t-40.06\t-0.22";

var tmp;
tmp = load(c);

document.write(SMC(tmp[0]) + '<br>');

var M = tmp[0];

var a=[];
```

(Continued)

Additional algorithm A.2 (Continued)

```javascript
    for(var j=1; j<M[0].length; j++){

        document.write('<hr>column=' +j);

        for(var k=0; k<M.length; k++){
            a[k]=[];
            a[k][0] = Number(M[k][j].split('|')[0]);
            a[k][1] = M[k][j].split('|')[1];
        }

        a=iSort(a);

        document.write(SMC(a)+ '<br>');

        for(var k=0; k<M.length; k++){
            M[k][j]=a[k][0] + '|' + a[k][1];
        }
    }


document.write(SMC(M)+ '<br>');

//SORT
function iSort(a) {
    var n = a.length;

    for (var i = 1; i < n; i++) {

        let n = a[i][0];
        let j = i-1;

        while ((j > -1) && (n < a[j][0])) {
            a[j+1][0] = a[j][0];

            var t = a[j+1][1];
            a[j+1][1] = a[j][1];
            a[j][1]=t;

            j--;
        }

        a[j+1][0] = n;
    }
    return a;
}

//LOAD MATRICES FROM STRINGS
function load(t){
    var n = [];
    var m = [];
    var L = [];

    m = t.split('|');
    var max = 0;
```

```
    for(var i=1; i<m.length; i++) {
        L[i-1]=[];
        n = m[i].split('\t');
        for(var j=0; j<n.length; j++){
            L[i-1][j]=n[j]; //Number(n[j]);
            if(j>0){L[i-1][j]+='|'+L[i-1][0];}
            if(max<=L[i-1][j]){max=L[i-1][j];}
        }
    }
    return [L, max];
}

// SHOW MATRIX CONTENT
function SMC(m) {
    var r = "<table border=1>";
    for(var i=0; i<m.length; i++) {
        r += "<tr>";
        for(var j=0; j<m[i].length; j++){
            r += "<td>"+m[i][j]+"</td>";
        }
        r += "</tr>";
    }
    r += "</table>";

    return r;
}

</script>


Output:

G -0.92|G -0.22|G  0.69|G   1.39|G -40.06|G   0.69|G  -0.92|G   1.16|G  -0.92|G
A  0.18|A  0.47|A -0.22|A -40.06|A -40.06|A   0.69|A   1.03|A  -0.92|A -40.06|A
T -0.22|T  0.18|T -0.22|T -40.06|T   1.39|T -40.06|T -40.06|T  -0.92|T   1.03|T
C  0.47|C -0.92|C -0.92|C -40.06|C -40.06|C -40.06|C  -0.22|C -40.06|C  -0.22|C

column=1
-0.92   G
-0.22   T
0.18    A
0.47    C

column=2
-0.92   C
-0.22   G
0.18    T
0.47    A

column=3
-0.92   C
-0.22   A
-0.22   T
0.69    G
```

(Continued)

Additional algorithm A.2 (Continued)

```
column=4
-40.06 A
-40.06 T
-40.06 C
 1.39   G

column=5
-40.06 G
-40.06 A
-40.06 C
 1.39   T

column=6
-40.06 T
-40.06 C
 0.69   G
 0.69   A

column=7
-40.06 T
-0.92   G
-0.22   C
 1.03   A

column=8
-40.06 C
-0.92   A
-0.92   T
 1.16   G

column=9
-40.06 A
-0.92   G
-0.22   C
 1.03   T
```

```
G -0.92|G -0.92|C -0.92|C -40.06|A -40.06|G -40.06|T -40.06|T -40.06|C -40.06|A
A -0.22|T -0.22|G -0.22|A -40.06|T -40.06|A -40.06|C  -0.92|G  -0.92|A  -0.92|G
T  0.18|A  0.18|T -0.22|T -40.06|C -40.06|C  0.69|G  -0.22|C  -0.92|T  -0.22|C
C  0.47|C  0.47|A  0.69|G  1.39|G  1.39|T  0.69|A  1.03|A  1.16|G  1.03|T
```

A.3 The Implementation of a Sequence Logo

Here, the implementation of a sequence logo is described (Additional algorithm A.3). Up to this point, a bidimensional array *M* has been loaded using the *load* function and the values on the columns have been sorted using an *iSort* function. What now? In a first instance, a PWM (probability weight matrix) is used as an input instead of the LLM (log-likelihood matrix). In the previous implementation, a matrix M contained all the values on the columns in a sorted order. However, to optimize this implementation, matrix M can be disregarded and variable a can be used directly on each step of the sequence logo construction. Since function *load* returns an unsorted matrix M, the number of columns is still known (t=M[0].length). By using nested loops, each column j in matrix M is traversed and the string values on column k are extracted and injected into the elements of variable a. Inside the nested loop, once variable a was loaded and sorted accordingly, it can be used to plot the letters of the sequence logo in the j area on the canvas. But what is the j area of the canvas? To calculate the length of the region in which the letters are to be plotted (iw) on the horizontal axis, the width (w) of the primary canvas object (ctx) is divided by the number of positions (columns) found in matrix M, minus the first column in matrix M, which contains only the letters (t-1). Then, the x coordinate on the primary canvas (the sequence logo canvas) is calculated by multiplying the current j position by the width of the letter region (iw). Once the width of the letter area is known, a third loop traverses the elements of variable a to collect the letters in the order dictated by the *iSort* function (a[u][1]). Inside this third loop, the plotting of the letters is made. Each letter to be plotted is first printed at a giant size on a secondary canvas object (ctl), then the image is copied and compressed to the x, y, and iw, ih coordinates. First, the u letter from variable a is printed at full size on the entire surface of the secondary canvas object using specific colors for DNA. Before the actual drawing of the letter (a[u][1]), a few calculations are made for centering the letter on the x-axis (to ensure a central position). Next, the y position of this letter is calculated for the primary canvas. The height of the primary canvas is divided by the maximum value found over matrix M. Note: since matrix M is a PWM, the maximum value possible will be 1. The result of this division is multiplied further by the value from element u of variable a (a[u][0]) plus the previous value from element $u - 1$ of variable a (pune). To reverse the axis, the entire result is subtracted from the height of the primary canvas. The height of each letter (ih) on the j area of the primary canvas object is computed. Again, the height of the canvas object is divided by the maximum value over matrix M (h/max), and then is multiplied by the value found in the element $u - 1$ of variable a (pune). From this result, the y value is further subtracted. Again, to reverse the axis, the result is further subtracted from the height of the primary canvas object (h-((h/max)*(pune))-y). Next, the entire surface of the secondary canvas object is copied to the primary canvas object at the following coordinates: x, y, iw, ih. The process continues with the next step of the loop where another column of matrix M is injected into variable a, then ordered, and then used to plot the next column of letters on the primary canvas object. This concludes the sequence logo approach.

Additional algorithm A.3 **Note that the source code is in context and works with copy/paste.**

```
<canvas id="bio" width="900" height="200"></canvas><hr>
<canvas id="letter" width="370" height="410" style="display: none;">
<div id="print"></div><br>

<script>
window.onload = function() {

    //MAKE LOGO FROM MATRICES
    c = '|G\t0.1\t0.2\t0.5\t1\t0\t0.5\t0.1\t0.8\t0.1' +
        '|A\t0.3\t0.4\t0.2\t0\t0\t0.5\t0.7\t0.1\t0' +
        '|T\t0.2\t0.3\t0.2\t0\t1\t0\t0\t0.1\t0.7' +
        '|C\t0.4\t0.1\t0.1\t0\t0\t0\t0.2\t0\t0.2';

    var msg;
    var tmp;
    tmp = load(c);

    msg = 'Unordered:<br>' + SMC(tmp[0])+'<br>';

    var M = tmp[0];
    var max = tmp[1];

    var a = [];
    var t = M[0].length;

    var canvas = document.getElementById('bio');
    var canvasl = document.getElementById('letter');
    var ctl = canvasl.getContext('2d');

    var w = canvas.width;
    var h = canvas.height;

    var wl = canvasl.width;
    var hl = canvasl.height;

    if (canvas.getContext) {

        var ctx = canvas.getContext('2d');

        for(var j=1; j<t; j++){

            //ORDER VALUES ON EACH COLUMN
            for(var k=0; k<M.length; k++){
                a[k]=[];
                a[k][0] = M[k][j].split('|')[0];
                a[k][1] = M[k][j].split('|')[1];
            }

            a = iSort(a);
```

```
            for(var k=0; k<M.length; k++){
                M[k][j] = a[k][0] + '|' + a[k][1];
            }

            // LOGO
            var iw = Math.floor(w/(t-1));
            var x = Math.floor(((j-1)*iw));

            var pune = 0;

            for (var u=0; u<a.length; u++)
            {
                ctl.imageSmoothingQuality = 'high';
                ctl.clearRect(0, 0, wl, hl);
                ctl.font = 'bold 540px Arial';

                var cl = 'black';
                //if(a[u][1]=='G'){cl='#f4244c';}
                //if(a[u][1]=='T'){cl='#607b46';}
                //if(a[u][1]=='A'){cl='#f4244c';}
                //if(a[u][1]=='C'){cl='#dcb655';}

                if(a[u][1]=='G'){cl='#fcaf07';}
                if(a[u][1]=='T'){cl='#d50000';}
                if(a[u][1]=='A'){cl='#07d607';}
                if(a[u][1]=='C'){cl='#0909c8';}

                ctl.fillStyle = cl;

                var ltr = ctl.measureText(a[u][1]).width;
                ctl.fillText(a[u][1], (wl/2)-(ltr/2), hl-5);

                var y = h-Math.floor((h/max)*(Number(a[u][0])+pune));

                if(u>0){var ih = h-((h/max)*(pune))-y;}
                if(u==0){var ih = h-y;}

                pune += Number(a[u][0]);

                ih = Math.floor(ih);

                ctx.imageSmoothingQuality = 'high';
                ctx.drawImage(canvasl, x, y, iw, ih);
            }
        }
    }

    msg += 'Ordered by columns:<br>' + SMC(M);
    document.getElementById('print').innerHTML = msg;
}

//SORT
```

(Continued)

Additional algorithm A.3 (Continued)

```javascript
function iSort(a) {
    var n = a.length;
    for (var i = 1; i < n; i++) {
        let n = a[i][0];
        let j = i-1;

        while ((j > -1) && (n < a[j][0])) {

            a[j+1][0] = a[j][0];

            var t = a[j+1][1];
            a[j+1][1] = a[j][1];
            a[j][1]=t;
            j--;
        }
        a[j+1][0] = n;
    }
    return a;
}

//LOAD MATRICES FROM STRINGS
function load(t){
    var n = [];
    var m = [];
    var L = [];

    m = t.split('|');
    var max = 0;

    for(var i=1; i<m.length; i++) {
        L[i-1]=[];
        n = m[i].split('\t');
        for(var j=0; j<n.length; j++){
            L[i-1][j]=n[j];
            if(j>0){
                if(max<=L[i-1][j]){max=L[i-1][j];}
                L[i-1][j]+='|'+L[i-1][0];
            }
        }
    }
    return [L, max];
}

// SHOW MATRIX CONTENT
function SMC(m) {
    var r = "<table border=1>";
    for(var i=0; i<m.length; i++) {
        r += "<tr>";
        for(var j=0; j<m[i].length; j++){
            r += "<td>"+m[i][j]+"</td>";
        }
```

```
        r += "</tr>";
    }
    r += "</table>";

    return r;
}

</script>
```

Output:

Unordered:

```
G  0.1|G  0.2|G  0.5|G   1|G   0|G  0.5|G  0.1|G  0.8|G  0.1|G
A  0.3|A  0.4|A  0.2|A   0|A   0|A  0.5|A  0.7|A  0.1|A  0.0|A
T  0.2|T  0.3|T  0.2|T   0|T   1|T  0.0|T  0.0|T  0.1|T  0.7|T
C  0.4|C  0.1|C  0.1|C   0|C   0|C  0.0|C  0.2|C  0.0|C  0.2|C
```

Ordered by columns

```
G  0.1|G  0.1|C  0.1|C   0|A   0|G  0.0|T  0.0|T  0.0|C  0.0|A
A  0.2|T  0.2|G  0.2|A   0|T   0|A  0.0|C  0.1|G  0.1|A  0.1|G
T  0.3|A  0.3|T  0.2|T   0|C   0|C  0.5|G  0.2|C  0.1|T  0.2|C
C  0.4|C  0.4|A  0.5|G   1|G   1|T  0.5|A  0.7|A  0.8|G  0.7|T
```

A.4 Sequence Logos Based on Maximum Values

This implementation slightly simplifies the approach (Additional algorithm A.4). One thing to notice is that *pune* variable is not used anymore. The height of the letters in this case is calculated based on the maximum value above the M matrix. However, compared to the previous implementation (Additional algorithm A.3) in which each letter was represented graphically in relation to the previous value found in element $u - 1$ of variable a, this time each letter is represented graphically in direct relation to the maximum value.

Additional algorithm A.4 **Note that the source code is in context and works with copy/paste.**

```
<canvas id="bio" width="900" height="200"></canvas>
<canvas id="letter" width="370" height="410" style="display: none;">
</canvas>
<div id="print"></div><br>

<script>
window.onload = function() {

    //MAKE LOGO FROM MATRICES
    c = '|G\t0.5\t0.2\t0.5\t1\t0\t0.5\t0.1\t0.8\t0.1' +
        '|A\t0.3\t0.4\t0.2\t0\t0\t0.5\t0.7\t0.1\t0' +
        '|T\t0.2\t0.3\t0.2\t0\t1\t0\t0\t0.1\t0.7' +
        '|C\t0.4\t0.1\t0.1\t0\t0\t0\t0.2\t0\t0.2';

    var msg;
    var tmp;
    tmp = load(c);

    msg = 'Unordered:<br>' + SMC(tmp[0])+'<br>';

    var M = tmp[0];
    var max = tmp[1];

    var a = [];
    var t = M[0].length;

    var canvas = document.getElementById('bio');
    var canvasl = document.getElementById('letter');
    var ctl = canvasl.getContext('2d');

    var w = canvas.width;
    var h = canvas.height - 5;

    var wl = canvasl.width;
    var hl = canvasl.height;

    if (canvas.getContext) {

        var ctx = canvas.getContext('2d');

        for(var j=1; j<t; j++){

            //ORDER VALUES ON EACH COLUMN
            for(var k=0; k<M.length; k++){
                a[k]=[];
                a[k][0] = M[k][j].split('|')[0];
                a[k][1] = M[k][j].split('|')[1];
            }

            a = iSort(a);
```

```
            for(var k=0; k<M.length; k++){
                M[k][j] = a[k][0] + '|' + a[k][1];
            }

            // LOGO
            var iw = w/(t-1);
            var x = (j-1)*iw;

            for (var u=0; u<a.length; u++)
            {
                ctl.imageSmoothingQuality = 'high';
                ctl.clearRect(0, 0, wl, hl);
                ctl.font = 'bold 540px Arial';

                var cl = 'black';
                if(a[u][1]=='G'){cl='#f4244c';}
                if(a[u][1]=='T'){cl='#607b46';}
                if(a[u][1]=='A'){cl='#3464ac';}
                if(a[u][1]=='C'){cl='#dcb655';}

                //if(a[u][1]=='G'){cl='#fcaf07';}
                //if(a[u][1]=='T'){cl='#d50000';}
                //if(a[u][1]=='A'){cl='#07d607';}
                //if(a[u][1]=='C'){cl='#0909c8';}

                ctl.fillStyle = cl;

                var ltr = ctl.measureText(a[u][1]).width;
                ctl.fillText(a[u][1], (wl/2)-(ltr/2), hl-5);

                var y = h-(h/max)*a[u][0];

                if(u>0){var ih = h-((h/max)*a[u-1][0])-y;}
                if(u==0){var ih = h-y;}

                ctx.imageSmoothingQuality = 'high';
                ctx.drawImage(canvas1, x, y, iw, ih);
            }
        }
    }

    msg += 'Ordered by columns:<br>' + SMC(M);
    document.getElementById('print').innerHTML = msg;
}

//SORT
function iSort(a) {
    var n = a.length;
    for (var i = 1; i < n; i++) {
        let n = a[i][0];
        let j = i-1;
```

Additional algorithm A.4 (Continued)

```
            while ((j > -1) && (n < a[j][0])) {

                a[j+1][0] = a[j][0];

                var t = a[j+1][1];
                a[j+1][1] = a[j][1];
                a[j][1]=t;
                j--;
            }
            a[j+1][0] = n;
        }
        return a;
}

//LOAD MATRICES FROM STRINGS
function load(t){
    var n = [];
    var m = [];
    var L = [];

    m = t.split('|');
    var max = 0;

    for(var i=1; i<m.length; i++) {
        L[i-1]=[];
        n = m[i].split('\t');
        for(var j=0; j<n.length; j++){
            L[i-1][j]=n[j];
            if(j>0){
                if(max<=L[i-1][j]){max=L[i-1][j];}
                L[i-1][j]+='|'+L[i-1][0];
            }
        }
    }
    return [L, max];
}

// SHOW MATRIX CONTENT
function SMC(m) {
    var r = "<table border=1>";
    for(var i=0; i<m.length; i++) {
        r += "<tr>";
        for(var j=0; j<m[i].length; j++){
            r += "<td>"+m[i][j]+"</td>";
        }
        r += "</tr>";
    }
    r += "</table>";

    return r;
}

</script>
```

```
Output:
```

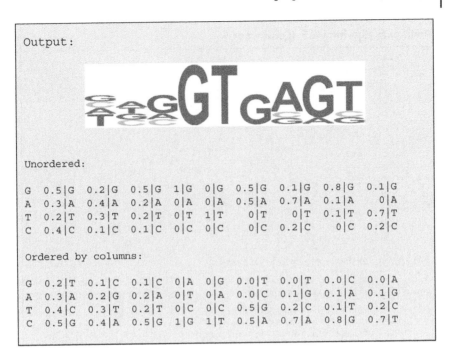

```
Unordered:

G  0.5|G  0.2|G  0.5|G  1|G  0|G  0.5|G  0.1|G  0.8|G  0.1|G
A  0.3|A  0.4|A  0.2|A  0|A  0|A  0.5|A  0.7|A  0.1|A    0|A
T  0.2|T  0.3|T  0.2|T  0|T  1|T    0|T    0|T  0.1|T  0.7|T
C  0.4|C  0.1|C  0.1|C  0|C  0|C    0|C  0.2|C    0|C  0.2|C

Ordered by columns:

G  0.2|T  0.1|C  0.1|C  0|A  0|G  0.0|T  0.0|T  0.0|C  0.0|A
A  0.3|A  0.2|G  0.2|A  0|T  0|A  0.0|C  0.1|G  0.1|A  0.1|G
T  0.4|C  0.3|T  0.2|T  0|C  0|C  0.5|G  0.2|C  0.1|T  0.2|C
C  0.5|G  0.4|A  0.5|G  1|G  1|T  0.5|A  0.7|A  0.8|G  0.7|T
```

A.5 Using Logarithms to Build Sequence Logos

The difference between this implementation (Additional algorithm A.5) and the
previous implementation (Additional algorithm A.4) is related to the use of the
logarithm to process the probability values from the PWM. This process takes
place directly in the *load* function (`-Log(2,n[j])`). Note that the maximum
and the minimum value over *M* are detected outside the *load* function. This is a
preparatory strategy for the next step in which loaded matrices are not used.

**Additional algorithm A.5 Note that the source code is in context and works
with copy/paste.**

```
<canvas id="bio" width="900" height="200"></canvas>
<canvas id="letter" width="370" height="410" style="display: none;">
</canvas>
<div id="print"></div><br>

<script>
window.onload = function() {
```

(Continued)

Additional algorithm A.5 (Continued)

```
//MAKE LOGO FROM MATRICES
c = '|G\t0.1\t0.2\t0.5\t1\t0\t0.5\t0.1\t0.8\t0.1' +
    '|A\t0.3\t0.4\t0.2\t0\t0\t0.5\t0.7\t0.1\t0' +
    '|T\t0.2\t0.3\t0.2\t0\t1\t0\t0\t0.1\t0.7' +
    '|C\t0.4\t0.1\t0.1\t0\t0\t0\t0.2\t0\t0.2';

var msg;

var M = load(c);
var max = 0;
var min = 0;

msg = 'Unordered:<br>' + SMC(M)+'<br>';

var a = [];
var t = M[0].length;

//FIND THE MAXIMUM SUM OVER COLUMNS
var fmax=0;
var fmin=0;
for(var j=1; j<t; j++){
    for(var k=0; k<M.length; k++){
        fmax += Number(M[k][j].split('|')[0]);
        fmin += Number(M[k][j].split('|')[0]);
    }
    if(max<=fmax){max=fmax;}
    if(min>=fmin){min=fmin;}
    fmax=0;
    fmin=0;
}

var canvas = document.getElementById('bio');
var canvasl = document.getElementById('letter');
var ctl = canvasl.getContext('2d');

var w = canvas.width;
var h = canvas.height-(max-min)-10;

var wl = canvasl.width;
var hl = canvasl.height;

if (canvas.getContext) {

    var ctx = canvas.getContext('2d');

    for(var j=1; j<t; j++){

        //ORDER VALUES ON EACH COLUMN
        for(var k=0; k<M.length; k++){
            a[k]=[];
            a[k][0] = M[k][j].split('|')[0];
            a[k][1] = M[k][j].split('|')[1];
        }
```

```
            a = iSort(a);

            for(var k=0; k<M.1length; k++){
                M[k][j] = a[k][0] + '|' + a[k][1];
            }

            // LOGO
            var iw = w/(t-1);
            var x = (j-1)*iw;

            var pune = 0;

            for (var u=0; u<a.length; u++)
            {
                ctl.imageSmoothingQuality = 'high';
                ctl.clearRect(0, 0, wl, hl);
                ctl.font = 'bold 540px Arial';

                var cl = 'black';
                if(a[u][1]=='G'){cl='#fcaf07';}
                if(a[u][1]=='T'){cl='#d50000';}
                if(a[u][1]=='A'){cl='#07d607';}
                if(a[u][1]=='C'){cl='#0909c8';}

                ctl.fillStyle = cl;

                var ltr = ctl.measureText(a[u][1]).width;
                ctl.fillText(a[u][1], (wl/2)-(ltr/2), hl-5);

                var y = h-(h/max)*(Number(a[u][0])+pune);

                if(u>0){var ih = h-((h/max)*(pune))-y;}
                if(u==0){var ih = h-y;}

                pune += Number(a[u][0]);

                ih = ih;

                ctx.imageSmoothingQuality = 'high';
                ctx.drawImage(canvasl, x, y, iw, ih);
            }
        }
    }

    msg += 'Ordered by columns:<br>' + SMC(M);
    document.getElementById('print').innerHTML = msg;
}


//SORT
function iSort(a) {
    var n = a.length;
    for (var i = 1; i < n; i++) {
        let n = a[i][0];
```

(Continued)

Additional algorithm A.5 (Continued)

```
        let j = i-1;

        while ((j > -1) && (n < a[j][0])) {

            a[j+1][0] = a[j][0];

            var t = a[j+1][1];
            a[j+1][1] = a[j][1];
            a[j][1]=t;
            j--;
        }
        a[j+1][0] = n;
    }
    return a;
}

//LOAD MATRICES FROM STRINGS
function load(t){
    var n = [];
    var m = [];
    var L = [];

    m = t.split('|');
    var maxc = 0;
    var max = 0;


    for(var i=1; i<m.length; i++) {
        L[i-1]=[];
        n = m[i].split('\t');
        for(var j=0; j<n.length; j++){
            L[i-1][j]=n[j];
            if(j>0){
                //if(L[i-1][j]>0){L[i-1][j]=n[j]*Log(2,(1/n[j]));}
                if(L[i-1][j]>0){L[i-1][j]=-Log(2,n[j]).toFixed(2);}
                L[i-1][j]+='|'+L[i-1][0];
            }
        }
    }
    return L;
}


// SHOW MATRIX CONTENT
function SMC(m) {
    var r = "<table border=1>";
    for(var i=0; i<m.length; i++) {
        r += "<tr>";
        for(var j=0; j<m[i].length; j++){
            r += "<td>"+m[i][j]+"</td>";
        }
        r += "</tr>";
    }
```

```
        r += "</table>";

        return r;
}

function Log(n, v) {
  return Math.log(v) / Math.log(n);
}

</script>
```

Output:

Unordered:

```
G  3.32|G  2.32|G  1.00|G  0|G  0|G  1|G  3.32|G  0.32|G  3.32|G
A  1.74|A  1.32|A  2.32|A  0|A  0|A  1|A  0.51|A  3.32|A  0.00|A
T  2.32|T  1.74|T  2.32|T  0|T  0|T  0|T  0.00|T  3.32|T  0.51|T
C  1.32|C  3.32|C  3.32|C  0|C  0|C  0|C  2.32|C  0.00|C  2.32|C
```

Ordered by columns:

```
G  1.32|C  1.32|A  1.00|G  0|G  0|G  0|T  0.00|T  0.00|C  0.00|A
A  1.74|A  1.74|T  2.32|A  0|A  0|A  0|C  0.51|A  0.32|G  0.51|T
T  2.32|T  2.32|G  2.32|T  0|T  0|T  1|G  2.32|C  3.32|A  2.32|C
C  3.32|G  3.32|C  3.32|C  0|C  0|C  1|A  3.32|G  3.32|T  3.32|G
```

A.6   From a Motif Set to a Sequence Logo

The last implementation uses the exact steps that were discussed in the chapter related to motifs, namely the implementation starts from a set of motif sequences and ends up printing a sequence logo. This latest implementation concludes the methodology used to plot sequence logo graphics. Note that online, on the book's website, there are a number of more dynamic implementations for sequence logo plots, which are based on the implementation from below (Additional algorithm A.6).

Additional algorithm A.6 Note that the source code is in context and works with copy/paste.

```
<canvas id="bio" width="900" height="200" style="border:1px solid #d3d3d3;">
</canvas>
<canvas id="letter" width="370" height="410" style="display: none;">
</canvas>
<div id="print"></div><br>

<script>

//RPN4
var c = 'GGGTGGCAAAC,' +
        'CGGTGGCAAAG,' +
        'GAGTGGCAAAA,' +
        'CAGTGGCAAAT,' +
        'GAGTAGCAAAC,' +
        'CGGCGGCAAAA,' +
        'AGGTGGCAAAA,' +
        'AGGTGGCGAAA,' +
        'TGGTGGCAAAA,' +
        'CGGTGGCAAAA,' +
        'TGGTGGCAAAT,' +
        'AGGTGGCAAAA,' +
        'GAGTGGCAATA,' +
        'CGGTGGCAAAA,' +
        'AGGTGGCTAGG,' +
        'GGGTGGCAAAA,' +
        'CGGTGGCGAAA,' +
        'CGGTGGCAAAA,' +
        'GGATGGCGACA,' +
        'TGATGGCAAAA,' +
        'AAGTGGCCTAC,' +
        'TAGTGGAAAAT,' +
        'TGGTGGCGAAA,' +
        'AAGTGGCGTCT,' +
        'GGGTGGCAAAC,' +
        'GGGTGGCAAAA';

var s = [];
var m = [];

m = c.split(',');
var n = m.length;

//THE ALIGNMENT MATRIX
for(var i=0; i<n; i++){
    s[i] = [];
    s[i]=m[i].split('');
}

document.write('<hr>Alignment matrix: ' + SMC(s));
```

```
//DETECT ALL LETTERS USING ARRAYS
var a = [];

var t = c.replace(/,/g, '').split('');
var k = t.length;

for(var i=0; i<=k; i++){
    var q = 1;
    for(var j=0; j<=a.length; j++){
        if (t[i] === a[j]) {q = 0;}
    }
    if (q === 1) {a.push(t[i]);}
}

// PROFILE MATRIX INITIALIZATION
var p = [];

for(var h=0; h<a.length; h++){
    p[h]=[];
    for(var i=0; i<=s[0].length; i++) {
        p[h][i]=0;
        p[h][0]=a[h];
    }
}

// THE POSITION FREQUENCY MATRIX
for(var i=0; i<s.length; i++) {

    for(var j=0; j<s[i].length; j++){

        for(var h=0; h<a.length; h++){

            if (s[i][j] === a[h]) {p[h][j+1]++;}
        }
    }
}

document.write('<hr>Position frequency matrix: ' + SMC(p));


// THE POSITION PROBABILITY MATRIX
var max = 0;
for(var i=0; i<p.length; i++) {
    for(var j=0; j<p[i].length-1; j++){

        p[i][j+1]=p[i][j+1]/s.length;
        p[i][j+1]=p[i][j+1].toFixed(2);

        if(max<=p[i][j+1]){max=p[i][j+1];}

        p[i][j+1]+='|'+p[i][0]
    }
}
```

(Continued)

Additional algorithm A.6 (Continued)

```
document.write('<hr>Position probability matrix: ' + SMC(p));

logo(p,max);

function logo(M, max) {

    //MAKE LOGO FROM MATRICES
    var msg;

    msg = 'Unordered:<br>' + SMC(M)+'<br>';

    var a = [];
    var t = M[0].length;

    var canvas = document.getElementById('bio');
    var canvasl = document.getElementById('letter');
    var ctl = canvasl.getContext('2d');

    var w = canvas.width;
    var h = canvas.height - 5;

    var wl = canvasl.width;
    var hl = canvasl.height;

    if (canvas.getContext) {

        var ctx = canvas.getContext('2d');

        for(var j=1; j<t; j++){

            //ORDER VALUES ON EACH COLUMN
            for(var k=0; k<M.length; k++){
                a[k]=[];
                a[k][0] = M[k][j].split('|')[0];
                a[k][1] = M[k][j].split('|')[1];
            }

            a = iSort(a);

            for(var k=0; k<M.length; k++){
                M[k][j] = a[k][0] + '|' + a[k][1];
            }

            // LOGO
            var iw = w/(t-1);
            var x = (j-1)*iw;

            for (var u=0; u<a.length; u++)
            {
                ctl.imageSmoothingQuality = 'high';
                ctl.clearRect(0, 0, wl, hl);
                ctl.font = 'bold 540px Arial';

                var cl = 'black';
```

```
            //if(a[u][1]=='G'){cl='#f4244c';}
            //if(a[u][1]=='T'){cl='#607b46';}
            //if(a[u][1]=='A'){cl='#f4244c';}
            //if(a[u][1]=='C'){cl='#dcb655';}

            if(a[u][1]=='G'){cl='#fcaf07';}
            if(a[u][1]=='T'){cl='#d50000';}
            if(a[u][1]=='A'){cl='#07d607';}
            if(a[u][1]=='C'){cl='#0909c8';}

            ctl.fillStyle = cl;

            var ltr = ctl.measureText(a[u][1]).width;
            ctl.fillText(a[u][1], (wl/2)-(ltr/2), hl-5);

            var y = h-(h/max)*a[u][0];

            if(u>0){var ih = h-((h/max)*a[u-1][0])-y;}
            if(u==0){var ih = h-y;}

            ctx.imageSmoothingQuality = 'high';
            ctx.drawImage(canvas1, x, y, iw, ih);
        }
    }
}

    msg += 'Ordered by columns:<br>' + SMC(M);
    document.getElementById('print').innerHTML = msg;
}

//SORT
function iSort(a) {
    var n = a.length;
    for (var i = 1; i < n; i++) {
        let n = a[i][0];
        let j = i-1;

        while ((j > -1) && (n < a[j][0])) {

            a[j+1][0] = a[j][0];

            var t = a[j+1][1];
            a[j+1][1] = a[j][1];
            a[j][1]=t;
            j--;
        }
        a[j+1][0] = n;
    }
    return a;
}

// SHOW MATRIX CONTENT
function SMC(m) {
    var r = "<table border=1>";
```

(Continued)

Additional algorithm A.6 (Continued)

```
    for(var i=0; i<m.length; i++) {
        r += "<tr>";
        for(var j=0; j<m[i].length; j++){
            r += "<td>"+m[i][j]+"</td>";
        }
        r += "</tr>";
    }
    r += "</table>";

    return r;
}

</script>
```

Output:

Unordered:

```
G 0.3|G 0.7|G 0.9|G 0.00|G 0.96|G 1|G 0.00|G 0.19|G 0.00|G 0.04|G 0.08|G
T 0.2|T 0.0|T 0.0|T 0.96|T 0.00|T 0|T 0.00|T 0.04|T 0.08|T 0.04|T 0.15|T
C 0.3|C 0.0|C 0.0|C 0.04|C 0.00|C 0|C 0.96|C 0.04|C 0.00|C 0.08|C 0.15|C
A 0.2|A 0.3|A 0.1|A 0.00|A 0.04|A 0|A 0.04|A 0.73|A 0.92|A 0.85|A 0.62|A
```

Ordered by columns:

```
G 0.2|T 0.0|T 0.0|T 0.00|G 0.00|T 0|T 0.00|G 0.04|T 0.00|G 0.04|G 0.08|G
T 0.2|A 0.0|C 0.0|C 0.00|A 0.00|C 0|C 0.00|T 0.04|C 0.00|C 0.04|T 0.15|T
C 0.3|C 0.3|A 0.1|A 0.04|C 0.04|A 0|A 0.04|A 0.19|G 0.08|T 0.08|C 0.15|C
A 0.3|G 0.7|G 0.9|G 0.96|T 0.96|G 1|G 0.96|C 0.73|A 0.92|A 0.85|A 0.62|A
```

Alignment matrix:

```
G  G  G  T  G  G  C  A  A  A  C
C  G  G  T  G  G  C  A  A  A  G
G  A  G  T  G  G  C  A  A  A  A
C  A  G  T  G  G  C  A  A  A  T
G  A  G  T  A  G  C  A  A  A  C
C  G  G  C  G  G  C  A  A  A  A
A  G  G  T  G  G  C  A  A  A  A
A  G  G  T  G  G  C  G  A  A  A
T  G  G  T  G  G  C  A  A  A  A
C  G  G  T  G  G  C  A  A  A  A
T  G  G  T  G  G  C  A  A  A  T
```

```
A  G  G  T  G  G  C  A  A  A  A
G  A  G  T  G  G  C  A  A  T  A
C  G  G  T  G  G  C  A  A  A  A
A  G  G  T  G  G  C  T  A  G  G
G  G  G  T  G  G  C  A  A  A  A
C  G  G  T  G  G  C  G  A  A  A
C  G  G  T  G  G  C  A  A  A  A
G  G  A  T  G  G  C  G  A  C  A
T  G  A  T  G  G  C  A  A  A  A
A  A  G  T  G  G  C  C  T  A  C
T  A  G  T  G  G  A  A  A  A  T
T  G  G  T  G  G  C  G  A  A  A
A  A  G  T  G  G  C  G  T  C  T
G  G  G  T  G  G  C  A  A  A  C
G  G  G  T  G  G  C  A  A  A  A
```

Position frequency matrix:

```
G   8   19   24    0   25   26    0    5    0    1    2
T   5    0    0   25    0    0    0    1    2    1    4
C   7    0    0    1    0    0   25    1    0    2    4
A   6    7    2    0    1    0    1   19   24   22   16
```

Position probability matrix:

```
G 0.3|G 0.7|G 0.9|G 0.00|G 0.96|G 1|G 0.00|G 0.19|G 0.00|G 0.04|G 0.08|G
T 0.2|T 0.0|T 0.0|T 0.96|T 0.00|T 0|T 0.00|T 0.04|T 0.08|T 0.04|T 0.15|T
C 0.3|C 0.0|C 0.0|C 0.04|C 0.00|C 0|C 0.96|C 0.04|C 0.00|C 0.08|C 0.15|C
A 0.2|A 0.3|A 0.1|A 0.00|A 0.04|A 0|A 0.04|A 0.73|A 0.92|A 0.85|A 0.62|A
```

References

1 Wilde, S.A., Valley, J.W., Peck, W.H., and Graham, C.M. (2001). Evidence from detrital zircons for the existence of continental crust and oceans on the Earth 4.4 Gyr ago. *Nature* 409 (6817): 175–178.

2 Zahnle, K., Schaefer, L., and Fegley, B. (2010). Earth's earliest atmospheres. *Cold Spring Harbor Perspectives in Biology* 2 (10): a004895.

3 Cleaves, H.J. II, (2012). Prebiotic chemistry: what we know, what we don't. *Evolution: Education and Outreach* 5: 342–360.

4 Cleaves, H.J. II, (2013). Prebiotic chemistry: geochemical context and reaction screening. *Life (Basel)* 3 (2): 331–345.

5 Altermann, W. and Kazmierczak, J. (2003). Archean microfossils: a reappraisal of early life on Earth. *Research in Microbiology* 154 (9): 611–617.

6 Noffke, N., Christian, D., Wacey, D., and Hazen, R.M. (2013). Microbially induced sedimentary structures recording an ancient ecosystem in the *ca.* 3.48 billion-year-old Dresser Formation, Pilbara, Western Australia. *Astrobiology* 13 (12): 1103–1124.

7 Mojzsis, S.J., Arrhenius, G., McKeegan, K.D. et al. (1996). Evidence for life on Earth before 3,800 million years ago. *Nature* 384 (6604): 55–59.

8 Bell, E.A., Boehnke, P., Harrison, T.M., and Mao, W.L. (2015). Potentially biogenic carbon preserved in a 4.1 billion-year-old zircon. *Proceedings of the National Academy of Sciences of the United States of America* 112 (47): 14518–14521.

9 Demoulin, C.F., Lara, Y.J., Cornet, L. et al. (2019). Cyanobacteria evolution: insight from the fossil record. *Free Radical Biology and Medicine* 140: 206–223.

10 Schopf, J.W. (2014). Geological evidence of oxygenic photosynthesis and the biotic response to the 2400-2200 ma "great oxidation event". *Biochemistry (Moscow)* 79 (3): 165–177.

Algorithms in Bioinformatics: Theory and Implementation, First Edition. Paul A. Gagniuc.
© 2021 John Wiley & Sons, Inc. Published 2021 by John Wiley & Sons, Inc.
Companion website: www.wiley.com/go/gagniuc/algorithmsinbioinformatics

11 Albani, A.E., Bengtson, S., Canfield, D.E. et al. (2010). Large colonial organisms with coordinated growth in oxygenated environments 2.1 Gyr ago. *Nature* 466: 100–104.

12 Douzery, E.J.P., Snell, E.A., Bapteste, E. et al. (2004). The timing of eukaryotic evolution: does a relaxed molecular clock reconcile proteins and fossils? *Proceedings of the National Academy of Sciences of the United States of America* 101 (43): 15386–15391.

13 Doolittle, R.F., Feng, D.-F., Tsang, S. et al. (1996). Determining divergence times of the major kingdoms of living organisms with a protein clock. *Science* 271 (5248): 470–477.

14 Parfrey, L.W. and Lahr, D.J.G. (2013). Multicellularity arose several times in the evolution of eukaryotes. *BioEssays* 35 (4): 339–347.

15 Evans, S.D., Hughes, I.V., Gehling, J.G., and Droser, M.L. (2020). Discovery of the oldest bilaterian from the Ediacaran of South Australia. *Proceedings of the National Academy of Sciences of the United States of America* 117 (14): 7845–7850.

16 Gooday, A., Holzmann, M., Caulle, C. et al. (2017). Giant protists (xenophyophores, Foraminifera) are exceptionally diverse in parts of the abyssal eastern Pacific licensed for polymetallic nodule exploration. *Biological Conservation* 207: 106–116.

17 Freeman, G. (2009). The rise of bilaterians. *Historical Biology* 21 (1–2): 99–114.

18 Cunningham, J.A., Liu, A.G., Bengtson, S., and Donoghue, P.C.J. (2017). The origin of animals: can molecular clocks and the fossil record be reconciled? *BioEssays* 39 (1): 1–12.

19 Monahan-Earley, R., Dvorak, A.M., and Aird, W.C. (2013). Evolutionary origins of the blood vascular system and endothelium. *Journal of Thrombosis and Haemostasis* 11 (s1): 46–66.

20 Chen, Z., Zhou, C., Yuan, X., and Xiao, S. (2019). Death march of a segmented and trilobate bilaterian elucidates early animal evolution. *Nature* 573 (7774): 412–415.

21 Shu, D.-G., Luo, H.-L., Morris, S.C. et al. (1999). Lower Cambrian vertebrates from south China. *Nature* 402: 42–46.

22 Dunn, C.W. (2013). Evolution: out of the ocean. *Current Biology* 23 (6): R241–R243.

23 Daeschler, E.B., Shubin, N.H., and Jr, F.A.J. (2006). A Devonian tetrapod-like fish and the evolution of the tetrapod body plan. *Nature* 440 (7085): 757–763.

24 Novacek, M.J. (1997). Mammalian evolution: an early record bristling with evidence. *Current Biology* 7 (8): PR489–R491.

25 Brusatte, S.L., Butler, R.J., Prieto-Márquez, A., and Norell, M.A. (2012). Dinosaur morphological diversity and the end-Cretaceous extinction. *Nature Communications* 3: 804.

26 Brusatte, S.L., Butler, R.J., Barrett, P.M. et al. (2015). The extinction of the dinosaurs. *Biological Reviews of the Cambridge Philosophical Society* 90 (2): 628–642.

27 Soligo, C., Will, O.A., Tavaré, S. et al. (2007). New light on the dates of primate origins and divergence. In: *PRIMATE ORIGINS: Adaptations and Evolution* (eds. M.J. Ravosa and M. Dagosto), 29–49. Boston, MA: Springer.

28 Brocks, J.J., Logan, G.A., Buick, R., and Summons, R.E. (1999). Archean molecular fossils and the early rise of eukaryotes. *Science* 285 (5430): 1033–1036.

29 Koonin, E.V. (2010). The origin and early evolution of eukaryotes in the light of phylogenomics. *Genome Biology* 11 (5): 209.

30 Hübner, M.R., Eckersley-Maslin, M.A., and Spector, D.L. (2013). Chromatin organization and transcriptional regulation. *Current Opinion in Genetics & Development* 23 (2): 89–95.

31 Gagniuc, P. and Ionescu-Tirgoviste, C. (2013). Gene promoters show chromosome-specificity and reveal chromosome territories in humans. *BMC Genomics* 14: 278.

32 Mirny, L.A. (2011). The fractal globule as a model of chromatin architecture in the cell. *Chromosome Research* 19 (1): 37–51.

33 Maeshima, K., Ide, S., and Babokhov, M. (2019). Dynamic chromatin organization without the 30-nm fiber. *Current Opinion in Cell Biology* 58: 95–104.

34 Ashwin, S.S., Nozaki, T., Maeshima, K., and Sasai, M. (2019). Organization of fast and slow chromatin revealed by single-nucleosome dynamics. *Proceedings of the National Academy of Sciences of the United States of America* 116 (40): 19939–19944.

35 Krebs, J.E., Kuo, M.-H., Allis, C.D., and Peterson, C.L. (1999). Cell cycle-regulated histone acetylation required for expression of the yeast HO gene. *Genes & Development* 13 (11): 1412–1421.

36 Eberharter, A. and Becker, P.B. (2002). Histone acetylation: a switch between repressive and permissive chromatin. *EMBO Reports* 3 (3): 224–229.

37 Grunstein, M. (1997). Histone acetylation in chromatin structure and transcription. *Nature* 389: 349–352.

38 Gilkerson, R., Bravo, L., Garcia, I. et al. (2013). The mitochondrial nucleoid: integrating mitochondrial DNA into cellular homeostasis. *Cold Spring Harbor Perspectives in Biology* 5 (5): a011080.

39 Farge, G. and Falkenberg, M. (2019). Organization of DNA in mammalian mitochondria. *International Journal of Molecular Sciences* 20 (11): 2770.

40 Osbourn, A.E. and Field, B. (2009). Operons. *Cellular and Molecular Life Sciences* 66 (23): 3755–3775.

41 Yang, X., Coulombe-Huntington, J., Kang, S. et al. (2016). Widespread expansion of protein interaction capabilities by alternative splicing. *Cell* 164 (4): 805–817.

42 Poss, Z.C., Ebmeier, C.C., and Taatjes, D.J. (2013). The Mediator complex and transcription regulation. *Critical Reviews in Biochemistry and Molecular Biology* 48 (6): 575–608.

43 Harris, L.B. and Rogers, S.O. (2011). Evolution of small putative group I introns in the SSU rRNA gene locus of *Phialophora* species. *BMC Research Notes* 4: 258.

44 Lambowitz, A.M. and Zimmerly, S. (2011). Group II introns: mobile ribozymes that invade DNA. *Cold Spring Harbor Perspectives in Biology* 3 (8): a003616.

45 Zimmerly, S. and Semper, C. (2015). Evolution of group II introns. *Mobile DNA* 6: 7.

46 Yoshihisa, T. (2014). Handling tRNA introns, archaeal way and eukaryotic way. *Frontiers in Genetics* 5: 213.

47 Pelley, J.W. (2012). Protein synthesis and degradation. In: *Elsevier's Integrated Review Biochemistry*, 2e (ed. J.W. Pelley), 149–160. Elsevier.

48 Kearse, M.G. and Wilusz, J.E. (2017). Non-AUG translation: a new start for protein synthesis in eukaryotes. *Genes & Development* 31 (17): 1717–1731.

49 Noller, H.F., Lancaster, L., Zhou, J., and Mohan, S. (2017). The ribosome moves: RNA mechanics and translocation. *Nature Structural & Molecular Biology* 24 (12): 1021–1027.

50 Belardinelli, R., Sharma, H., Peske, F. et al. (2016). Translocation as continuous movement through the ribosome. *RNA Biology* 13 (12): 1197–1203.

51 Schmeing, T.M., Huang, K.S., Strobel, S.A., and Steitz, T.A. (2005). An induced-fit mechanism to promote peptide bond formation and exclude hydrolysis of peptidyl-tRNA. *Nature* 438: 520–524.

52 Davey, C.A., Sargent, D.F., Luger, K. et al. (2002). Solvent mediated interactions in the structure of the nucleosome core particle at 1.9 Å resolution. *Journal of Molecular Biology* 319: 1097–1113.

53 Sehnal, D., Rose, A., Kovca, J. et al. (2018). Mol*: towards a common library and tools for web molecular graphics. *MolVA/EuroVis Proceedings*: 29–33.

54 Yusupova, G.Z., Yusupov, M.M., Cate, J.H.D., and Noller, H.F. (2001). The path of messenger RNA through the ribosome. *Cell* 106: 233–241.

55 Rould, M.A., Perona, J.J., and Steitz, T.A. (1991). Structural basis of anticodon loop recognition by glutaminyl-tRNA synthetase. *Nature* 352: 213–218.

56 Belinky, F., Babenko, V.N., Rogozin, I.B., and Koonin, E.V. (2018). Purifying and positive selection in the evolution of stop codons. *Scientific Reports* 8 (1): 9260.

57 Fournier, G.P., Andam, C.P., Alm, E.J., and Gogarten, J.P. (2011). Molecular evolution of aminoacyl tRNA synthetase proteins in the early history of life. *Origins of Life and Evolution of the Biosphere* 41 (6): 621–632.

58 Kaiser, F., Krautwurst, S., Salentin, S. et al. (2020). The structural basis of the genetic code: amino acid recognition by aminoacyl-tRNA synthetases. *Scientific Reports* 10: 12647.

59 Ognjenović, J. and Simonović, M. (2018). Human aminoacyl-tRNA synthetases in diseases of the nervous system. *RNA Biology* 15 (4–5): 623–634.

60 Melnikov, S.V., van den Elzen, A., Stevens, D.L. et al. (2018). Loss of protein synthesis quality control in host-restricted organisms. *Proceedings of the National Academy of Sciences of the United States of America* 115 (49): E11505–E11512.

61 Melnikov, S.V. and Söll, D. (2019). Aminoacyl-tRNA synthetases and tRNAs for an expanded genetic code: what makes them orthogonal? *International Journal of Molecular Sciences* 20 (8): 1929.

62 Woese, C.R., Olsen, G.J., Ibba, M., and Sö, D. (2000). Aminoacyl-tRNA synthetases, the genetic code, and the evolutionary process. *Microbiology and Molecular Biology Reviews* 64 (1): 202–236.

63 Zhang, C. (2009). Novel functions for small RNA molecules. *Current Opinion in Molecular Therapeutics* 11 (6): 641–651.

64 Wahid, F., Shehzad, A., Khan, T., and Kim, Y.Y. (2010). MicroRNAs: synthesis, mechanism, function, and recent clinical trials. *Biochimica et Biophysica Acta* 1803 (11): 1231–1243.

65 MacFarlane, L.-A. and Murphy, P.R. (2010). MicroRNA: biogenesis, function and role in cancer. *Current Genomics* 11 (7): 537–561.

66 Groot, M. and Lee, H. (2020). Sorting mechanisms for microRNAs into extracellular vesicles and their associated diseases. *Cell* 9 (4): 1044.

67 Li, J., Jiang, X., and Wang, K. (2019). Exosomal miRNA: an alternative mediator of cell-to-cell communication. *ExRNA* 1: 31.

68 Mitchell, P.S., Parkin, R.K., Kroh, E.M. et al. (2008). Circulating microRNAs as stable blood-based markers for cancer detection. *Proceedings of the National Academy of Sciences of the United States of America* 105 (30): 10513–10518.

69 Ramsköld, D., Wang, E.T., Burge, C.B., and Sandberg, R. (2009). An abundance of ubiquitously expressed genes revealed by tissue transcriptome sequence data. *PLoS Computational Biology* 5 (12): e1000598.

70 Brown, P.O. and Botstein, D. (1999). Exploring the new world of the genome with DNA microarrays. *Nature Genetics* 21 (1): 33–37.

71 Jones, K.W. and Robertson, F.W. (1970). Localisation of reiterated nucleotide sequences in *Drosophila* and mouse by in situ hybridisation of complementary RNA. *Chromosoma* 31 (3): 331–345.

72 Rao, M.S., Vleet, T.R.V., Ciurlionis, R. et al. (2019). Comparison of RNA-Seq and microarray gene expression platforms for the toxicogenomic evaluation of liver from short-term rat toxicity studies. *Frontiers in Genetics* 9: 636.

73 García-Ortega, L.F. and Martínez, O. (2015). How many genes are expressed in a transcriptome? Estimation and results for RNA-Seq. *PLoS One* 10 (6): e0130262.

74 Mora, C., Tittensor, D.P., Adl, S. et al. (2011). How many species are there on Earth and in the ocean? *PLoS Biology* 9 (8): e1001127.

75 Louca, S., Mazel, F., Doebeli, M., and Parfrey, L.W. (2019). A census-based estimate of Earth's bacterial and archaeal diversity. *PLoS Biology* 17 (2): e3000106.

76 Amann, R. and Rosselló-Móra, R. (2016). After all, only millions? *MBio* 7 (4): e00999–e00916.

77 Schloss, P.D., Girard, R.A., Martin, T. et al. (2016). Status of the archaeal and bacterial census: an update. *MBio* 7 (3): e00201.

78 Williams, S.L. and Grosholz, E.D. (2002). Preliminary reports from the *Caulerpa taxifolia* invasion. *Marine Ecology Progress Series* 233: 307–310.

79 Tendal, O. and Lewis, K.B. (1978). New Zealand xenophyophores: upper bathyal distribution, photographs of growth position, and a new species. *New Zealand Journal of Marine and Freshwater Research* 12 (2): 197–203.

80 Eslick, E.M., Beilby, M.J., and Moon, A.R. (2014). A study of the native cell wall structures of the marine alga *Ventricaria ventricosa* (Siphonocladales, Chlorophyceae) using atomic force microscopy. *Microscopy* 63 (2): 131–140.

81 Lemieux, C., Otis, C., and Turmel, M. (2014). Six newly sequenced chloroplast genomes from prasinophyte green algae provide insights into the relationships among prasinophyte lineages and the diversity of streamlined genome architecture in picoplanktonic species. *BMC Genomics* 15 (1): 857.

82 Chrétiennot-Dinet, M.-J., Courties, C., Vaquer, A. et al. (1995). A new marine picoeucaryote: ostreococcus tauri gen. et sp. nov. (Chlorophyta, Prasinophyceae). *Phycologia* 34 (4): 285–292.

83 Árnason, Ú., Lammers, F., Kumar, V. et al. (2018). Whole-genome sequencing of the blue whale and other rorquals finds signatures for introgressive gene flow. *Science Advances* 4 (4): eaap9873.

84 Laursen, L. and Bekoff, M. (1978). Loxodonta africana. *Mammalian Species* 92: 1–8.

85 Withers, P.C. (1983). Energy, water, and solute balance of the ostrich Struthio camelus. *Physiological Zoology* 56 (4): 568–579.

86 Razin, S. (1992). Peculiar properties of mycoplasmas: the smallest self-replicating prokaryotes. *FEMS Microbiology Letters* 100 (1–3): 423–431.

87 Atalla, H., Lysnyansky, I., Raviv, Y., and Rottem, S. (2015). *Mycoplasma gallisepticum* inactivated by targeting the hydrophobic domain of the membrane preserves surface lipoproteins and induces a strong immune response. *PLoS One* 10 (3): e0120462.

88 Schulz, H.N., Brinkhoff, T., Ferdelman, T.G. et al. (1999). Dense populations of a giant sulfur bacterium in Namibian shelf sediments. *Science* 284 (5413): 493–495.

89 Bresler, V., Montgomery, W.L., Fishelson, L., and Pollak, P.E. (1998). Gigantism in a bacterium, *Epulopiscium fishelsoni*, correlates with complex patterns in arrangement, quantity, and segregation of DNA. *Journal of Bacteriology* 180 (21): 5601–5611.

90 Perlmutter, J.D., Qiao, C., and Hagan, M.F. (2013). Viral genome structures are optimal for capsid assembly. *eLife* 2: e00632.

91 Abergel, C., Legendre, M., and Claverie, J.-M. (2015). The rapidly expanding universe of giant viruses: Mimivirus, Pandoravirus, Pithovirus and Mollivirus. *FEMS Microbiology Reviews* 39 (6): 779–796.

92 Legendre, M., Bartoli, J., Shmakova, L. et al. (2014). Thirty-thousand-year-old distant relative of giant icosahedral DNA viruses with a pandoravirus morphology. *Proceedings of the National Academy of Sciences of the United States of America* 111 (11): 4274–4279.

93 Tischer, I., Gelderblom, H., Vettermann, W., and Koch, M.A. (1982). A very small porcine virus with circular single-stranded DNA. *Nature* 295: 64–66.

94 Zhai, S.-L., Lu, S.-S., Wei, W.-K. et al. (2019). Reservoirs of porcine circoviruses: a mini review. *Frontiers in Veterinary Science* 6: 319.

95 Weinbauer, M.G. (2004). Ecology of prokaryotic viruses. *FEMS Microbiology Reviews* 28 (2): 127–181.

96 Martin, W. and Kowallik, K.V. (1999). Annotated English translation of Mereschkowsky's 1905 paper 'Über Natur und Ursprung der Chromatophoren im Pflanzenreiche'. *European Journal of Phycology* 34 (3): 287–295.

97 Mereschkowski, K. (1905). Über Natur und Ursprung der Chromatophoren im Pflanzenreiche. *Biologisches Centralblatt* 25: 593–604.

98 Glansdorff, N., Xu, Y., and Labedan, B. (2008). The last universal common ancestor: emergence, constitution and genetic legacy of an elusive forerunner. *Biology Direct* 3: 29.

99 Eme, L., Sharpe, S.C., Brown, M.W., and Roger, A.J. (2014). On the age of eukaryotes: evaluating evidence from fossils and molecular clocks. *Cold Spring Harbor Perspectives in Biology* 6: a016139.

100 Chernikova, D., Motamedi, S., Csürös, M. et al. (2011). A late origin of the extant eukaryotic diversity: divergence time estimates using rare genomic changes. *Biology Direct* 6: 26.

101 McInerney, J.O. and O'Connell, M.J. (2017). Mind the gaps in cellular evolution. *Nature* 541: 297–299.

102 Spang, A., Saw, J.H., Jørgensen, S.L. et al. (2015). Complex archaea that bridge the gap between prokaryotes and eukaryotes. *Nature* 521 (7551): 173–179.

103 Bock, R. and Timmis, J.N. (2008). Reconstructing evolution: gene transfer from plastids to the nucleus. *BioEssays* 30 (6): 556–566.

104 Michalovova, M., Vyskot, B., and Kejnovsky, E. (2013). Analysis of plastid and mitochondrial DNA insertions in the nucleus (NUPTs and NUMTs) of six plant species: size, relative age and chromosomal localization. *Heredity (Edinb)* 111 (4): 314–320.

105 Roark, L.M., Hui, A.Y., Donnelly, L. et al. (2010). Recent and frequent insertions of chloroplast DNA into maize nuclear chromosomes. *Cytogenetic and Genome Research* 129 (1–3): 17–23.

106 Asada, K. (2006). Production and scavenging of reactive oxygen species in chloroplasts and their functions. *Plant Physiology* 141 (2): 391–396.

107 Murphy, M.P. (2009). How mitochondria produce reactive oxygen species. *Biochemical Journal* 417 (1): 1–13.

108 Daley, D.O. and Whelan, J. (2005). Why genes persist in organelle genomes. *Genome Biology* 6 (5): 110.

109 Woodson, J.D. and Chory, J. (2008). Coordination of gene expression between organellar and nuclear genomes. *Nature Reviews Genetics* 9 (5): 383–395.

110 Murat, D., Byrne, M., and Komeili, A. (2010). Cell biology of prokaryotic organelles. *Cold Spring Harbor Perspectives in Biology* 2 (10): a000422.

111 Mauriello, E. (2019). How bacteria arrange their organelles. *eLife* 8: e43777.

112 MacCready, J.S., Hakim, P., Young, E.J. et al. (2018). Protein gradients on the nucleoid position the carbon-fixing organelles of cyanobacteria. *eLife* 7: e39723.

113 Andersson, S.G.E., Zomorodipour, A., Andersson, J.O. et al. (1998). The genome sequence of *Rickettsia prowazekii* and the origin of mitochondria. *Nature* 396: 133–140.

114 Emelyanov, V.V. (2001). Evolutionary relationship of *Rickettsiae* and mitochondria. *FEBS Letters* 501: 11–18.

115 Hedges, S.B., Blair, J.E., Venturi, M.L., and Shoe, J.L. (2004). A molecular timescale of eukaryote evolution and the rise of complex multicellular life. *BMC Evolutionary Biology* 4: 2.

116 Cole, L.W. (2016). The evolution of per-cell organelle number. *Frontiers in Cell and Development Biology* 4: 85.

117 Yarlett, N. and Hackstein, J.H.P. (2005). Hydrogenosomes: one organelle, multiple origins. *BioScience* 55 (8): 657–668.

118 Martin, W. (2005). The missing link between hydrogenosomes and mitochondria. *Trends in Microbiology* 13 (10): 457–459.

119 Embley, T.M., van der Giezen, M., Horner, D.S. et al. (2003). Mitochondria and hydrogenosomes are two forms of the same fundamental organelle. *Philosophical Transactions of the Royal Society of London, Series B: Biological Sciences* 358 (1429): 191–202.

120 Akhmanova, A., Voncken, F., van Alen, T. et al. (1998). A hydrogenosome with a genome. *Nature* 396: 527–528.

121 Kraus, F. and Ryan, M.T. (2017). The constriction and scission machineries involved in mitochondrial fission. *Journal of Cell Science* 130 (18): 2953–2960.

122 Kamerkar, S.C., Kraus, F., Sharpe, A.J. et al. (2018). Dynamin-related protein 1 has membrane constricting and severing abilities sufficient for mitochondrial and peroxisomal fission. *Nature Communications* 9 (1): 5239.

123 Wexler-Cohen, Y., Stevens, G.C., Barnoy, E. et al. (2014). A dynamin-related protein contributes to *Trichomonas vaginalis* hydrogenosomal fission. *The FASEB Journal* 28 (3): 1113–1121.

124 Miyagishima, S.-y., Nishida, K., Mori, T. et al. (2003). A plant-specific dynamin-related protein forms a ring at the chloroplast division site. *Plant Cell* 15 (3): 655–665.

125 Jilly, R., Khan, N.Z., Aronsson, H., and Schneider, D. (2018). Dynamin-like proteins are potentially involved in membrane dynamics within chloroplasts and cyanobacteria. *Frontiers in Plant Science* 9: 206.

126 Wu, Z., Sloan, D.B., Brown, C.W. et al. (2017). Mitochondrial retroprocessing promoted functional transfers of rpl5 to the nucleus in grasses. *Molecular Biology and Evolution* 34 (9): 2340–2354.

127 Smith, J.J. and Aitchison, J.D. (2013). Peroxisomes take shape. *Nature Reviews Molecular Cell Biology* 14: 803–817.

128 Sugiura, A., Mattie, S., Prudent, J., and McBride, H.M. (2017). Newly born peroxisomes are a hybrid of mitochondrial and ER-derived pre-peroxisomes. *Nature* 542 (7640): 251–254.

129 Daubin, V. and Szöllősi, G.J. (2016). Horizontal gene transfer and the history of life. *Cold Spring Harbor Perspectives in Biology* 8 (4): a018036.

130 Griffith, F. (1928). The significance of pneumococcal types. *Journal of Hygiene (London)* 27 (2): 113–159.

131 Lacroix, B. and Citovsky, V. (2016). Transfer of DNA from bacteria to eukaryotes. *MBio* 7 (4): e00863–e00816.

132 Schell, J. and Montagu, M.V. (1977). The Ti-plasmid of *Agrobacterium tumefaciens*, a natural vector for the introduction of nif genes in plants? *Basic Life Sciences* 9: 159–179.

133 Sieber, K.B., Bromley, R.E., and Hotopp, J.C.D. (2017). Lateral gene transfer between prokaryotes and eukaryotes. *Experimental Cell Research* 358 (2): 421–426.

134 Gilbert, C. and Cordaux, R. (2017). Viruses as vectors of horizontal transfer of genetic material in eukaryotes. *Current Opinion in Virology* 25: 16–22.

135 Ma, P.-F., Zhang, Y.-X., Guo, Z.-H., and Li, D.-Z. (2015). Evidence for horizontal transfer of mitochondrial DNA to the plastid genome in a bamboo genus. *Scientific Reports* 5: 11608.

136 Qiu, Y., Filipenko, S.J., Darracq, A., and Adams, K.L. (2014). Expression of a transferred nuclear gene in a mitochondrial genome. *Current Plant Biology* 1: 68–72.

137 Soucy, S.M., Huang, J., and Gogarten, J.P. (2015). Horizontal gene transfer: building the web of life. *Nature Reviews Genetics* 16 (8): 472–482.

138 Keeling, P.J. and Palmer, J.D. (2008). Horizontal gene transfer in eukaryotic evolution. *Nature Reviews Genetics* 9 (8): 605–618.

139 Ginzburg, L.R., Bingham, P.M., and Yoo, S. (1984). On the theory of speciation induced by transposable elements. *Genetics* 107 (2): 331–341.

140 Serrato-Capuchina, A. and Matute, D.R. (2018). The role of transposable elements in speciation. *Gene* 9: 254.

141 Baidouri, M.E., Carpentier, M.-C., Cooke, R. et al. (2014). Widespread and frequent horizontal transfers of transposable elements in plants. *Genome Research* 24 (5): 831–838.

142 McClintock, B. (1950). The origin and behavior of mutable loci in maize. *Proceedings of the National Academy of Sciences of the United States of America* 36 (6): 344–355.

143 Hales, K.G., Korey, C.A., Larracuente, A.M., and Roberts, D.M. (2015). Genetics on the fly: a primer on the *Drosophila* model system. *Genetics* 201 (3): 815–842.

144 Arakaki, Y., Kawai-Toyooka, H., Hamamura, Y. et al. (2013). The simplest integrated multicellular organism unveiled. *PLoS One* 8 (12): e81641.

145 Featherston, J., Arakaki, Y., Hanschen, E.R. et al. (2018). The 4-celled *Tetrabaena socialis* nuclear genome reveals the essential components for genetic control of cell number at the origin of multicellularity in the volvocine lineage. *Molecular Biology and Evolution* 35 (4): 855–870.

146 Woznica, A., Cantley, A.M., Beemelmanns, C. et al. (2016). Bacterial lipids activate, synergize, and inhibit a developmental switch in choanoflagellates. *Proceedings of the National Academy of Sciences of the United States of America* 113 (28): 7894–7899.

147 Alegado, R.A. and King, N. (2014). Bacterial influences on animal origins. *Cold Spring Harbor Perspectives in Biology* 6 (11): a016162.

148 Alegado, R.A., Brown, L.W., Cao, S. et al. (2012). A bacterial sulfonolipid triggers multicellular development in the closest living relatives of animals. *eLife* 1: e00013.

149 King, N. (2004). The unicellular ancestry of animal development. *Developmental Cell* 7 (3): 313–325.

150 King, N., Hittinger, C.T., and Carroll, S.B. (2003). Evolution of key cell signaling and adhesion protein families predates animal origins. *Science* 301 (5631): 361–363.

151 Sebé-Pedrós, A., Roger, A.J., Lang, F.B. et al. (2010). Ancient origin of the integrin-mediated adhesion and signaling machinery. *Proceedings of the National Academy of Sciences of the United States of America* 107 (22): 10142–10147.

152 Fairclough, S.R., Dayel, M.J., and King, N. (2010). Multicellular development in a choanoflagellate. *Current Biology* 20 (20): R875–R876.

153 Laundon, D., Larson, B.T., McDonald, K. et al. (2019). The architecture of cell differentiation in choanoflagellates and sponge choanocytes. *PLoS Biology* 17 (4): e3000226.

154 Sogabe, S., Hatleberg, W.L., Kocot, K.M. et al. (2019). Pluripotency and the origin of animal multicellularity. *Nature* 570: 519–522.

155 Durston, A. (2013). Dictyostelium: the mathematician's organism. *Current Genomics* 14 (6): 355–360.

156 Du, Q., Kawabe, Y., Schilde, C. et al. (2015). The evolution of aggregative multicellularity and cell–cell communication in the dictyostelia. *Journal of Molecular Biology* 427 (23): 3722–3733.

157 Gaudet, P., Williams, J.G., Fey, P., and Chisholm, R.L. (2008). An anatomy ontology to represent biological knowledge in *Dictyostelium discoideum*. *BMC Genomics* 9: 130.

158 Brückner, S. and Mösch, H.-U. (2012). Choosing the right lifestyle: adhesion and development in *Saccharomyces cerevisiae*. *FEMS Microbiology Reviews* 36 (1): 25–58.

159 Mináriková, L., Kuthan, M., Ricicová, M. et al. (2001). Differentiated gene expression in cells within yeast colonies. *Experimental Cell Research* 271 (2): 296–304.

160 Vopálenská, I., Hůlková, M., Janderová, B., and Palková, Z. (2005). The morphology of *Saccharomyces cerevisiae* colonies is affected by cell adhesion and the budding pattern. *Research in Microbiology* 156 (9): 921–923.

161 Bourque, G., Burns, K.H., Gehring, M. et al. (2018). Ten things you should know about transposable elements. *Genome Biology* 19 (1): 199.

162 Bodea, G.O., McKelvey, E.G.Z., and Faulkner, G.J. (2018). Retrotransposon-induced mosaicism in the neural genome. *Open Biology* 8 (7): 180074.

163 Hultén, M.A., Jonasson, J., Nordgren, A., and Iwarsson, E. (2010). Germinal and somatic trisomy 21 mosaicism: how common is it, what are the implications for individual carriers and how does it come about? *Current Genomics* 11 (6): 409–419.

164 Papavassiliou, P., York, T.P., Gursoy, N. et al. (2009). The phenotype of persons having mosaicism for trisomy 21/Down syndrome reflects the percentage of trisomic cells present in different tissues. *American Journal of Medical Genetics. Part A* 149A (4): 573–583.

165 Frank, M.H. and Chitwood, D.H. (2016). Plant chimeras: the good, the bad, and the 'Bizzaria'. *Developmental Biology* 419 (1): 41–53.

166 Smith, J.J., Timoshevskaya, N., Timoshevskiy, V.A. et al. (2019). A chromosome-scale assembly of the axolotl genome. *Genome Research* 29: 317–324.

167 Stevens, K.A., Wegrzyn, J.L., Zimin, A. et al. (2016). Sequence of the sugar pine megagenome. *Genetics* 204 (4): 1613–1626.

168 Gonzalez-Ibeas, D., Martinez-Garcia, P.J., Famula, R.A. et al. (2016). Assessing the gene content of the megagenome: sugar pine (*Pinus lambertiana*). *G3 (Bethesda)* 6 (12): 3787–3802.

169 Scott, A.D., Stenz, N.W.M., Ingvarsson, P.K., and Baum, D.A. (2016). Whole genome duplication in coast redwood (*Sequoia sempervirens*) and its implications for explaining the rarity of polyploidy in conifers. *The New Phytologist* 211 (1): 186–193.

170 Garcia, R., Gemperlein, K., and Müller, R. (2014). *Minicystis rosea* gen. nov., sp. nov., a polyunsaturated fatty acid-rich and steroid-producing soil myxobacterium. *International Journal of Systematic and Evolutionary Microbiology* 64: 3733–3742.

171 Han, K., Li, Z.-f., Peng, R. et al. (2013). Extraordinary expansion of a *Sorangium cellulosum* genome from an alkaline milieu. *Scientific Reports* 3: 2101.

172 Martínez-Cano, D.J., Reyes-Prieto, M., Martínez-Romero, E. et al. (2014). Evolution of small prokaryotic genomes. *Frontiers in Microbiology* 5: 742.

173 Bennett, G.M. and Moran, N.A. (2013). Small, smaller, smallest: the origins and evolution of ancient dual symbioses in a phloem-feeding insect. *Genome Biology and Evolution* 5 (9): 1675–1688.

174 Corradi, N., Pombert, J.-F., Farinelli, L. et al. (2010). The complete sequence of the smallest known nuclear genome from the microsporidian *Encephalitozoon intestinalis*. *Nature Communications* 1: 77.

175 Derelle, E., Ferraz, C., Rombauts, S. et al. (2006). Genome analysis of the smallest free-living eukaryote *Ostreococcus tauri* unveils many unique features. *Proceedings of the National Academy of Sciences of the United States of America* 103 (31): 11647–11652.

176 Doležel, J. and Bartoš, J. (2005). Plant DNA flow cytometry and estimation of nuclear genome size. *Annals of Botany* 95 (1): 99–110.

177 Dolezel, J., Bartos, J., Voglmayr, H., and Greilhuber, J. (2003). Nuclear DNA content and genome size of trout and human. *Cytometry A* 51 (2): 127–128.

178 Pellicer, J., Fay, M.F., and Leitch, I.J. (2010). The largest eukaryotic genome of them all? *Botanical Journal of the Linnean Society* 164 (1): 10–15.

179 Seeman, N.C. (2003). DNA in a material world. *Nature* 421: 427–431.

180 Schakenraad, K., Biebricher, A.S., Sebregts, M. et al. (2017). Hyperstretching DNA. *Nature Communications* 8: 2197.

181 Larsson, J., Nylander, J.A., and Bergman, B. (2011). Genome fluctuations in cyanobacteria reflect evolutionary, developmental and adaptive traits. *BMC Evolutionary Biology* 11: 187.

182 Krupovic, M., Ravantti, J.J., and Bamford, D.H. (2009). Geminiviruses: a tale of a plasmid becoming a virus. *BMC Evolutionary Biology* 9: 112.

183 Kazlauskas, D., Varsani, A., Koonin, E.V., and Krupovic, M. (2019). Multiple origins of prokaryotic and eukaryotic single-stranded DNA viruses from bacterial and archaeal plasmids. *Nature Communications* 10: 3425.

184 Erdmann, S., Tschitschko, B., Zhong, L. et al. (2017). A plasmid from an *Antarctic haloarchaeon* uses specialized membrane vesicles to disseminate and infect plasmid-free cells. *Nature Microbiology* 2: 1446–1455.

185 Xiao-Ming, Z., Junrui, W., Li, F. et al. (2017). Inferring the evolutionary mechanism of the chloroplast genome size by comparing whole-chloroplast genome sequences in seed plants. *Scientific Reports* 7 (1): 1555.

186 Daniell, H., Lin, C.-S., Yu, M., and Chang, W.-J. (2016). Chloroplast genomes: diversity, evolution, and applications in genetic engineering. *Genome Biology* 17 (134): 1–29.

187 Rivas, J.D.L., Lozano, J.J., and Ortiz, A.R. (2002). Comparative analysis of chloroplast genomes: functional annotation, genome-based phylogeny, and deduced evolutionary patterns. *Genome Research* 12 (4): 567–583.

188 Egea, N. and Lang-Unnasch, N. (1995). Phylogeny of the large extrachromosomal DNA of organisms in the phylum Apicomplexa. *The Journal of Eukaryotic Microbiology* 42 (6): 679–684.

189 Lim, L. and McFadden, G.I. (2010). The evolution, metabolism and functions of the apicoplast. *Philosophical Transactions of the Royal Society of London, Series B: Biological Sciences* 365 (1541): 749–763.

190 Arisue, N. and Hashimoto, T. (2015). Phylogeny and evolution of apicoplasts and apicomplexan parasites. *Parasitology International* 64 (3): 254–259.

191 McFaddena, G.I. and Yeh, E. (2016). The apicoplast: now you see it, now you don't. *International Journal for Parasitology* 47 (2–3): 137–144.

192 Lhee, D., Ha, J.-S., Kim, S. et al. (2019). Evolutionary dynamics of the chromatophore genome in three photosynthetic *Paulinella species*. *Scientific Reports* 9 (1): 2560.

193 Nowack, E.C.M., Melkonian, M., and Glöckner, G. (2008). Chromatophore genome sequence of *Paulinella sheds* light on acquisition of photosynthesis by eukaryotes. *Current Biology* 18 (6): 410–418.

194 Löffelhardt, W. and Bohnert, H.J. (1994). Structure and function of the cyanelle genome. *International Review of Cytology* 151: 29–65.

195 Trench, R.K., Pool, R.R., Logan, M., and Engelland, A. (1978). Aspects of the relation between *Cyanophora paradoxa* (Korschikoff) and its endosymbiotic cyanelles *Cyanocyta korschikoffiana* (Hall & Claus) - I. Growth, ultrastructure, photosynthesis and the obligate nature of the association. *Proceedings of the Royal Society of London* 202 (1149): 423–443.

196 Steiner, J.M. and Löffelhardt, W. (2002). Protein import into cyanelles. *Trends in Plant Science* 7 (2): 72–77.

197 Yusa, F., Steiner, J.M., and Löffelhardt, W. (2008). Evolutionary conservation of dual Sec translocases in the cyanelles of *Cyanophora paradoxa. BMC Evolutionary Biology* 8: 304.

198 Wise, R.R. and Hoober, J.K. (2006). *The Structure and Function of Plastids.* Dordrecht: Springer.

199 Shlomai, J. (2004). The structure and replication of kinetoplast DNA. *Current Molecular Medicine* 4 (6): 623–647.

200 Ogbadoyi, E.O., Robinson, D.R., and Gull, K. (2003). A high-order trans-membrane structural linkage is responsible for mitochondrial genome positioning and segregation by flagellar basal bodies in trypanosomes. *Molecular Biology of the Cell* 14 (5): 1769–1779.

201 Lukeš, J., Guilbride, D.L., Votýpka, J. et al. (2002). Kinetoplast DNA network: evolution of an improbable structure. *Eukaryotic Cell* 1 (4): 495–502.

202 Jensen, R.E. and Englund, P.T. (2012). Network news: the replication of kinetoplast DNA. *Annual Review of Microbiology* 66 (1): 473–491.

203 Liu, Y. and Englund, P.T. (2007). The rotational dynamics of kinetoplast DNA replication. *Molecular Microbiology* 64 (3): 676–690.

204 Lane, N. and Martin, W. (2010). The energetics of genome complexity. *Nature* 467 (7318): 929–934.

205 Friedman, J.R. and Nunnari, J. (2014). Mitochondrial form and function. *Nature* 505 (7483): 335–343.

206 Kühlbrandt, W. (2015). Structure and function of mitochondrial membrane protein complexes. *BMC Biology* 13 (1): 89.

207 Jang, J.Y., Blum, A., Liu, J., and Finkel, T. (2018). The role of mitochondria in aging. *The Journal of Clinical Investigation* 128 (9): 3662–3670.

208 Gray, M.W., Lang, B.F., Cedergren, R. et al. (1998). Genome structure and gene content in protist mitochondrial DNAs. *Nucleic Acids Research* 26 (4): 865–878.

209 de la Cueva-Méndez, G. and Pimentel, B. (2007). Gene and cell survival: lessons from prokaryotic plasmid R1. *EMBO Reports* 8 (5): 458–464.

210 Smillie, C., Garcillán-Barcia, M.P., Francia, M.V. et al. (2010). Mobility of plasmids. *Microbiology and Molecular Biology Reviews* 74 (3): 434–452.

211 Lederberg, J. (1952). Cell genetics and hereditary symbiosis. *Physiological Reviews* 32 (4): 403–430.

212 Esser, K., Kück, U., Lang-Hinrichs, C. et al. (1986). *Plasmids of Eukaryotes, Fundamentals and Applications*. Berlin Heidelberg: Springer-Verlag.

213 Salanoubat, M., Genin, S., Artiguenave, F. et al. (2002). Genome sequence of the plant pathogen *Ralstonia solanacearum*. *Nature* 415: 497–502.

214 Medema, M.H., Trefzer, A., Kovalchuk, A. et al. (2010). The sequence of a 1.8-mb bacterial linear plasmid reveals a rich evolutionary reservoir of secondary metabolic pathways. *Genome Biology and Evolution* 2: 212–224.

215 Wolf, Y.I., Kazlauskas, D., Iranzo, J. et al. (2018). Origins and evolution of the global RNA virome. *MBio* 9 (6): e02329–e02318.

216 Duffy, S. (2018). Why are RNA virus mutation rates so damn high? *PLoS Biology* 16 (8): e3000003.

217 Legendre, M., Fabre, E., Poirot, O. et al. (2018). Diversity and evolution of the emerging *Pandoraviridae family*. *Nature Communications* 9: 2285.

218 Sun, C., Feschotte, C., Wu, Z., and Mueller, R.L. (2015). DNA transposons have colonized the genome of the giant virus *Pandoravirus salinus*. *BMC Biology* 13: 38.

219 Fux, R., Söckler, C., Link, E.K. et al. (2018). Full genome characterization of porcine circovirus type 3 isolates reveals the existence of two distinct groups of virus strains. *Virology Journal* 15: 25.

220 Chen, F., Pan, Y., Liao, C. et al. (2012). Complete genome sequence of porcine circovirus 2d strain GDYX. *Journal of Virology* 86 (22): 12457–12458.

221 Wang, K.-S., Choo, Q.-L., Weiner, A.J. et al. (1987). Structure, sequence and expression of the hepatitis delta (δ) viral genome. *Nature* 328: 456.

222 Sureau, C., Guerra, B., and Lanford, R.E. (1993). Role of the large hepatitis B virus envelope protein in infectivity of the hepatitis delta virion. *Journal of Virology* 67 (1): 366–372.

223 Gudima, S., He, Y., Meier, A. et al. (2007). Assembly of hepatitis delta virus: particle characterization, including the ability to infect primary human hepatocytes. *Journal of Virology* 81 (7): 3608–3617.

224 Jeudy, S., Rigou, S., Alempic, J.-M. et al. (2020). The DNA methylation landscape of giant viruses. *Nature Communications* 11: 2657.

225 Rout, S.K., Friedmann, M.P., Riek, R., and Greenwald, J. (2018). A prebiotic template-directed peptide synthesis based on amyloids. *Nature Communications* 9: 234.

226 Diener, T.O. (1971). Potato spindle tuber "virus". IV. A replicating, low molecular weight RNA. *Virology* 45 (2): 411–428.

227 Diener, T.O. (1989). Circular RNAs: relics of precellular evolution? *Proceedings of the National Academy of Sciences of the United States of America* 86 (23): 9370–9374.

228 Maurel, M.-C., Leclerc, F., Vergne, J., and Zaccai, G. (2019). RNA back and forth: looking through ribozyme and viroid motifs. *Viruses* 11 (3): 283.

229 Dinter-Gottlieb, G. (1986). Viroids and virusoids are related to group I introns. *Proceedings of the National Academy of Sciences of the United States of America* 83 (17): 6250–6254.

230 Hausner, G., Hafez, M., and Edgell, D.R. (2014). Bacterial group I introns: mobile RNA catalysts. *Mobile DNA* 5: 8.

231 Landthaler, M. and Shub, D.A. (1999). Unexpected abundance of self-splicing introns in the genome of bacteriophage Twort: introns in multiple genes, a single gene with three introns, and exon skipping by group I ribozymes. *Proceedings of the National Academy of Sciences of the United States of America* 96 (12): 7005–7010.

232 Diener, T.O. (1981). Are viroids escaped introns? *Proceedings of the National Academy of Sciences of the United States of America* 78 (8): 5014–5015.

233 Hedberg, A. and Johansen, S.D. (2013). Nuclear group I introns in self-splicing and beyond. *Mobile DNA* 4: 17.

234 Willyard, C. (2018). New human gene tally reignites debate. *Nature* 558 (7710): 354–355.

235 Hatje, K., Mühlhausen, S., Simm, D., and Kollmar, M. (2019). The protein-coding human genome: annotating high-hanging fruits. *BioEssays* 41 (11): e1900066.

236 Mische, S.M., Fisher, N.C., Meyn, S.M. et al. (2020). A review of the scientific rigor, reproducibility, and transparency studies conducted by the ABRF research groups. *Journal of Biomolecular Techniques* 31 (1): 11–26.

237 Pavesi, A., Vianelli, A., Chirico, N. et al. (2018). Overlapping genes and the proteins they encode differ significantly in their sequence composition from non-overlapping genes. *PLoS One* 13 (10): e0202513.

238 Chen, C.-H., Pan, C.-Y., and Lin, W.-c. (2019). Overlapping protein-coding genes in human genome and their coincidental expression in tissues. *Scientific Reports* 9: 13377.

239 Uesaka, M., Nishimura, O., Go, Y. et al. (2014). Bidirectional promoters are the major source of gene activation-associated non-coding RNAs in mammals. *BMC Genomics* 15: 35.

240 Levenshtein, V.I. (1966). Binary codes capable of correcting deletions, insertions, and reversals. *Soviet Physics Doklady* 10 (8): 707–710.

241 Needleman, S.B. and Wunsch, C.D. (1970). A general method applicable to the search for similarities in the amino acid sequence of two proteins. *Journal of Molecular Biology* 48 (3): 443–453.

242 Smith, T.F. and Waterman, M.S. (1981). Identification of common molecular subsequences. *Journal of Molecular Biology* 147 (1): 195–197.

243 Czarnul, P. (2018). *Parallel Programming for Modern High Performance Computing Systems*. CRC Press.

244 Lubbers, P., Albers, B., and Salim, F. (2010). Creating HTML5 offline web applications. In: *Pro HTML5 Programming* (ed. C. Andres), 243–257. Apress.

245 Pagni, M. and Jongeneel, C.V. (2001). Making sense of score statistics for sequence alignments. *Briefings in Bioinformatics* 2 (1): 51–67.

246 Altschul, S.F., Bundschuh, R., Olsen, R., and Hwa, T. (2001). The estimation of statistical parameters for local alignment score distributions. *Nucleic Acids Research* 29 (2): 351–361.

247 Kneusel, R. (2018). *Random Numbers and Computers*, 1–260. Springer International Publishing.

248 Bonilla, L.L., Alvaro, M., and Carretero, M. (2016). Chaos-based true random number generators. *Journal of Mathematics in Industry* 7 (1): 1–17.

249 Gagniuc, P.A. (2017). *Markov Chains: From Theory to Implementation and Experimentation*. Hoboken, NJ: Wiley.

250 Gagniuc, P.A., Ionescu-Tirgoviste, C., Gagniuc, E. et al. (2020). Spectral forecast: a general purpose prediction model as an alternative to classical neural networks. *Chaos* 30: 033119–033116.

251 Sayle, R.A. and Milner-White, E.J. (1995). RASMOL: biomolecular graphics for all. Trends Biochem. *Trends in Biochemical Sciences* 20 (9): 374.

252 Eger, S. (2013). Sequence alignment with arbitrary steps and further generalizations, with applications to alignments in linguistics. *Information Sciences* 237: 287–304.

253 Jäger, G. (2015). Support for linguistic macrofamilies from weighted sequence alignment. *Proceedings of the National Academy of Sciences of the United States of America* 112 (41): 12752–12757.

254 Gatenby, R.A. and Frieden, B.R. (2007). Information theory in living systems, methods, applications, and challenges. *Bulletin of Mathematical Biology* 69 (2): 635–657.

255 Frieden, B.R. and Gatenby, R.A. (2011). Information dynamics in living systems: prokaryotes, eukaryotes, and cancer. *PLoS One* 6 (7): e22085.

256 Rojas, W.A.G., McMorrow, J.J., Geier, M.L. et al. (2017). Solution-processed carbon nanotube true random number generator. *Nano Letters* 17 (8): 4976–4981.

257 Pironio, S. (2018). The certainty of randomness. *Nature* 556 (7700): 176–177.

258 Johnson, J. (1928). Thermal agitation of electricity in conductors. *Physical Review* 32 (97): 97–109.

259 Nyquist, H. (1928). Thermal agitation of electric charge in conductors. *Physical Review* 32 (110): 110–113.

260 Huffman, D. (1952). A method for the construction of minimum-redundancy codes. *Proceedings of the IRE* 40 (9): 1098–1101.

261 Gagniuc, P.A. and Ionescu-Tirgoviste, C. (2016). Compression up to singularity: a model for lossless compression and decompression of information. *Proceedings of the Romanian Academy, Series B* 18 (3): 169–174.

262 Reanney, D.C., MacPhee, D.G., and Pressing, J. (1983). Intrinsic noise and the design of the genetic machinery. *Australian Journal of Biological Sciences* 36 (1): 77–90.

263 Lu, B., Fleming, S., Szalay, T., and Golovchenko, J. (2015). Thermal motion of DNA in an MspA pore. *Biophysical Journal* 109 (7): 1439–1445.

264 Reanney, D.C. and Pressing, J. (1984). Temperature as a determinative factor in the evolution of genetic systems. *Journal of Molecular Evolution* 21 (1): 72–75.

265 Rao, C.V., Wolf, D.M., and Arkin, A.P. (2002). Control, exploitation and tolerance of intracellular noise. *Nature* 420 (6912): 231–237.

266 Turaeva, N. and Brown-Kennerly, V. (2015). Marcus model of spontaneous point mutation in DNA. *Chemical Physics* 461: 106–110.

267 Bancaud, A., Huet, S., Daigle, N. et al. (2009). Molecular crowding affects diffusion and binding of nuclear proteins in heterochromatin and reveals the fractal organization of chromatin. *The EMBO Journal* 28 (24): 3785–3798.

268 Jeziorska, D.M., Murray, R.J.S., Gobbi, M.D. et al. (2017). DNA methylation of intragenic CpG islands depends on their transcriptional activity during differentiation and disease. *Proceedings of the National Academy of Sciences of the United States of America* 14 (36): E7526–E7535.

269 Gagniuc, P. and Ionescu-Tirgoviste, C. (2012). Eukaryotic genomes may exhibit up to 10 generic classes of gene promoters. *BMC Genomics* 13: 512.

270 Deaton, A.M. and Bird, A. (2011). CpG islands and the regulation of transcription. *Genes & Development* 5 (10): 1010–1022.

271 Reik, W. (2007). Stability and flexibility of epigenetic gene regulation in mammalian development. *Nature* 447: 425–432.

272 Suelves, M., Carrió, E., Núñez-Álvarez, Y., and Peinado, M.A. (2016). DNA methylation dynamics in cellular commitment and differentiation. *Briefings in Functional Genomics* 15 (6): 443–453.

273 Morselli, M., Pastor, W.A., Montanini, B. et al. (2015). In vivo targeting of de novo DNA methylation by histone modifications in yeast and mouse. *eLife* 4: e06205.

274 Kelley, K. and Lin, S.-L. (2012). Induction of somatic cell reprogramming using the microRNA miR-302. In: *Progress in Molecular Biology and Translational Science*, 83–107. Elsevier.

275 Ehrlich, M. and Lacey, M. (2013). DNA methylation and differentiation: silencing, upregulation and modulation of gene expression. *Epigenomics* 5 (5): 553–568.

276 Bartels, A., Han, Q., Nair, P. et al. (2018). Dynamic DNA methylation in plant growth and development. *International Journal of Molecular Sciences* 19 (7): 2144.

277 Cheng, X., Virk, N., Chen, W. et al. (2013). CpG usage in RNA viruses: data and hypotheses. *PLoS One* 8 (9): e74109.

278 Jang, H.S., Shin, W.J., Lee, J.E., and Do, J.T. (2017). CpG and non-CpG methylation in epigenetic gene regulation and brain function. *Genes (Basel)* 8 (6): 148.

279 Gagniuc, P., Cimponeriu, D., Panduru, N.M. et al. (2011). A sensitive method for detecting dinucleotide islands and clusters through depth analysis. *Romanian Journal of Diabetes Nutrition and Metabolic Diseases* 18 (2): 165–170.

280 Ionescu-Tîrgovişte, C., Gagniuc, P.A., and Guja, C. (2015). Structural properties of gene promoters highlight more than two phenotypes of diabetes. *PLoS One* 10 (9): e0137950.

281 Stormo, G.D. (2000). DNA binding sites: representation and discovery. *Bioinformatics* 16 (1): 16–23.

282 Luscombe, N.M., Austin, S.E., Berman, H.M., and Thornton, J.M. (2000). An overview of the structures of protein-DNA complexes. *Genome Biology* (1): 1, reviews001.1–reviews001.37.

283 Siddharthan, R. (2010). Dinucleotide weight matrices for predicting transcription factor binding sites: generalizing the position weight matrix. *PLoS One* 5 (3): e9722.

284 Wu, Q. and Krainer, A.R. (1999). AT-AC Pre-mRNA splicing mechanisms and conservation of minor introns in voltage-gated ion channel genes. *Molecular and Cellular Biology* 19 (5): 3225–3236.

285 Bitton, D.A., Rallis, C., Jeffares, D.C. et al. (2014). LaSSO, a strategy for genome-wide mapping of intronic lariats and branch points using RNA-seq. *Genome Research* 24 (7): 1169–1179.

286 Shepard, P.J. and Hertel, K.J. (2008). Conserved RNA secondary structures promote alternative splicing. *RNA* 14 (8): 1463–1469.

287 Athavale, S.S., Petrov, A.S., Hsiao, C. et al. (2012). RNA folding and catalysis mediated by iron(II). *PLoS One* 7 (5): e38024.

288 Olson, K.E., Dolan, G.F., and Müller, U.F. (2014). In vivo evolution of a catalytic RNA couples trans-splicing to translation. *PLoS One* 9 (1): e86473.

289 Wachutka, L., Caizzi, L., Gagneur, J., and Cramer, P. (2019). Global donor and acceptor splicing site kinetics in human cells. *eLife* 8: e45056.

290 Zhu, L., Zhang, Y., Zhang, W. et al. (2009). Patterns of exon-intron architecture variation of genes in eukaryotic genomes. *BMC Genomics* 47: 10.

291 Frey, K. and Pucker, B. (2020). Animal, fungi, and plant genome sequences harbor different non-canonical splice sites. *Cell* 9 (2): 458.

292 Pucker, B. and Brockington, S.F. (2018). Genome-wide analyses supported by RNA-Seq reveal non-canonical splice sites in plant genomes. *BMC Genomics* 19: 980.

293 Modrek, B., Resch, A., Grasso, C., and Lee, C. (2001). Genome-wide detection of alternative splicing in expressed sequences of human genes. *Nucleic Acids Research* 29 (13): 2850–2859.

294 Gao, K., Masuda, A., Matsuura, T., and Ohno, K. (2008). Human branch point consensus sequence is yUnAy. *Nucleic Acids Research* 36 (7): 2257–2267.

295 Konarska, M., Grabowski, P.J., Padgett, R.A., and Sharp, P.A. (1985). Characterization of the branch site in lariat RNAs produced by splicing of mRNA precursors. *Nature* 313 (6003): 552–557.

296 Cartegni, L., Chew, S.L., and Krainer, A.R. (2002). Listening to silence and understanding nonsense: exonic mutations that affect splicing. *Nature Reviews Genetics* 3 (4): 285–298.

297 Carmel, I., Tal, S., Vig, I., and Ast, A.G. (2004). Comparative analysis detects dependencies among the 5' splice-site positions. *RNA* 10 (5): 828–840.

298 Sibley, C.R., Blazquez, L., and Ule, J. (2016). Lessons from non-canonical splicing. *Nature Reviews Genetics* 17 (7): 407–421.

299 Ast, G. (2004). How did alternative splicing evolve? *Nature Reviews Genetics* 5 (10): 773–782.

300 Shmelkov, S.V., Jun, L., Clair, R.S. et al. (2004). Alternative promoters regulate transcription of the gene that encodes stem cell surface protein AC133. *Blood* 103 (6): 2055–2061.

301 Ruskin, B., Greene, J., and Green, M. (1985). Cryptic branch point activation allows accurate in vitro splicing of human beta-globin intron mutants. *Cell* 41: 833–844.

302 Boudreault, S., Martenon-Brodeur, C., Caron, M. et al. (2016). Global profiling of the cellular alternative RNA splicing landscape during virus-host interactions. *PLoS One* 11 (9): e0161914.

303 Panahi, B. and Hejazi, M.A. (2020). Integrative analysis of gene expression and alternative splicing in microalgae grown under heterotrophic condition. *PLoS One* 15 (6): e0234710.

304 Suzuki, H., Osaki, K., Sano, K. et al. (2011). Comprehensive analysis of alternative splicing and functionality in neuronal differentiation of P19 cells. *PLoS One* 6 (2): e16880.

305 Naftelberg, S., Schor, I.E., Ast, G., and Kornblihtt, A.R. (2015). Regulation of alternative splicing through coupling with transcription and chromatin structure. *Annual Review of Biochemistry* (84): 165–198.

306 Zhang, Y. and Ding, Y. (2020). The simultaneous coupling of transcription and splicing in plants. *Molecular Plant* 13 (2): 184–186.

307 Kornblihtt, A.R., Mata, M.D.L., Fededa, J.P. et al. (2004). Multiple links between transcription and splicing. *RNA* 10 (10): 1489–1498.

308 Stormo, G.D., Schneider, T.D., Gold, L., and Ehrenfeucht, A. (1982). Use of the 'Perceptron' algorithm to distinguish translational initiation sites in *E. coli*. *Nucleic Acids Research* 10 (9): 2997–3011.

309 Fredericks, A.M., Cygan, K.J., Brown, B.A., and Fairbrother, W.G. (2015). RNA-binding proteins: splicing factors and disease. *Biomolecules* 5 (2): 893–909.

310 Rio, D.C. (1992). RNA binding proteins, splice site selection, and alternative pre-mRNA splicing. *Gene Expression* 2 (1): 1–5.

311 Martelly, W., Fellows, B., Senior, K. et al. (2019). Identification of a non-canonical RNA binding domain in the U2 snRNP protein SF3A1. *RNA* 25 (11): 1509–1521.

312 Zhang, S., Samocha, K.E., Rivas, M.A. et al. (2018). Base-specific mutational intolerance near splice sites clarifies the role of nonessential splice nucleotides. *Genome Research* 28 (7): 968–974.

Index

Algorithms in Bioinformatics: Theory and Implementation, First Edition. Paul A. Gagniuc.
© 2021 John Wiley & Sons, Inc. Published 2021 by John Wiley & Sons, Inc.
Companion website: www.wiley.com/go/gagniuc/algorithmsinbioinformatics

Printed and bound by CPI Group (UK) Ltd, Croydon, CR0 4YY